OXFORD MATHEMATICAL MONOGRAPHS

Editors

POLYHARMONIC FUNCTIONS

by

NACHMAN ARONSZAJN
The University of Kansas

THOMAS M. CREESE
The University of Kansas

LEONARD J. LIPKIN
The University of North Florida

CLARENDON PRESS · OXFORD
1983

Oxford University Press, Walton Street, Oxford OX2 6DP

London Glasgow New York Toronto
Delhi Bombay Calcutta Madras Karachi
Kuala Lumpur Singapore Hong Kong Tokyo
Nairobi Dar es Salaam Cape Town
Melbourne Wellington

and associate companies in
Beirut Berlin Ibadan Mexico City

Published in the United States by
Oxford University Press, New York

British Library Cataloguing in Publication Data

Aronszajn, N.
Polyharmonic functions.—(Oxford mathematical monographs)
1. Functions, Subharmonic
I. Title   II. Creese, T. M.
III. Lipkin, L. J.
515′.53   QA405

ISBN 0-19-853906-1

Printed in Northern Ireland at The Universities Press (Belfast) Ltd.

PREFACE

A solution of the equation $\Delta^m u = 0$, where Δ is the Laplace operator in $\mathbb{R}^n$, is usually called 'polyharmonic of order m'. Such functions form a subclass of a much larger class of analytic functions which share many important properties, particularly the tendency to exhibit globally properties which analytic functions generally exhibit only locally. This larger class was identified by Aronszajn [6], and was called the class of functions 'harmonic of infinite order'. In this monograph we develop the theory and applications of these functions, which we now call 'polyharmonic' functions (Chapter II, §1).

In our terminology, solutions of the equation $\Delta^m u = 0$ are called 'polyharmonic of degree m'. The term 'order' has been reserved because the order of an entire function is closely related to a property of polyharmonic functions which we call 'Laplacian order'. This relationship is revealed in Chapter II, §1, Chapter V, §2, and Chapter VI, §§1 and 2. Laplacian order also extends the ordering of the Gevrey classes (Chapter VI, §4).

Among the polyharmonic functions are found the sections $u(x, t_0)$ of analytic solutions of the heat equation in n space variables, as well as the solutions of the partial differential equation $P(\Delta)u + Au = 0$, where P is a polynomial of degree m and A is any linear differential operator of order less than $2m$ with constant coefficients. All entire functions of one or more complex variables are polyharmonic; many rational functions are not. (Following an equivalent definition of polyharmonicity by Lelong [59], Avanissian and Fernique [15] defined distributions 'harmonic of infinite order' and proved that these distributions are actually polyharmonic functions.)

One of the interesting features of the polyharmonic functions is that some of their properties are basically complex while others are basically real. The harmonicity hulls (Chapter II, §§3 and 4), for example, are complex domains. The nature of singularities (Chapter III), convexity (Chapter IV, §2), and Almansi coefficients (Chapter V, §2) are exhibited in real space. In Chapter II, §2 an ostensibly real phenomenon is seen to play an important role in complex space also. This kind of interplay is worthy of attention.

Preliminary definitions and results are drawn together in Chapter I. In Chapter II, §1 the concepts of Laplacian order and type are set forth for functions defined on domains in $\mathbb{R}^n$. The fundamental serio-integral representation is developed in §2, at which point it is observed that the polyharmonic functions can be continued analytically into well-determined domains in $\mathbb{C}^n$. This point is taken up in §§3 and 4 with a study of 'harmonicity hulls' in $\mathbb{C}^n$.

Even though holomorphic functions do not have isolated singularities in $\mathbb{C}^n$, $n \geq 2$, analytic functions in $\mathbb{R}^n$ can exhibit such singular behaviour. In Chapter III, §1, a study is made of the behaviour of polyharmonic functions near isolated singularities in $\mathbb{R}^n$, and in Chapter III, §2, a theorem is proved on additive separation of singularities (occurring on arbitrary closed sets without interior) for polyharmonic functions. A fundamental tool in the latter development is an approximation lemma of the Runge type.

Local and pointwise orders and types are defined in Chapter IV and are shown to satisfy certain maximum and minimum principles. The most important consequence is that polyharmonicity is preserved by analytic continuation. Other applications appear in Chapter VI, §1. Convexity properties for local order and type are developed in Chapter IV, §2.

A series expansion for holomorphic functions is derived in Chapter V, §1. This is a generalization of the finite Almansi expansion for polyharmonic functions of finite degree in $\mathbb{R}^n$. The generalized Almansi expansion plays a central role in the study of polyharmonic functions, but it also applies locally to the whole class of holomorphic functions. Formulas for the coefficients in the series are studied in Chapter V, §2, and are used in §3 of this chapter to construct examples illuminating all the previous material.

Additional applications and extensions of the theory of polyharmonic functions appear in Chapter VI. These include harmonic extension from $\mathbb{R}^n$ to $\mathbb{R}^{n+1}$, a theorem on bounded polyharmonic functions, and several topics relating to partial differential equations.

A theory of holomorphic capacity in $\mathbb{C}^n$ is developed in the Appendix. Its applications to the study of polyharmonic functions appear in Chapters IV and VI, but the Appendix can be read independently since it deals with holomorphic functions in general. After the definitions are given, the most important examples are given and lower bounds on the capacity function are derived (§§A3 and A4). These are needed in §A5, which is the bridge to applications in the study of polyharmonic functions. Results on sequences and series of functions of several complex variables are given there, some of which are reminiscent of Hartogs' lemma on uniform convergence of series. A notion of completeness of sets in the sense of capacity is discussed in §A6, and remarks about capacity on complex manifolds appear in §A7.

ACKNOWLEDGMENTS

Professor N. Aronszajn died in February 1980, before this monograph was completed. Some of his earliest and latest work involved the subject presented here. After his initial contributions in 1935–1938 and the later work of other authors, Aronszajn again took up this topic in the early 1960s. In 1963–1964 he delivered a series of lectures at the University of Chicago. By that time he had discovered Laplacian order and type for analytic functions and had developed three central tools: the serio-integral representation, a preliminary version of holomorphic capacity, and a significant generalization of the classical Almansi expansion of polyharmonic functions of finite degree. The last two of these have application beyond the theory of polyharmonic functions to the general subject of analytic functions of several variables. The expansion theorem in Chapter V, which Aronszajn always refered to as the Almansi expansion, really should be known as the Almansi–Aronszajn expansion for holomorphic functions.

In 1969, Aronszajn showed us the notes from his Chicago lectures. He suggested that the three of us collaborate in polishing them for publication. Improvements led us to new results and eventually to a complete revision of the whole work. Since the Chicago lectures, virtually all of the old material has been significantly extended or illustrated with examples.

The notes taken by E. Gerlach from Aronszajn's Chicago lectures were the most complete written version of the subject at the time. We made extensive use of them.

Research support was given by several sources, notably: Office of the Naval Research, NONR-583(13); University of Kansas General Research Fund, 3093-5038, 4381-2703, 4381-2570, and 4381-6570; National Science Foundation, GP-8110, GP-16292, MPS73-05227, and MC 877-02683; College of Arts and Sciences and the Department of Mathematical Sciences at the University of North Florida; Oregon State University.

Mrs Jeanne Aronszajn, taking a great interest in seeing the project concluded properly, has given us cordial and timely encouragement. Professor Kennan Smith went through the mathematical papers left at Aronszajn's death, kindly sorting out for us those pertaining to the monograph.

Mrs Jeanne Steffey presided over typing and other aspects of the work, quietly overcoming obstacles of time and distance. Her dedication to the project was crucial to its completion.

The surviving authors dedicate their work to their families and gratefully acknowledge their profound debt to Professor Aronszajn.

Kansas T. M. C.
Florida 1982 L. J. L.

CONTENTS

CHAPTER I. PRELIMINARIES

1. Polyharmonic functions of finite degree 1
2. Fundamental solutions, Green's formula and Green's function 7
3. Estimates 12
4. Analytic and holomorphic functions of several variables 16

CHAPTER II. POLYHARMONICITY AND ANALYTIC CONTINUATION

1. Polyharmonicity: Laplacian order and type 25
2. Serio-integral representation 34
3. Harmonicity hulls of domains in $\mathbb{R}^n$, $n \geq 2$ 55
4. Monodromy groups of harmonicity hulls 69

CHAPTER III. SINGULARITIES IN $\mathbb{R}^n$

1. Principal parts 77
2. Separation of singularities

CHAPTER IV. LOCAL LAPLACIAN ORDER AND TYPE

1. Basic properties and extremal principles 98
2. Convexity 112

CHAPTER V. THE ALMANSI EXPANSION

1. The Almansi expansion for holomorphic functions 122
2. Characterization of order and type by the Almansi coefficients 131
3. Examples 148

CHAPTER VI. EXTENSIONS AND APPLICATIONS

1. Harmonic extension to $\mathbb{R}^{n+1}$ 154
2. Relations between Laplacian order and exponential order of entire functions 164

3. Bounded polyharmonic functions 169
4. Comparison with Gevrey classes 173
5. Solutions of metaharmonic and related partial differential equations 174
6. Boundary value problems for polyharmonic functions 180
7. Analytic solutions of the heat equation 197

APPENDIX. HOLOMORPHIC CAPACITY

A1. Basic properties 208
A2. Some examples 210
A3. Polar and non-polar sets 220
A4. Estimates and continuity properties 223
A5. Applications to sequences of holomorphic functions 230
A6. Completeness 245
A7. Holomorphic capacity on manifolds 250

REFERENCES 252

NON-ALPHABETIC SYMBOLS 257

INDEX 258

I

PRELIMINARIES

1. Polyharmonic functions of finite degree

The real n-dimensional Euclidean space will be denoted $\mathbb{R}^n$ and will usually be considered to be the real part of complex n-dimensional space $\mathbb{C}^n$. A point in $\mathbb{R}^n$ or $\mathbb{C}^n$ will be written $x=(x_1,\ldots,x_n)$ with the space to which the point belongs being clear from the context. A multi-index is an n-tuple $q=(q_1,\ldots,q_n)$ of non-negative integers. In the usual notation

$$|q|=\sum_{j=1}^{n} q_j, \quad q!=q_1!\,q_2!\ldots q_n!,$$

$$x^q=x_1^{q_1}\ldots x_n^{q_n},$$

and

$$D^q=\frac{\partial^{|q|}}{\partial x_1^{q_1}\ldots\partial x_n^{q_n}}$$

in $\mathbb{R}^n$ or in $\mathbb{C}^n$. The Laplace operator in $\mathbb{R}^n$ or $\mathbb{C}^n$ is

$$\Delta=\sum_{j=1}^{n}\frac{\partial^2}{\partial x_j^2},$$

and its iterates are denoted Δ^p, $p=1,2,\ldots$. The identity operator is Δ^0. *It is important to point out that the operator Δ so defined will be called the Laplace operator in both real and complex space, with the x_j representing the complex co-ordinates in $\mathbb{C}^n$.* Thus, in $\mathbb{C}^n$ this operator is not the same as the usual Laplace operator in $2n$ real co-ordinates. Our convention is natural, since we shall be concerned with analytic continuation of functions from open sets in $\mathbb{R}^n$ into functions defined on open sets in $\mathbb{C}^n$ (see §4). We shall denote by $\mathscr{C}^k(D)$, $k=0,1,\ldots,\infty$, the space of k times continuously differentiable (real or complex valued) functions on an open set D, where D can be in either $\mathbb{R}^n$ or $\mathbb{C}^n$. The term *domain* will mean open connected set. The open ball with centre x and radius R in $\mathbb{R}^n$ (or in $\mathbb{C}^n$) will be denoted $B_R^n(x)$ (or $\mathscr{B}_R^n(x)$), with the superscript n being suppressed when the dimension is clear from the context. If x is the origin we shall often write B_R^n (or $\mathscr{B}_R^n$). The symbol $|x|$ means the non-negative value of

$$\left(\sum_{j=1}^{n}|x_j|^2\right)^{\frac{1}{2}}$$

in both $\mathbb{R}^n$ and $\mathbb{C}^n$. The function $r(x)$ is defined by

$$r(x)=\left(\sum_{j=1}^{n}x_j^2\right)^{\frac{1}{2}}.$$

In $\mathbb{R}^n$ we shall usually choose the non-negative values of this function, but in $\mathbb{C}^n$ we shall be careful to state the choice of branch. In $\mathbb{R}^n$ we shall often use $r \equiv r(x) \equiv |x|$, and therefore use r as the first co-ordinate in polar (spherical) co-ordinates. Finally, the terms *real analytic function, complex analytic function,* and *holomorphic function* are used in the standard way, with the term holomorphic being reserved exclusively for the complex case. Some remarks on these terms will be made in §4.

In the rest of this section, and in §§2 and 3, we shall work in $\mathbb{R}^n$.

Definition 1.1. A function $u(x) \in \mathscr{C}^{2p}(D)$ is called *polyharmonic of finite degree p* in the open set $D \subset \mathbb{R}^n$ if $\Delta^p u(x) = 0$ for all $x \in D$. The function $u(x)$ is *polyharmonic of least degree p* in D if it is polyharmonic of finite degree p but not polyharmonic of any smaller degree. For $p = 1$ the function is called *harmonic.*

It is well known (and will be proved in the next section) that polyharmonic functions of finite degree are real analytic and therefore polyharmonicity of finite degree p is preserved by analytic continuation.

Lemma 1.1. If $u, v \in \mathscr{C}^{2p}(D)$, then

$$\Delta^p(uv) = \sum_{k+l+|q|=p} \frac{2^{|q|}p!}{k!\,l!\,q!}(D^q\Delta^k u)(D^q\Delta^l v). \tag{1.1}$$

Proof. The proof is by induction on p. ‖

As a consequence of (1.1) we obtain

$$\Delta^p \sum_{j=1}^n x_j \frac{\partial u}{\partial x_j} = 2p\Delta^p u + \sum_{j=1}^n x_j \frac{\partial}{\partial x_j}\Delta^p u. \tag{1.2}$$

For $r > 0$ we have

$$\frac{\partial u}{\partial r} = \sum_{j=1}^n \frac{x_j}{r}\frac{\partial u}{\partial x_j}, \tag{1.3}$$

so that if u depends only on r, (1.2) can be written

$$\Delta^p \sum_{j=1}^n x_j \frac{\partial u}{\partial x_j} \equiv \Delta^p r\frac{\partial u}{\partial r} = 2p\Delta^p u + r\frac{\partial}{\partial r}\Delta^p u. \tag{1.4}$$

By induction on p we obtain

$$\begin{aligned} \Delta^p r^k &= \{k(k-2)\ldots(k-2p+2)(k-2+n) \\ &\quad \times (k-4+n)\ldots(k-2p+n)\}r^{k-2p} \\ &= 2^{2p}\frac{\Gamma(k/2+1)\Gamma((k+n)/2)}{\Gamma(k/2-p+1)\Gamma((k+n)/2-p)}\,r^{k-2p}. \end{aligned} \tag{1.5}$$

We denote by $A_{p,k}$ the coefficient in (1.5):

$$A_{p,k} = 2^{2p} \frac{\Gamma(k/2+1)\Gamma((k+n)/2)}{\Gamma(k/2-p+1)\Gamma((k+n)/2-p)}. \tag{1.6}$$

If k is even and $k \leqslant -2$ or if $k+n$ is even and $k+n \leqslant 0$, we use the first form of the coefficient in (1.5).

Next we consider the functions $r^k \log r$. It is easy to verify that

$$\Delta r^k \log r = k(k+n-2)r^{k-2} \log r + (2k+n-2)r^{k-2},$$

and by induction on p we obtain

$$\Delta^p(r^k \log r) = A_{p,k} r^{k-2p} \log r + r^{k-2p} \sum_{i=1}^{p} A_{i-1,k} A_{p-i,k-2i}(2k+n+2-4i).$$

By (1.6) the preceding formula is transformed into

$$\Delta^p (r^k \log r) = A_{p,k} r^{k-2p} \left\{ \log r + \sum_{i=1}^{p} \left(\frac{1}{k-2i+2} + \frac{1}{k+n-2i} \right) \right\}. \tag{1.7}$$

Remark 1.1. By writing the Laplace operator in the spherical form

$$\Delta = \frac{\partial^2}{\partial r^2} + \frac{n-1}{r} \frac{\partial}{\partial r} + \frac{1}{r^2} \Delta_S$$

where Δ_S is the Laplace–Beltrami operator on the unit sphere, we can obtain the formulae involving Laplacians of r^k and $r^k \log r$ given above, as well as the proofs of Lemma 1.2 and Propositions 1.1 and 1.2 below, in a slightly different way. Moreover, the proof of Proposition 1.3 uses the spherical co-ordinates $x = r\theta$, θ on the unit sphere, and the spherical form of the Laplace operator is perhaps the most natural one to use.

Lemma 1.2. If $\varphi \in \mathscr{C}^1(B_R(0))$ and if

$$\sum_{j=1}^{n} x_j \frac{\partial \varphi}{\partial x_j} + 2p\varphi + c = 0$$

in $B_R(0)$ with p any positive integer and c a constant, then $\varphi \equiv -c/2p$ in $B_R(0)$ and hence $\sum_{j=1}^{n} x_j(\partial\varphi/\partial x_j) = 0$ there.

Proof. For $r > 0$ we have, by (1.3),

$$r \frac{\partial \varphi}{\partial r} + 2p\varphi + c = 0.$$

Multiplying both sides by r^{2p-1} we obtain

$$\frac{\partial}{\partial r}\left(r^{2p}\varphi + \frac{c}{2p} r^{2p} \right) = 0.$$

Since φ is continuous at the origin,

$$r^{2p}\varphi+\frac{c}{2p}r^{2p}=0$$

and the lemma follows. ‖

Proposition 1.1. If a function u is polyharmonic of degree p in a domain D containing the origin then $\sum x_j\, \partial u/\partial x_j$ *is also polyharmonic of degree p, and conversely. If u is polyharmonic of least degree p then so is* $\sum x_j\, \partial u/\partial x_j$, *and conversely.*

Proof. Since u is analytic in D it is enough to consider instead of D a small ball $B_R(0)\subset D$. There (1.2) and Lemma 1.2 (with $c=0$) give the result. ‖

Proposition 1.2. If u is polyharmonic of degree p then $r^{2k}u$ *is polyharmonic of degree* $k+p$ *for every non-negative integer k. If u is polyharmonic of least degree p then* $r^{2k}u$ *is of least degree* $k+p$.

Proof. Consider first the case $p=1$ and $k=1$. We have

$$\Delta r^2 u = 2nu + 4\sum_{j=1}^{n} x_j \frac{\partial u}{\partial x_j}.$$

Hence $\Delta r^2 u$ is harmonic and $r^2 u$ is polyharmonic of degree 2. In this case, the part of the assertion pertaining to least degree is trivial. By induction on p we obtain our assertion for $p=2, 3,\ldots$ and $k=1$. By using Proposition 1.1, induction on k proves the proposition. ‖

Definition 1.2. A domain $D\subset\mathbb{R}^n$ is a *star domain with centre* 0 if $x\in D$ and $0\leq t\leq 1$ imply that $tx\in D$. A *star domain with centre* x^0 is the result of translating by x^0 a star domain with centre 0.

The following result is due to Almansi [3]. A proof is given by Nicolescu [68], together with a treatment of polyharmonic functions of finite degree using this expansion as a principal tool. The Almansi expansion in the form given below, as well as other forms obtained by changes of coordinates, has been used extensively in the study of solutions of the equation $\Delta^p u(x)=0$ (see the Notes at the end of this section).

Proposition 1.3. (*Finite Almansi expansion*). *If u is polyharmonic of finite degree p in a star domain D with centre* 0, *then there exist unique functions* $h_0,\ldots, h_{p-1}$, *each harmonic in D such that*

$$u(x) = \sum_{k=0}^{p-1} r^{2k}(x)h_k(x) \quad \textit{for} \quad x\in D. \tag{1.8}$$

Proof. We use induction on p. For $p=1$ the proposition is obvious.

Suppose that it is proved for $p = p_0 - 1$. Then there are unique harmonic functions $g_0, \ldots, g_{p_0-2}$ with

$$\Delta u = \sum_{k=0}^{p_0-2} r^{2k} g_k = \sum_{k=1}^{p_0-1} r^{2k-2} g_{k-1}.$$

For the desired harmonic functions $h_0, \ldots, h_{p-1}$ we must have

$$(1.9) \qquad \Delta u = \sum_{k=1}^{p_0-1} r^{2k-2} g_{k-1} = \sum_{k=1}^{p_0-1} \Delta(r^{2k} h_k)$$
$$= \sum_{k=1}^{p_0-1} r^{2k-2} \left\{ 4k\left(k - 1 + \frac{n}{2}\right) h_k + 4kr \frac{\partial h_k}{\partial r} \right\}.$$

By the induction hypothesis the Almansi expansion for Δu is unique, so it is necessary and sufficient to find unique solutions, harmonic on D, for the equations

$$4k\left(k - 1 + \frac{n}{2}\right) h_k + 4kr \frac{\partial h_k}{\partial r} = g_{k-1}, \qquad k = 1, \ldots, p_0 - 1.$$

Multiplying by the integrating factor $r^{k-2+n/2}$, integrating, and noticing that at the origin we have $r^{k-1+n/2} h_k = 0$, we obtain

$$r^{k-1+n/2} h_k(r\theta) = \frac{1}{4k} \int_0^r \rho^{k-2+n/2} g_{k-1}(\rho\theta)\, d\rho$$

where $x = r\theta \in D$, θ being a point on the unit sphere about the origin. Hence

$$h_k(x) = \frac{1}{4k} \int_0^1 \tau^{k-2+n/2} g_{k-1}(\tau x)\, d\tau,$$

and h_k is harmonic since the last integral implies that $\Delta h_k = 0$. Finally, setting

$$h_0(x) = u(x) - \sum_{k=1}^{p_0-1} r^{2k}(x) h_k(x)$$

we obtain by (1.9) that $\Delta h_0 = 0$. $\|$

Proposition 1.4. If h is polyharmonic of finite degree p on the ball $B_R(0) \subset \mathbb{R}^n$, and if

$$(1.10) \qquad g(y) = r^{2p-n}(y) h\left(\frac{y}{|y|^2}\right),$$

then g is polyharmonic of degree p on $\mathbb{R}^n \setminus \overline{B_{1/R}(0)}$.

Proof. The well-known special case $p = 1$, the Kelvin formula, follows at once from the identity

$$\sum_{j=1}^{n} \frac{\partial^2}{\partial x_j^2} h(x) = |y|^{n+2} \sum_{j=1}^{n} \frac{\partial^2}{\partial y_j^2} \left\{ |y|^{2-n} h\left(\frac{y}{|y|^2}\right) \right\}, \tag{1.11}$$

$$y = \frac{x}{|x|^2}.$$

For the general case, let the Almansi expansion of h be

$$h(x) = \sum_{j=0}^{p-1} r^{2k}(x) h_k(x).$$

For $y \in \mathbb{R}^n \backslash \overline{B_{1/R}(0)}$

$$\begin{aligned} g(y) &= |y|^{2p-n} \sum_{k=0}^{p-1} |y|^{-2k} h_k\left(\frac{y}{|y|^2}\right) \\ &= \sum_{j=0}^{p-1} |y|^{2j} g_j(y) \end{aligned} \tag{1.12}$$

where

$$g_j(y) = |y|^{2-n} h_{p-1-j}\left(\frac{y}{|y|^2}\right)$$

is harmonic on $\mathbb{R}^n \backslash \overline{B_{1/R}(0)}$ by the special case $p=1$. By Proposition 1.2, $|y|^{2j} g_j(y)$ is polyharmonic of degree $j+1 \leqslant p$. therefore $g(y)$ is polyharmonic of degree p. ‖

Remark 1.2: The transformation (1.10) can be considered as an extension of the Kelvin transformation to polyharmonic functions of degree p, but, whereas the Kelvin transform changes a function polyharmonic of least degree 1 into a function of the same kind, the extended Kelvin transform (1.10) does not preserve degree. An extreme example is the case of $r^{2p-2}h(x)$ with h harmonic and not identically zero, which is polyharmonic of least degree p. The extended Kelvin transform gives the function

$$g(y) = |y|^{2-n} h\left(\frac{y}{|y|^2}\right)$$

which is harmonic.

Remark 1.3. The inverse of the extended Kelvin transform (as in the case of the usual Kelvin transform) is of the same form (1.10) except for the interchange of x and y and of $B_R(0)$ and $\mathbb{R}^n \backslash \overline{B_{1/R}(0)}$.

Remark 1.4. The usual Kelvin transform is often used to define regularity of a harmonic function at infinity. In fact, for a harmonic function $g(y)$

in a neighbourhood of infinity, we apply the inverse Kelvin transform and, if the resulting function is regular at the origin, we say that the harmonic function g is regular at infinity. For functions polyharmonic of degree p in a neighbourhood of infinity, we can apply the inverse of transform (1.10) and see whether the resulting function has a regular extension to a whole neighbourhood of the origin. If it has, we can say that the original function $g(y)$ is regular at infinity.

Notes

There is an extensive literature on the theory of polyharmonic functions of finite degree. Much of the motivation for the study stems from boundary value problems of mathematical physics, particularly from elasticity theory. Nicolescu's monograph [68] summarizes much of the work done (or at least the ideas proposed) up to the time of its writing. Although it is not our intention to study polyharmonic functions of finite degree specifically (see Chapter II, §1, for our classes of functions), we give here an indication of some of the work which is related in spirit to the material presented in this monograph.

Work related to the Almansi expansion (Proposition 1.3) is discussed by Bremekampt [23], Colucci [28], Fichera [31, 32], Pascali [74], Teleman [85], and Tolotti [87–89], as well as in Nicolescu's monograph [68]. Various characterizations of polyharmonic functions of finite degree in terms of means are given by Cheng [24], Bramble and Payne [22], Friedman [35], Ghermanescu [40, 43], Kampé de Fériet [55], Pčelin [75], and Reade [79]. Convergence and growth properties are studied by Duffin and Nehari [30], Gagaeff [39], Nicolescu [69–71], Ovčarenko [73], Picone [76], and Privaloff and Pčelin [77, 78]. Reflection principles (using the Almansi expansion as a tool) are studied by Armitage [4], Bramble [21], Duffin [29], Huber [52], and Kraft [57]. The literature on boundary value problems is too large to describe here and the interested reader should consult books on partial differential equations.

2. Fundamental solutions, Green's formula and Green's function

To each operator Δ^l there corresponds a *characteristic singular solution* of the equation $\Delta^l u = 0$ which is polyharmonic of least degree l in $\mathbb{R}^n \backslash \{0\}$ and is given by

$$r^{2l-n}/\gamma_{l-1} \text{ for odd } n \text{ and } l = 1, 2, \ldots \tag{2.1}$$

$$\begin{cases} r^{2l-n}/\gamma'_{l-1} \text{ for even } n \text{ and } l = 1, 2, \ldots, n/2 - 1 \\ -r^{2l-n}(\log r)/\gamma'_{l-1} \text{ for even } n \text{ and } l = n/2,\ n/2 + 1, \ldots . \end{cases} \tag{2.1'}$$

The constants γ_l and γ'_l are given in the notation of (1.6) as follows:

(2.2) $$\text{for odd } n,\ \gamma_l = A_{l,2l+2-n} \equiv 2^{2l} l! \frac{\Gamma(l+2-n/2)}{\Gamma(2-n/2)}, \qquad l=0,1,2,\ldots$$

(2.2′) $$\text{for even } n=2m\geq 4, \begin{cases} \gamma'_l = A_{l,2l+2-n} = (-1)^l 2^{2l} l! \dfrac{(m-2)!}{(m-l-2)!} \\ \qquad \text{for } l=0,1,\ldots,(m-2) \\ \gamma'_l = (-1)^{m+1} 2^{2l-1}(m-2)!\, l!\, (l+1-m)! \\ \qquad \text{for } l=(m-1), m,\ldots \end{cases}$$

(2.2″) $$\text{for } n=2,\ \gamma'_l = 2^{2l}(l!)^2.$$

A *fundamental solution* for the equation $\Delta^l u = 0$ is the corresponding characteristic singular solution plus any polyharmonic function of degree l in the whole space $\mathbb{R}^n$. Among the fundamental solutions we choose one, which will be denoted by $\mathscr{r}_l(x)$, and which is defined by

(2.3) $$\mathscr{r}_l(x) = \frac{r^{2l-n}(x)}{\gamma_{l-1}}, \qquad l=1,2,\ldots, \quad \text{for odd } n$$

(2.3′) $$\begin{cases} \mathscr{r}_l(x) = \dfrac{r^{2l-n}(x)}{\gamma'_{l-1}} \quad \text{for} \quad l=1,2,\ldots,(m-1) \\ \mathscr{r}_l(x) = \dfrac{-r^{2l-n}(x)\log r(x)}{\gamma'_{l-1}} + \dfrac{c_{l-m}}{\gamma'_{l-1}} r^{2l-n}(x) \quad \text{for } l\geq m, \text{ for } n=2m, \end{cases}$$

where the constants c_l, $l=0,1,\ldots$, in the second line of (2.3′) are determined as follows:

(2.4) $$c_0 = 0, \qquad c_l = \sum_{k=1}^{l} \frac{1}{2k} + \sum_{k=m}^{l+m-1} \frac{1}{2k} \quad \text{for} \quad l>0.$$

With this choice of the constants c_l it follows immediately that for all dimensions $n=1,2,\ldots$, and all degrees $l=1,2,\ldots$, we have

(2.5) $$\Delta^p \mathscr{r}_l(x) = \mathscr{r}_{l-p}(x) \quad \text{for} \quad p<l.$$

In order to use Green's formulae, we shall have to integrate over $(n-1)$-dimensional manifolds which are sufficiently regular. We define a *Lipschitzian graph manifold* to be a topological submanifold $\mathscr{M}$ in $\mathbb{R}^n$ such that for each of its points y there exists a neighbourhood U of y on the manifold and an $(n-1)$-dimensional plane Π in $\mathbb{R}^n$ such that the orthogonal projection of U on Π is a bi-Lipschitzian homeomorphism, that is there exists a constant $M\geq 1$ such that if P_Π is the projection and y_1 and y_2 are points in U, then

(2.6) $$\frac{1}{M} \leq \frac{|P_\Pi y_1 - P_\Pi y_2|}{|y_1 - y_2|} \leq 1.$$

On each Lipschitzian submanifold $\mathcal{M}$ of $\mathbb{R}^n$ there exists an intrinsic measure $\sigma \equiv \sigma_{\mathcal{M}}$ which is defined as follows: on the neighbourhood U projecting homeomorphically on the hyperplane Π, we consider the projection $P_\Pi(U)$ and the inverse homeomorphism P_Π^{-1}, and for any Borel set $E \subset U$ we put

$$\sigma(E) = \int_{P_\Pi(E)} \frac{D(P_\Pi^{-1}(x))}{D(x)}\, \mathrm{d}x \tag{2.7}$$

where $D(P_\Pi^{-1}(x))/D(x)$ is the Jacobian of the Lipschitzian mapping P_Π^{-1} at the point x and $\mathrm{d}x$ is Lebesgue measure on the hyperplane Π. The Jacobian exists almost everywhere and is uniformly bounded by M^{n-1} in view of (2.6). The measure σ is independent of the choice of the hyperplane Π and therefore defines an intrinsic measure on the whole manifold $\mathcal{M}$.

At almost every point of $\mathcal{M}$ (relative to the measure σ) there exists a uniquely determined tangent hyperplane.

All manifolds which we consider will be Lipschitzian graph manifolds. When a manifold is part of the boundary of a region $D \subset \mathbb{R}^n$, the boundary ∂D will be orientable and we fix on it the orientation consistent with the orientation of $\mathbb{R}^n$. At almost every point of ∂D, there is an inner normal vector on the normal line (orthogonal to the tangent hyperplane), and we denote by ν the *unit inner normal.*

We shall use manifolds only in the two following classes:

(1) bounded orientable piecewise smooth Lipschitzian manifolds of dimension $n-1$;

(2) bounded orientable piecewise flat manifolds of dimension $n-1$.

The second class is clearly a subclass of the first, but we emphasize it because it will play an important role in later constructions.

The following lemma is obvious.

Lemma 2.1. For any region $D \subset \mathbb{R}^n$ and every compact set $K \subset D$ there exists an open bounded set U with $K \subset U \subset \bar{U} \subset D$ such that ∂U belongs to either of the two classes (1) or (2).

We can now write the first Green formula

$$\int_{\partial D} \left\{u(x)\frac{\partial v(x)}{\partial \nu} - v(x)\frac{\partial u(x)}{\partial \nu}\right\} \mathrm{d}\sigma(x) + \int_D \{u(x)\Delta v(x) - v(x)\Delta u(x)\}\, \mathrm{d}x = 0 \tag{2.8}$$

where ∂D is a Lipschitzian graph, $\partial/\partial\nu$ is the inner normal derivative, and $\mathrm{d}\sigma(x)$ is the element of the intrinsic measure σ on ∂D. This formula is valid for all dimensions $n=1, 2, \ldots$.† In (2.8) we replace u by $\Delta^l u$ and v by $\Delta^{p-1-l}v$ for $l=0, 1, \ldots, p-1$, add the p resulting formulas and obtain

$$(2.9)\quad \sum_{l=0}^{p-1}\int_{\partial D}\left(\Delta^l u\frac{\partial\Delta^{p-1-l}v}{\partial\nu}-\Delta^{p-1-l}v\frac{\partial\Delta^l u}{\partial\nu}\right)d\sigma+\int_D(u\Delta^p v-v\Delta^p u)\,dx=0.$$

Here the functions u, v are in the class $\mathscr{C}^{2p}$ in an open neighbourhood of $\bar{D}$ and ∂D is in one of the classes (1) or (2) defined above.

We now consider the second Green formula. Denote by ω_n the area of the unit sphere in $\mathbb{R}^n$ where

$$\omega_n=\frac{2\pi^{n/2}}{\Gamma(n/2)},$$

and put

$$(2.10)\qquad \Omega_n=\begin{cases}1/(n-2)\omega_n & \text{for}\quad n\neq 2\\ 1/\omega_n & \text{for}\quad n=2.\end{cases}$$

By the standard procedure, if $y\in\mathbb{R}^n\backslash\bar{D}$ we replace $v(x)$ in (2.9) by $\imath_p(x-y)$, and if $y\in D$ we take a small ball $B_R(y)\subset\overline{B_R(y)}\subset D$ and replace D in (2.9) by $D'=D\backslash\overline{B_R(y)}$. In the second case, taking the limit as R approaches zero, we obtain by a standard argument the second Green formula

$$(2.11)\quad \Omega_n\sum_{l=0}^{p-1}\int_{\partial D}\left(\Delta^l u(x)\frac{\partial\imath_{l+1}(x-y)}{\partial\nu_x}-\imath_{l+1}(x-y)\frac{\partial\Delta^l u(x)}{\partial\nu}\right)d\sigma(x)$$
$$-\Omega_n\int_D\imath_p(x-y)\Delta^p u(x)\,dx=\begin{cases}u(y) & \text{if}\quad y\in D\\ 0 & \text{if}\quad y\notin\bar{D}.\end{cases}$$

Remark 2.1. The usual second Green formula is (2.11) for $p=1$.

Remark 2.2. If u is polyharmonic of degree p in D, (2.11) shows immediately that u is analytic in D, as was stated in §1.

† When $n=1$ and $D=(a, b)$, we adopt the conventions that

$$\int_{\partial D}f(x)\,d\sigma(x)=f(b)+f(a),$$

$$\left.\frac{\partial}{\partial\nu}\right|_{x=a}=\left.\frac{d}{dx}\right|_{x=a},\quad\text{and}\quad\left.\frac{\partial}{\partial\nu}\right|_{x=b}=-\left.\frac{d}{dx}\right|_{x=b}.$$

Formula (2.8) and all succeeding formulas in this section and the next are valid for $n=1$. Note that the area of the unit 'sphere' is $\omega_1=2$, and formula (2.10) below gives $\Omega_1=-\frac{1}{2}$.

The classical Green function $G(x, y)$ for the ball $B_R(a)$ is†

$$G(x,y)=\imath_1(x-y)-\imath_1\left(\frac{|y-a|}{R}(x-y^*)\right) \tag{2.12}$$

$$=\imath_1(x-y)-\imath_1\left(\frac{|x-a|}{R}(x^*-y)\right)$$

where

$$x^*=\frac{R^2}{|x-a|^2}(x-a)+a.$$

For details see, for example, Helms [47]. This function is defined and analytic for all x and y in $\mathbb{R}^n$ except for $y=x$ and $y=x^*$. Both expressions in (2.12) can be used to calculate $G(x, y)$ except when one of the points x or y is the centre a, in which case one expression is undetermined and the other gives $G(x, y)$. From (2.12) it follows immediately that

(2.13) $G(x, y)=G(y, x)$, G is harmonic in x and y

(2.14) $G(x, y)=0$ for $x\in\partial B_R(a)$ and $G(x, y)>0$ for $x, y\in B_R(a)$‡

$$\begin{cases}\dfrac{\partial G(x,y)}{\partial\nu_x}=0 & \text{for}\quad x,y\in\partial B_R(a),\qquad x\neq y\\[2ex] \dfrac{\partial G(x,y)}{\partial\nu_x}>0 & \text{for}\quad x\in\partial B_R(a),\qquad y\in B_R(a)\end{cases} \tag{2.15}$$

where $\partial/\partial\nu_x$ is the derivative with respect to x in the direction of the inner normal.

Now consider (2.9) for $p=1$, $D=B_R(a)$,

$$v=\imath_1\left(\frac{|y-a|}{R}(x-y^*)\right)$$

and y fixed in $B_R(a)$. Subtracting (2.9) from formula (2.11) with $p=1$, $D=B_R(a)$, we obtain from (2.14)

$$u(y)=\Omega_n\int_{\partial B_R(a)}\frac{\partial G(x,y)}{\partial\nu_x}u(x)\,d\sigma(x) \tag{2.16}$$

$$-\Omega_n\int_{B_R(a)}G(x,y)\Delta u(x)\,dx.$$

This is *Green's formula* for a ball.

† In Chapter VI, §6, we shall define and use more general Green functions.

‡ For $n=1$, $G(x, y)<0$ for x and y in $B_R(a)$. However, formulae (2.16) and (2.16′) are valid because $\Omega_1<0$.

By calculating $\partial G(x, y)/\partial \nu_x$ for $x \in \partial B_R(a)$ and by using (2.3) and (2.3′) we obtain

$$u(y) = \frac{1}{\omega_n} \int_{\partial B_R(a)} \frac{R^2 - |y-a|^2}{R\,|x-y|^n}\, u(x)\, d\sigma(x) \tag{2.16′}$$
$$- \Omega_n \int_{B_R(a)} G(x, y) \Delta u(x)\, dx.$$

The function $(R^2 - |y-a|^2)/R\,|x-y|^n$ is the *Poisson kernel* of $B_R(a)$.

3. Estimates

We shall derive local estimates for the derivatives of the functions $\Delta^l u$. Let us first assume that $u \in \mathscr{C}^2(\overline{B_R(a)})$ and that

$$|u(x)| \le M_0, \ |\Delta u(x)| \le M_1, \quad \text{for} \quad x \in \overline{B_R(a)}.$$

From the formula for Green's function and from Green's formula (equations (2.12) and (2.16′)), we see that the first derivatives $\partial/\partial y_k$ of the integrands in (2.16′) are absolutely integrable. Hence we can differentiate under the integral signs. For the Poisson kernel we obtain

$$\frac{\partial}{\partial y_k} \frac{R^2 - |y-a|^2}{R\,|x-y|^n}\bigg|_{y=a} = \frac{n(x_k - a_k)}{R^{n+1}}.$$

For Green's function we calculate the derivatives of the summands separately and obtain

$$\frac{\partial}{\partial y_k} \Omega_n \mathscr{e}_1(x-y)\bigg|_{y=a} = \frac{1}{\omega_n} \frac{x_k - a_k}{|x-a|^n}$$

$$\frac{\partial}{\partial y_k} \Omega_n \mathscr{e}_1\left(\frac{|x-a|}{R}(x^* - y)\right)\bigg|_{y=a} = \frac{1}{\omega_n} \frac{|x-a|^2 (x_k^* - a_k)}{R^{2+n}} = \frac{1}{\omega_n} \frac{(x_k - a_k)}{R^n}.$$

Since the derivative in the direction of any unit vector τ is given by

$$\frac{\partial}{\partial \tau} = \sum_{j=1}^{n} \tau_j \frac{\partial}{\partial y_j},$$

substituting the above expressions in (2.16′) and using

$$\sum_{j=1}^{n} \tau_j (x_j - a_j) \le |x-a|$$

show that

$$\left|\frac{\partial u(y)}{\partial \tau}\right|_{y=a} \le \frac{n}{R} M_0 + \frac{n}{n+1} R M_1. \tag{3.1}$$

Denote by $d(y, A)$ the minimum distance from the point y to the closed set A.

Proposition 3.1. Let $u \in \mathscr{C}^2(D)$ and suppose that $\|u\|_D \leqslant M_0$, $\|\Delta u\|_D \leqslant M_1$.† For any point $y \in D$ and any unit vector τ we have

$$\left|\frac{\partial u(y)}{\partial \tau}\right| \leqslant \frac{n}{d(y, \partial D)} M_0 + \frac{n}{n+1} d(y, \partial D) M_1. \tag{3.2}$$

Proof. For any R such that $\overline{B_R(y)} \subset D$ we can apply (3.1) and then let R increase to $d(y, \partial D)$. ‖

Proposition 3.2. Let $u \in \mathscr{C}^{2p}(D)$, set $M_l = \|\Delta^l u\|_D$, $l = 0, 1, \ldots, p$, and let q be a non-negative integer with $l + q \leqslant p$. For fixed $a \in D$ put $\rho = d(a, \partial D)/(q+1)$. If $y \in B_\rho(a)$ then

$$\left|\frac{\partial^q \Delta^l u(y)}{\partial y_{k_1} \ldots \partial y_{k_q}}\right| \leqslant \sum_{j=0}^{q} \frac{n^q M_{l+j}}{(n+1)^j \rho^{q-2j}} A'(q, j) \tag{3.3}$$

where $A'(q, j)$ is the coefficient of ξ^{q-j} in the Taylor expansion

$$(\xi + q + 1)(\xi + q) \ldots (\xi + 2) = \sum_{j=0}^{q} A'(q, j) \xi^{q-j}.$$

Proof. For the sequence of concentric balls $B_{\rho(q+1-i)}(a)$, $i = 1, 2, \ldots, q$, note that if $y \in B_{\rho(q+1-i)}(a)$ then $\rho < d(y, \partial B_{\rho(q+2-i)}(a)) < \rho(q+2-i)$. From Proposition 3.1 applied to the function $\Delta^l u$ in $B_{\rho(q+1)}(a)$ we obtain

$$\left|\frac{\partial \Delta^l u(y)}{\partial y_{k_1}}\right| \leqslant \frac{n}{\rho} M_l + \frac{n}{n+1} (q+1) \rho M_{l+1}, \quad y \in B_{\rho q}(a). \tag{3.4}$$

We shall prove that for $i = 1, 2, \ldots, q$ we have

$$\left|\frac{\partial^i \Delta^l u(y)}{\partial y_{k_1} \ldots \partial y_{k_i}}\right| \leqslant \sum_{j=0}^{i} \frac{n^i}{(n+1)^j} \rho^{2j-i} M_{l+j} A''(q, i, j), \quad y \in B_{\rho(q+1-i)} \tag{3.5}$$

where $A''(q, i, j)$ is the coefficient of ξ^{i-j} in the Taylor expansion

$$(\xi + q + 1)(\xi + q) \ldots (\xi + q + 2 - i) = \sum_{j=0}^{i} A''(q, i, j) \xi^{i-j}.$$

For $i = 1$, (3.5) is true by virtue of (3.4). Assuming now that (3.5) holds for i, we shall prove it for $i+1$. Using (3.4) with $y \in B_{\rho(q-i)}(a)$ we obtain

$$\left|\frac{\partial^{i+1} \Delta^l u(y)}{\partial y_{k_1} \ldots \partial y_{k_{i+1}}}\right| \leqslant \frac{n}{\rho} \left\|\frac{\partial^i \Delta^l u}{\partial y_{k_1} \ldots \partial y_{k_i}}\right\|_{B_{\rho(q+1-i)}(a)} + \frac{n}{n+1} (q+1-i) \rho \left\|\frac{\partial^i \Delta^{l+1} u}{\partial y_{k_1} \ldots \partial y_{k_i}}\right\|_{B_{\rho(q+1-i)}(a)}. \tag{3.6}$$

† The symbol $\|f\|_D$ means the supremum of f on D.

Using (3.5) to estimate the norms on the right-hand side of (3.6) we obtain (3.5) for $i+1$ with coefficients

$$A''(q, i+1, 0) = 1, A''(q, i+1, j) = A''(q, i, j) + (q+1-i)A''(q, i, j-1)$$

for $1 \leqslant j \leqslant i$, and

$$A''(q, i+1, i+1) = (q+1-i)A''(q, i, i).$$

Thus the coefficients $A''(q, i+1, j)$ are the Taylor coefficients in the expansion of $(\xi+q+1)(\xi+q) \dots (\xi+q+1-i)$. Therefore the inequality (3.5) is valid for $i \leqslant q$, and setting $i = q$ and noting that $A''(q, q, j) = A'(q, j)$, we obtain the conclusion of the proposition. ||

The next task is to derive estimates for the fundamental solutions $\imath_l$, defined by formulas (2.3) and (2.3′), and for their derivatives. Here and in later calculations we use three formulas involving the gamma function, often without making explicit reference:

(3.7) $\Gamma(x) = x^{x-\frac{1}{2}}e^{-x}(2\pi)^{\frac{1}{2}}e^{\theta(x)/12x}$, $x > 0$, $0 < \theta(x) < 1$ (Stirling's formula)

(3.7′) $\Gamma(z)\Gamma(z+\frac{1}{2}) = (2\pi)^{\frac{1}{2}}2^{\frac{1}{2}-2z}\Gamma(2z)$, for all complex z such that the arguments are not poles of Γ (Legendre's formula)

(3.7″) $\Gamma(z)\Gamma(-z) = -\dfrac{\pi}{z \sin \pi z}$, for complex z with $z \neq 0, \pm 1, \pm 2, \dots$.

From (2.2), (2.2′), and (2.2″), and using Stirling's formula (3.7) and (3.7″) we find that there exist positive constants C_1 and C_2 such that

$$(3.8) \qquad 0 < C_1 \frac{|l+2-n/2|^{(n-3)/2}+1}{(2l)!} \leqslant \begin{Bmatrix} 1/|\gamma_l| \\ 1/|\gamma_l'| \end{Bmatrix} \leqslant C_2 \frac{|l+2-n/2|^{(n-3)/2}+1}{(2l)!}.$$

In most of our applications, only an upper estimate is needed and the inequality

$$(3.9) \qquad \begin{Bmatrix} 1/|\gamma_l| \\ 1/|\gamma_l'| \end{Bmatrix} \leqslant C_n \frac{(l+1)^{n/2}}{(2l)!}, \qquad l \geqslant 1$$

suffices, where C_n is a positive constant depending only on the dimension n.

The constants c_{l-m} in equations (2.3′) and (2.4), and $c_{l-m} - 1/2(l-m)$

which will appear below, can be estimated as follows. By (2.4) we have

$$\frac{1}{2(l-m)} \leqslant c_{l-m} \leqslant \sum_{k=1}^{l-1} \frac{1}{k}, \quad \text{for} \quad l > m.$$

This last sum is majorized by $1+\log l$, and hence

$$0 \leqslant c_{l-m} - \frac{1}{2(l-m)} \leqslant c_{l-m} < 1 + \log l, \qquad l > m. \tag{3.10}$$

The fundamental solutions $\mathscr{k}_l$ are linear combinations of r^{2l-n} and $r^{2l-n} \log r$. In order to calculate arbitrary derivatives, first observe that

$$\frac{\partial}{\partial x_j} r^k = k x_j r^{k-2}$$

$$\frac{\partial}{\partial x_j} (r^k \log r) = x_j r^{k-2}(k \log r + 1)$$

for every real number k. If P_0 and Q_0 are homogeneous polynomials of degree λ, then

$$\frac{\partial}{\partial x_j} (P_0 r^k \log r + Q_0 r^k) = P r^{k-2} \log r + Q r^{k-2}$$

where P and Q are homogeneous polynomials of degree $\lambda + 1$. Any such polynomial has at most $n^{\lambda+1}$ distinct terms. Moreover, if every coefficient of P_0 and Q_0 is bounded in modulus by M, then every coefficient of P and Q is bounded in modulus by $M(2\lambda + |k| + 1)$. Now if q is a multi-index and if P_0 and Q_0 are constants, it follows that

$$D^q(P_0 r^k \log r + Q_0 r^k) = (P_q \log r + Q_q) r^{k-2|q|} \tag{3.11}$$

where P_q and Q_q are homogeneous polynomials of degree $|q|$, each with at most $n^{|q|}$ distinct terms, and each coefficient bounded in modulus by

$$M(|k|+1)(|k|+3) \ldots (|k|+2\,|q|-1).$$

Using Legendre's formula (3.7′), we can write this bound

$$M \frac{2^{|q|} \Gamma(|q| + (|k|+1)/2)}{\Gamma((|k|+1)/2)}. \tag{3.12}$$

Since the power of r in $\mathscr{k}_l$ is $k = 2l - n$, so that $|k| \leqslant 2l + n$, we can replace $|k|$ in (3.12) by $2l + n$ and obtain an upper estimate. Combining this with formulae (3.8)–(3.12) we have proved

Proposition 3.3. There exists a constant $C>0$ such that for all multi-indices q, all $l \geq 1$, and all $x \in \mathbb{R}^n \setminus \{0\}$,

$$(3.13) \quad |D^q \imath_l| \leq C \frac{\Gamma(|q|+l+(n+1)/2)}{\Gamma(l+(n+1)/2)} (2n)^{|q|} \frac{|l+1-n/2|^{(n-3)/2}+1}{(2l-2)!} \times r^{2l-n-|q|}(|\log r|+1+\log l)$$

$$\leq C \frac{\Gamma(|q|+l+(n+1)/2)}{\Gamma(l+(n+1)/2)} (2n)^{|q|} \frac{|l+1-n/2|^{(n-3)/2}+1}{(2l-2)!} \times r^{2l-n-|q|-1}(|\log r|+1)(1+\log l)$$

$$\leq C \frac{\Gamma(|q|+l+(n+1)/2)}{\Gamma(l+(n+1)/2)} (2n)^{|q|} \frac{|l+1-n/2|^{(n-3)/2}+1}{(2l-2)!} \times r^{2l-n-|q|-1}(r+1)^2 l$$

$$\leq \frac{\pi_q(l)}{(2l-2)!} r^{2l-n-|q|-1}(r^2+1)$$

where $\pi_q(l)$ is a polynomial in l of degree less than $|q|+n/2$, the coefficients of which depend only on $|q|$ and n. If x is replaced by $x-y$ then some or all of the differentiations may be with respect to the components of y. If n is odd, the factor $(|\log r|+1)(1+\log l)$ can be omitted. In $\mathbb{C}^n \setminus \{x : r(x)=0\}$ the inequality

$$(3.14) \qquad |D^q \imath_l| < \frac{\pi_q(l)}{(2l-2)!} |r|^{2l-n-|q|}(|\log r|+1)(1+\log l)$$

can be applied to the analytic continuations of $\imath_l$ and r.

Corollary 3.1. If τ is any unit vector then $|\partial \imath_l/\partial \tau|$ has the bound given in (3.13) *or* (3.14) *with $|q|=1$.*

The proof follows from the fact that $\imath_l(x)$ depends only on $|x|$, which is invariant under rotations about the origin. Thus an estimate in terms of $|x|$ in any one direction will hold in every direction.

4. Analytic and holomorphic functions of several variables

We begin this section by reviewing terminology and some basic facts concerning analytic and holomorphic functions of several variables (see, for example, Fuks [38], Grauert and Fritzsche [45], Hörmander [50], or Narasimhan [67]). We use the term *analytic* to refer to functions defined (initially) on domains in $\mathbb{R}^n$, and *holomorphic* to refer to those in $\mathbb{C}^n$. Next we take up questions of analytic continuation, and finally we review some properties of entire functions.

Definition 4.1. A (real or complex valued) function u defined on an open set $D \subset \mathbb{R}^n$ is *analytic* on D if for each $x^0 \in D$

$$u(x) = \sum_{|q|=0}^{\infty} a_q (x - x^0)^q \tag{4.1}$$

where the series is uniformly absolutely convergent on a neighbourhood of x^0 in D. Equivalently, u is analytic on D if $u \in \mathscr{C}^\infty(D)$ and for every compact set $K \subset D$ there exist positive constants C_K and R_K such that

$$|D^q u(x)| \leqslant C_K q!\, R_K^{-|q|} \tag{4.2}$$

for all $x \in K$ and all multi-indices q.

It follows that the coefficients a_q in (4.1) are given by

$$a_q = \frac{D^q u(x^0)}{q!}.$$

By (4.2), at every point $x^0 \in K$ the Taylor series (4.1) is uniformly absolutely convergent on $\overline{B_R(x^0)}$ for every $R < \min(R_K^{-1}, d(x^0, \partial D))$.

Definition 4.2. A complex valued function u defined on an open set $D \subset \mathbb{C}^n$ is *holomorphic* on D if the equivalent conditions (4.1) and (4.2) hold. In (4.2) the differentiation is with respect to the complex variables.

For the complex case, holomorphy in each variable separately is equivalent to holomorphy by a classical theorem of Hartogs, a fact which is not true in the analytic case. Another very useful property of holomorphic functions which is not shared by analytic functions is the following.

Proposition 4.1. If a sequence $\{u_n\}$ of holomorphic functions converges uniformly on every compact subset of an open set D, then the limit function is holomorphic on D.

We denote by $\mathscr{D}(x; R)$ the polydisc in $\mathbb{C}^n$ with centre $x = (x_1, \ldots, x_n)$ and polyradius $R = (R_1, \ldots, R_n)$, that is

$$\mathscr{D}(x; R) = \{z \in \mathbb{C}^n : |z_j - x_j| < R_j,\ j = 1, 2, \ldots, n\}.$$

In case $R_1 = R_2 = \ldots = R_n = R$, we denote the polydisc by $P(x; R)$.

Cauchy Theorem

(a) (*Cauchy integral formula*) *If u is holomorphic in $\mathscr{D}(x^0; R)$ and continuous on $\overline{\mathscr{D}(x^0; R)}$ then for all $z \in \mathscr{D}(x^0; R)$*

$$u(z) = \frac{1}{(2\pi i)^n} \int_{|\zeta_1 - x_1^0| = R_1} \cdots \int_{|\zeta_n - x_n^0| = R_n} \frac{u(\zeta)}{(\zeta_1 - x_1^0) \ldots (\zeta_n - x_n^0)} \, d\zeta_1 \ldots d\zeta_n$$

where $\zeta = (\zeta_1, \ldots, \zeta_n) \in \mathbb{C}^n$.

(b) (*Cauchy bounds*) *Under the hypothesis of* (a),

$$|D^q u(x^0)| \leq q!\, R^{-q} \max_{z \in \mathscr{D}(x^0, R)} |u(z)|$$

for every multi-index q.

(c) *If u is holomorphic on a neighbourhood of* x^0 *and if* $|D^q u(x^0)| \leq Cq!\, R^{-q}$ *holds for all multi-indices q with the constant C independent of q, then u is holomorphic on* $\mathscr{D}(x^0; R)$ *and the Taylor series*

$$\sum_q \frac{D^q u(x^0)}{q!} (z - x^0)^q$$

converges uniformly absolutely on every compact subset of $\mathscr{D}(x^0; R)$ *to the function u.*

Definition 4.3.

(a) A function u which is holomorphic on $\mathbb{C}^n$ is called an *entire* (holomorphic) *function.*

(b) A function u is *meromorphic* in a domain $D \subset \mathbb{C}^n$ if there exist holomorphic functions v and w on D, $w \not\equiv 0$, such that $u = v/w$ on $D \backslash N$, where N is the zero set of w.

(c) A meromorphic function $u = v/w$ is a *rational function* if v and w are polynomials.

Remark 4.1. A function analytic on all of $\mathbb{R}^n$ is not necessarily the restriction of an entire holomorphic function (for example, $u(x) = 1/(1+|x|^2)$, $x \in \mathbb{R}^n$). For this reason, the term *entire* will always refer to holomorphic functions.

An analytic (or holomorphic) function u in a domain $D \subset \mathbb{R}^n$ (or $\mathbb{C}^n$) presents at a point $x' \in \partial D$ a *non-removable singularity* if there is no Taylor development centred at x' which is valid in some neighbourhood U of x' and which gives the function u on $D \cap U$. All other points of ∂D are *removable singularities.* We can always extend a function u to a domain $D' \supset D$ such that u is analytic (holomorphic) on D' and D' contains all the removable singular points of ∂D. A non-removable singularity x' of an analytic (holomorphic) function u in D is called *polar* if there exists a non-identically vanishing analytic (holomorphic) function v in D such that the product uv has at x' a removable singularity.

If u is a holomorphic function in a domain $D \subset \mathbb{C}^n$, we can consider its analytic continuation outside D, if such continuation exists. To do this we first define analytic continuation along a path. An *admissible path* Γ is a continuous mapping of an interval $[\alpha, \beta] \subset \mathbb{R}^1$ into $\mathbb{C}^n$ such that

(1) $\Gamma(\alpha) \in D$;

(2) At each point $\Gamma(\tau)$, $\tau \in [\alpha, \beta]$, there exists a Taylor development

$\sum_q a_q(\tau)\{x-\Gamma(\tau)\}^q$, valid in some complex ball $\mathscr{B}_{R(\tau)}(\Gamma(\tau))$, such that at $\tau=\alpha$ this is the Taylor development of u about $\Gamma(\alpha)$;

(3) There exists a positive number R_1 (depending on the path and on u) such that for any τ_1 and τ_2 in $[\alpha, \beta]$ with $0<\tau_2-\tau_1<R_1$ the Taylor developments for τ with $\tau_1 \leqslant \tau \leqslant \tau_2$ define a (well-determined) single-valued function on $\bigcup_{\tau\in[\tau_1,\tau_2]} \mathscr{B}_{R(\tau)}(\Gamma-\tau))$.

Under these conditions we obtain a well-determined analytic function in $\mathscr{B}_{R(\beta)}(\Gamma(\beta))$ which will be called *the function obtained by analytic continuation of u from D to $\Gamma(\beta)$ along the path* Γ. It is easily shown that for each admissible path we can choose a constant R such that $0<R\leqslant R(\tau)$.

All Taylor developments† obtained by analytic continuation along all admissible paths will be considered as points forming the *analytic Weierstrassian manifold of u*. If the centre of such a Taylor development is x^0, the corresponding point in the manifold will be denoted $\hat{x}^0$. A fundamental system of neighbourhoods of $\hat{x}^0$ is defined as follows. If the Taylor development $\sum_q a_q(x-x^0)^q$ corresponding to $\hat{x}^0$ converges in a ball $\mathscr{B}_R(x^0)$, then all Taylor developments of the function $\sum_q a_q(x-x^0)^q$ with centres $x\in B_R(x^0)$ will form a neighbourhood of $\hat{x}^0$. The analytic Weierstrassian manifold is then a topological complex n-dimensional connected manifold. The projection P, assigning to $\hat{x}$ the centre x of its Taylor development, is an injection locally (but in general, not globally), and defines the analytic structure. The projection P assigns the same projection to at most enumerably many points. The whole domain D can be considered as a subset of the manifold. The original function u can be extended from D to the manifold by defining $\hat{u}(\hat{x})=a_0$, where a_0 is the zero-coefficient of the Taylor development determined by $\hat{x}$. In the analytic structure of the manifold, $\hat{u}$ is the analytic extension of u to the whole manifold.

If u is an analytic function on a domain $D\subset\mathbb{R}^n$ it can be extended canonically to a holomorphic function on a domain $D'\subset\mathbb{C}^n$ such that $D'\cap\mathbb{R}^n=D$. We construct this extension as follows. For each $x^0\in D$, let R^0 be the largest radius such that $B_{R^0}(x^0)\subset D$ and such that the Taylor series

$$\sum_q \frac{D^q u(x^0)}{q!}(x-x^0)^q$$

equals $u(x)$ in $B_{R^0}(x^0)$, the convergence being absolute and uniform on compact sets. (If $R^0=\infty$, replace the real variable x by the complex variable z and the extension of u to an entire holomorphic function is accomplished without further discussion.) In the Taylor series, if we

† Two Taylor developments are considered the same if they have the same coefficients a_q and the same centres x^0.

replace x by $z \in \mathbb{C}^n$, the resulting Taylor series is absolutely and uniformly convergent on compact subsets of $\mathscr{B}_{R^0}(x^0) \subset \mathbb{C}^n$ and defines a holomorphic function there. If x^0 varies in D, the complex Taylor developments define a single-valued holomorphic function on

$$D' = \bigcup_{x^0 \in D} \mathscr{B}_{R^0}(x^0).$$

To see this, suppose that

$$\sum_q \frac{D^q u(x')}{q!} (x - x')^q$$

and

$$\sum_q \frac{D^q u(x'')}{q!} (x - x'')^q$$

are two such developments valid in the complex balls $\mathscr{B}_{R'}(x')$ and $\mathscr{B}_{R''}(x'')$ respectively. If

$$z^0 = x^0 + iy^0 \in \mathscr{B}_{R'}(x') \cap \mathscr{B}_{R''}(x''),$$

then

$$x^0 \in B_{R'}(x') \cap B_{R''}(x'')$$

and

$$|y^0|^2 < \min\{R'^2 - |x^0 - x'|^2, R''^2 - |x^0 - x''|^2\} \leq (R^0)^2,$$

where R^0 is the radius corresponding to x^0. Thus the values given by each of the Taylor developments about x' and x'' at the point z^0 are the same as the Taylor development

$$\sum_q \frac{D^q u(x^0)}{q!} (x - x^0)^q \quad \text{at} \quad x = z^0.$$

The function so defined is the *canonical holomorphic extension* of u and it will be denoted by the same letter. This construction has two important uses in our work. First, if u is analytic on a domain $D \subset \mathbb{R}^n$, it can be automatically considered as a single-valued holomorphic function on a domain $D' \subset \mathbb{C}^n$, $D' \cap \mathbb{R}^n = D$, where D' is well determined and the extended function is the only holomorphic function on D' whose restriction to D is the given function u, that is, no branching can occur in D'. This means that analytic continuation, discussed before, can be considered to be continuation from the real domain $D \subset \mathbb{R}^n$ into the space $\mathbb{C}^n$. Thus the concept of analytic Weierstrassian manifold has a meaning for such a function u, as does the notion of unlimited analytic continuation (to be defined below). Second, suppose that A is a linear differential operator in the real variables $x_1,\ldots, x_n$, with analytic coefficients defined on a domain $D \subset \mathbb{R}^n$. Each of the coefficients can be extended to its canonical domain, and then all coefficients are uniquely determined on the intersection $G \subset \mathbb{C}^n$ of these domains. Moreover, since the formal rules of differentiation hold when the real variables are replaced by complex variables, making this replacement we obtain a well-determined extension A' of the differential operator A to the domain G. Now suppose that u is a function analytic on D and that D' is its domain of canonical extension. The extension of Au to $D' \cap G$ is $A'u$. It is precisely for this reason that we use

$$\Delta = \sum_{j=1}^{n} \frac{\partial^2}{\partial z_j^2}$$

as the Laplace operator in the space $\mathbb{C}^n$. This operator is, of course, quite different from the Laplace operator in $2n$ real variables.

Let u be analytic on $D \subset \mathbb{R}^n$ (or holomorphic on $D \subset \mathbb{C}^n$). We shall call the analytic extension of u to the corresponding analytic Weierstrassian manifold the *complete analytic extension* of u. If the projection P is a homeomorphism on a domain $\hat{G}$ on the corresponding Weierstrassian manifold, we obtain a *branch* of the complete function u. Identifying the points $\hat{x} \in \hat{G}$ with their projections on $\mathbb{R}^n$ (or $\mathbb{C}^n$), we obtain an analytic (or holomorphic) function on $G = P\hat{G}$. If $D \subset G$, we shall call the resulting function a *branch of u extended to G*.

As an example of the preceding constructions consider the function

$$r(x) = \left(\sum_{j=1}^{n} x_j^2\right)^{\frac{1}{2}}.$$

This symbol actually gives two distinct analytic functions on $\mathbb{R}^n \backslash \{0\}$ which differ by sign. For each choice of sign, the canonical holomorphic extension is obtained by using the domains $\mathcal{B}_{|x|/\sqrt{2}}(x)$ for each $x \in \mathbb{R}^n \backslash \{0\}$, and the extended function is

$$r(z) = \left(\sum_{j=1}^{n} z_j^2\right)^{\frac{1}{2}}$$

with the corresponding sign. The two determinations of $r(z)$ determine the same analytic Weierstrassian manifold and the same complete analytic function. The Weierstrassian manifold is projected by P onto the domain $G \subset \mathbb{C}^n$ defined by

$$\sum_{j=1}^{n} z_j^2 \neq 0,$$

and each point of G is the projection of exactly two points of the manifold. Thus the function is double-valued over G.

As another example, let

$$u(x)=\left(1-\sum_{j=1}^{n} x_j^2\right)^{-1}.$$

This formula defines two distinct analytic functions, one in $B_1(0)$ and the other in $\mathbb{R}^n\backslash\overline{B_1(0)}$. The holomorphic extensions of both these functions are part of the one holomorphic function

$$u(z)=\left(1-\sum_{j=1}^{n} z_j^2\right)^{-1}$$

defined on the domain

$$G=\left\{z\in\mathbb{C}^n:\ \sum_{j=1}^{n} z_j^2\neq 1\right\},$$

and this domain G is the analytic Weierstrassian manifold of u.

Let u be holomorphic on a domain $D\subset\mathbb{C}^n$ and let G be a domain with $D\subset G$. We say that u has *unlimited analytic continuation in* G if for every point $a\in D$ and every path in G starting at a, u has an analytic continuation along this path. In view of our discussion of the canonical holomorphic extension of an analytic function u defined in $D\subset\mathbb{R}^n$, it follows that if G is a domain in $\mathbb{C}^n$ with $D\subset G$, we can speak of unlimited analytic continuation in G of a function analytic in D. In most of our later work, this will be the point of view.

Finally we recall some properties of the growth of entire functions (for the case of one variable see Boas [18], and for more than one variable see Fuks [38] or Levin [64]). In our discussion, the results for one variable can be obtained from the $\mathbb{C}^n$ case by taking $n=1$. The definitions of order and type of an entire function, which follow, can be given using the maximum modulus of u on various kinds of domains, and in general the types (though not the orders) are dependent on the choice of domain (see Fuks [38]). For our purposes it is convenient to use the polydisc

$$P(0;R)=\{z\in\mathbb{C}^n:|z_j|<R,\ j=1,\ldots,n\}$$

as our domain.

For an entire function u, let

$$M(u;R)=\sup_{z\in P(0;R)}|u(z)|.$$

Definition 4.4. An entire holomorphic function u is of *exponential order* ρ, $0\leqslant\rho\leqslant\infty$, if

$$\limsup_{R\to\infty}\frac{\log\log M(u;R)}{\log R}=\rho.$$

Equivalently, u is of exponential order $\rho < \infty$ if for every $\varepsilon > 0$ and for no $\varepsilon < 0$ there exists a positive constant C_ε such that $M(u; R) \leqslant C_\varepsilon \exp(R^{\rho+\varepsilon})$.

Remark 4.2. We have used the term 'exponential order' rather than the standard term 'order' so that we can distinguish it from the 'Laplacian order' of an analytic function, to be defined in Chapter II, §1.

Definition 4.5. An entire holomorphic function u of positive finite exponential order ρ is of *order* ρ *and type* τ, $0 \leqslant \tau \leqslant \infty$, if

$$\limsup_{R\to\infty} R^{-\rho} \log M(u; R) = \tau.$$

Equivalently, u is of order ρ and type τ if for every $\varepsilon > 0$ and for no $\varepsilon < 0$ there exists a constant C_ε such that $M(u; R) \leqslant C_\varepsilon \exp\{(\tau+\varepsilon)R^\rho\}$.

Remark 4.3. We could define type for order $\rho = 0$, but then we would always have $\tau = \infty$ unless u is constant. For positive finite order ρ, functions of type τ are often said to be of maximal type if $\tau = \infty$, minimal type if $\tau = 0$, and medium type otherwise.

Theorem 4.1. Let u be entire with Taylor development

$$u(z) = \sum_q a_q z^q.$$

(a) *The exponential order of u is* ρ *if and only if*

$$\rho = \limsup_{|q|\to\infty} \frac{|q| \log |q|}{-\log |a_q|},$$

where the quotient is taken to be 0 *if* $a_q = 0$.

(b) *The exponential order of u is* ρ, $0 < \rho < \infty$, *if and only if for every* $\varepsilon > 0$ *and for no* $\varepsilon < 0$

$$\limsup_{|q|\to\infty} |a_q|^{1/|q|} \, |q|^{1/(\rho+\varepsilon)} = 0.$$

(c) *If u has positive finite exponential order* ρ, *then the type is* τ *if and only if*

$$(e\rho\tau)^{1/\rho} = \limsup_{|q|\to\infty} |a_q|^{1/|q|} \, |q|^{1/\rho}$$

where $\tau = \infty$ *if the* lim sup *is* ∞.

Remark 4.4. If u is an entire function and if we set $v(z) = u(z + z^0)$ for a fixed z^0, then $v(z)$ has the same exponential order and type as u. Letting

$$M(u; z^0; R) = \sup_{z\in P(z^0;R)} |u(z)|$$

where

$$P(z^0; R) = P(0; R) + z^0,$$

we have

$$M(v; R) = M(u; z^0; R).$$

If z^0 varies in a compact set $K \subset \mathbb{C}^n$, the constants C_ε in Definitions 4.4 and 4.5 need not depend on z^0, but only on K and ε. Similarly, the Taylor expansion in Theorem 4.1 can be centred at z^0 and the limit suprema are uniform on K.

II

POLYHARMONICITY AND ANALYTIC CONTINUATION

1. Polyharmonicity: Laplacian order and type

Until the mid-1930s, a function was said to be 'polyharmonic' if it was annihilated by some power of the Laplace operator. The monograph of Nicolescu [68] summarizes the situation at that time. The picture was changed substantially by the realization that important representations and analytic continuation properties were also valid in a much larger class of functions, which was then given the same name. This was identified in Aronszajn [6] as the class of 'harmonic functions of infinite order'.

Definition 1.1. A function u which is $\mathscr{C}^\infty$ on a domain D in $\mathbb{R}^n$ is *polyharmonic on* D if its sequence of Laplacians $\{\Delta^p u\}_{p=0}^\infty$ satisfies the condition that $\{[|\Delta^p u|/(2p)!]^{\frac{1}{2p}}\}_{p=0}^\infty$ approaches zero uniformly on every compact subset of D.

The members of the older smaller class are now called 'polyharmonic of finite degree' and are described in Chapter I, §§1 and 2. If $\Delta u = Cu$ for some constant C, then u is polyharmonic. Solutions of far more general differential equations are also polyharmonic, as are the restriction of entire functions on $\mathbb{C}^n$. The concept of Laplacian order and type, to be introduced now, creates a hierarchy in which many familiar classes of functions are located according to the growth of their sequences of Laplacians.

Definition 1.2. On a domain D in $\mathbb{R}^n$, a C^∞ function u is of *Laplacian order* ρ *and type* τ if for every compact $K \subset D$ and every $\varepsilon > 0$ there exists a constant $C_{K,\varepsilon}$ such that

$$\|\Delta^p u\|_K \leqslant C_{K,\varepsilon}(2p)!^{1-1/\rho}(\tau+\varepsilon)^{2p} \quad \text{for all } p,$$

that is

$$\limsup_{p\to\infty}\left(\frac{\|\Delta^p u\|_K}{(2p)!^{1-1/\rho}}\right)^{\frac{1}{2p}} \leqslant \tau.$$

Order $\rho=\infty$ is allowed by setting $1-1/\rho=1$. Thus, the allowable Laplacian orders with type are $0<\rho\leqslant\infty$, and the types so allowed are $0\leqslant\tau<\infty$. The classification is extended to allow Laplacian order ρ and type $\tau=\infty$ if for every compact $K\subset D$ and every $\varepsilon>0$ there exists a finite τ_K

and a constant $C_{K,\varepsilon}$ such that

$$\|\Delta^p u\|_K \leq C_{K,\varepsilon}(2p)!^{1-1/\rho}(\tau_K+\varepsilon)^{2p} \quad \text{for all } p.$$

In this case

$$\limsup_{p\to\infty}\left(\frac{\|\Delta^p u\|_K}{(2p)!^{1-1/\rho}}\right)^{\frac{1}{2p}} \leq \tau_K < \infty.$$

Clearly it is equivalent to say that u has Laplacian order ρ and type ∞ on D if every compact subdomain of D has a neighbourhood on which u has Laplacian order ρ and some finite type.

It is also said that u is simply of *Laplacian order* ρ on D if for every compact $K\subset D$ and every $\varepsilon>0$

$$\|\Delta^p u\|_K \leq C_{K,\varepsilon}(2p)!^{1-1/(\rho+\varepsilon)}.$$

The orders so allowed are $0 \leq \rho < \infty$.

The polyharmonic functions are recognizable as those functions of Laplacian order ∞ and type 0. Functions that are polyharmonic of finite degree have Laplacian order 0 (for which no type is allowed).

Remark 1.1. The constants $C_{K,\varepsilon}$ can be omitted from the criteria in Definition 1.2 by taking $C_{K,\varepsilon}=1$ in those applications where it suffices to consider only large values of p.

Example 1.1. Every function which is analytic on $D\subset\mathbb{R}^n$ is of Laplacian order ∞ and type ∞ there. (The converse, showing that every function having Laplacian order ∞ and type ∞ on D is necessarily analytic there, is proved in §2, Theorem 2.2.)

To see this, observe that such a function u has analytic continuation as an (unbranched) holomorphic function on the canonical domain $D'\subset\mathbb{C}^n$ (see Chapter I, §4). If K is a compact subset of D', then by the Cauchy bounds there exist numbers C_K and R_K such that

$$\|D^q u\|_K \leq C_K q!\, R_K^{-|q|} \quad \text{for each } q=(q_1,\ldots,q_n)$$

where $D^q u$ is any derivative of u.

$$\Delta^p u = \sum_{|q|=p}\binom{p}{q}D^{2q}u = \sum_{|q|=p}\frac{p!}{q_1!\ldots q_n!}D^{2q_1,\ldots,2q_n}u$$

and so

$$\begin{aligned}\|\Delta^p u\|_K &\leq \sum_{|q|=p}\binom{p}{q}C_K(2q)!\,R_K^{-2|q|}\\ &\leq C_K(2p)!\,R_K^{-2p}\sum_{|q|=p}\binom{p}{q}\\ &= C_K(2p)!\left(\frac{1}{R_K}\right)^{2p}n^p.\end{aligned}$$

In the definition of Laplacian order $\rho=\infty$ and type $\tau=\infty$, one could take $\tau_K=n^{\frac{1}{2}}/R_K$.

Example 1.2. If u satisfies $\Delta u=Cu$ on a domain $D\subset\mathbb{R}^n$, where C is some constant, then $\Delta^p u=C^p u$ and so

$$\|\Delta^p u\|_K\leq\|u\|_K\,(|C|^{\frac{1}{2}})^{2p},\quad \text{for all } p.$$

Thus u is of Laplacian order 1 and type $|C|^{\frac{1}{2}}$. More generally, solutions of the equation

$$\Delta^m u=Cu$$

are of Laplacian order 1 and type $|C|^{\frac{1}{2m}}$. This result is generalized in Chapter VI, §5. In Chapter VI, §7, it is shown that analytic solutions of the heat equation have 'sections' with Laplacian order 2 and some type.

Like Exx. 1.1 and 1.2, the examples which follow locate certain well-known classes of functions in the hierarchy of Laplacian order and type. Most of the proofs must be given in other sections, where techniques for defining, representing, or making estimates of functions are presented. The major collections of examples are concentrated in Chapter IV, §1, and Chapter V, §3. Chapter VI is also largely concerned with examples.

The radius function in $\mathbb{R}^n$ is $|x|$, that is

$$r(x)=\left(\sum_{j=1}^{n}x_j^2\right)^{\frac{1}{2}}.$$

The functions r^k and $r^k\log r$ will be used extensively, beginning in the next section. The evaluations made in Chapter I, §1, are the basis for the following summary.

Example 1.3. $\Delta^p r^k=0$ only in special cases. If k is a non-negative even integer then r^k is polyharmonic of finite degree $p=k/2+1$ on $\mathbb{R}^n$. If $k=2p-n$, then r^k is of finite degree p on $\mathbb{R}^n\backslash\{0\}$ provided that n is odd and p is a positive integer. If n is even, then p must be positive and $k=2p-n$ must be negative. In contrast, if n is even and p is a positive integer such that $k=2p-n\geq 0$, then $r^k\log r$ is polyharmonic of finite degree p on $\mathbb{R}^n\backslash\{0\}$.

These functions are fundamental solutions of the equation $\Delta^p u=0$. If v is a function of the form r^k or $r^k\log r$ but not one of those just described, then v is not polyharmonic of any finite degree on any domain in $\mathbb{R}^n$. In fact, it is easily seen from equations (1.5) and (1.7), Chapter I, that on any compact K in $\mathbb{R}^n\backslash\{0\}$ such a v has the property

$$\lim_{p\to\infty}\left(\frac{\|\Delta^p v\|_K}{(2p)!}\right)^{\frac{1}{2p}}=\frac{1}{R_K}$$

where R_k is the distance from the origin to the nearest point of K. Thus v will have Laplacian order ∞ and type ∞ of $\mathbb{R}^n\backslash\{0\}$ and it will not be polyharmonic on any domain in $\mathbb{R}^n$.

Definition 1.3. For a given domain D the following *function spaces* are defined: $\mathcal{P}(D)$ is the class of functions which are polyharmonic of finite degree on D, $\mathcal{P}^{\rho,\tau}(D)$ is the class of functions having Laplacian order ρ and type τ there, and $\mathcal{P}^{\rho}(D)$ is the class of functions having Laplacian order ρ (but perhaps no type). Often the reference to D will be omitted.

Proposition 1.1.

(a) *If* $0<\rho<\rho'<\infty$ *and* $0<\tau<\tau'<\infty$ *then*

$$\mathcal{P}\subset\mathcal{P}^{0}\subset\mathcal{P}^{\rho,0}\subset\mathcal{P}^{\rho,\tau}\subset\mathcal{P}^{\rho,\tau'}\subset\mathcal{P}^{\rho,\infty}\subset\mathcal{P}^{\rho}\subset\mathcal{P}^{\rho',0}\subset\mathcal{P}^{\infty,0}$$
$$\subset\mathcal{P}^{\infty,\tau}\subset\mathcal{P}^{\infty,\tau'}\subset\mathcal{P}^{\infty,\infty}.$$

(b) *If* $\rho<\infty$ *and* $0\leqslant\tau\leqslant\infty$ *then* $\mathcal{P}^{\rho}=\bigcap_{\sigma>\rho}\mathcal{P}^{\sigma,\tau}=\bigcap_{\sigma>\rho}\mathcal{P}^{\sigma}$.

If $\rho<\infty$ *and* $\tau<\infty$ *then* $\mathcal{P}^{\rho,\tau}=\bigcap_{\sigma>\tau}\mathcal{P}^{\rho,\sigma}$.

The proof consists only of applying Definitions 1.2 and 1.3.

Definition 1.4. On a domain D a function u has *least Laplacian order*

$$\lambda=\inf\{\rho;\, u\in\mathcal{P}^{\rho,\tau}(D) \text{ for some } \tau\}.$$

The least order λ is *unattainable* if u does not also have Laplacian order λ and some finite type τ on D. If the least order λ is attainable on D, then the *least type* is

$$\mu=\inf\{\tau;\, u\in\mathcal{P}^{\lambda,\tau}(D)\},$$

which is a finite number.

There is no need for a concept of 'unattainable' least type, but examples of functions with unattainable least order will be given in Chapter V, §3. It will also be shown that $\mathcal{P}^{\lambda}\backslash\mathcal{P}^{\lambda,\infty}$ is not empty, and some of its members turn out to be interesting. Similarly, in Chapter V, §3, it will be shown that there are functions of Laplacian order zero which are not polyharmonic of finite degree. For example,

$$u(x)=\sum_{k=0}^{\infty}\{r(x)\}^{2k}(2k)!^{-k}$$

can be shown to have this property (see Chapter V, §3, Ex. 3.3). Thus $\mathcal{P}^{0}\backslash\mathcal{P}$ is not empty either.

Example 1.4. If D is an interval in $\mathbb{R}^1$ then a function u is polyharmonic on D if and only if it is the restriction of an entire function v on $\mathbb{C}^1$, and the least Laplacian order of u is exactly the (exponential) order of v as an entire function. This biunique correspondence is the origin of the terms Laplacian order and type, although it holds only in dimension 1. The situation in $\mathbb{R}^n$ is different if $n\geqslant 2$, there being only inequalities in one direction (see Chapter VI, §2). For $n=1$ the full picture is given in the following theorem.

Theorem 1.1. Let u be defined on an interval D in $\mathbb{R}^1$.

(a) *u has Laplacian order ∞ and type $\tau<\infty$ on D (that is $u\in\mathcal{P}^{\infty,\tau}(D)$) if and only if u has analytic continuation to the domain in $\mathbb{C}^1$ consisting of all points lying within $1/\tau$ of D.*

(b) *u is polyharmonic on D (that is $u\in\mathcal{P}^{\infty,0}(D)$) if and only if it is in the restriction of an entire function on $\mathbb{C}^1$.*

(c) *u has finite least Laplacian order ρ (and finite least type τ) on D if and only if u is the restriction of an entire function with order ρ (and type τ^ρ/ρ).*

(d) *u is polyharmonic with finite degree p if and only if it is a polynomial of degree at most $2p-1$.*

Proof. Since $\Delta^p u=\mathrm{d}^{2p}u/\mathrm{d}x^{2p}$, the proof of (d) is obvious.

For each subset $K\subset D$ and each $R>0$ let

$$K(R)=\{z\in\mathbb{C};|z-x|<R \text{ for some } x\in K\}.$$

By the Cauchy bounds, if v is any function which is holomorphic on a neighbourhood of $\overline{K(R)}$ then

$$\|\Delta^p v\|_K=\left\|\frac{\mathrm{d}^{2p}v}{\mathrm{d}x^{2p}}\right\|_K\leqslant(2p)!\,\|v\|_{K(R)}\,R^{-2p},\quad\text{for all } p. \tag{1.1}$$

(a) It follows that the restriction of v to D has Laplacian order ∞ and type $\tau\leqslant 1/R$ if v is holomorphic on $D(R)$, that is u has type τ on D if v is holomorphic on $D(1/\tau)$ as described in the hypotheses.

For the converse, let K be a compact subset of D. By Proposition 3.1, Chapter I, if y lies in the interior of K then

$$\left|\frac{\mathrm{d}^{2p+1}u(y)}{\mathrm{d}y^{2p+1}}\right|\leqslant\frac{1}{d(y,\partial K)}\|\Delta^p u\|_K+\frac{d(y,\partial K)}{2}\|\Delta^{p+1}u\|_K$$

for all p. If u is of Laplacian order ρ and type τ on D, then for each $\varepsilon>0$

$$\left|\frac{\mathrm{d}^{2p+1}u(y)}{\mathrm{d}y^{2p+1}}\right|\leqslant C_1(2p)!^{1-1/\rho}(\tau+\varepsilon)^{2p}+C_2(2p+2)!^{1-1/\rho}(\tau+\varepsilon)^{2p+2}$$
$$\leqslant(2p+1)!^{1-1/\rho}(\tau+2\varepsilon)^{2p+1}\text{ for all large } p.$$

It follows from this and from the definition of Laplacian order and type that

$$\limsup_{k\to\infty}\left|\frac{1}{k!^{1-1/\rho}}u^{(k)}(y)\right|^{1/k}\leqslant\tau$$

for each $y\in K$, and hence for each $y\in D$. For $\rho=\infty$, the expression on the left is the reciprocal of the radius of convergence R of Taylor's series for

u about y. Hence $1/R \leq \tau$ or $R \geq 1/\tau$. Thus u has analytic continuation to $D(1/\tau)$. This proves (a), of which (b) is the special case $\tau = 0$.

(c) If u has least Laplacian order $\rho < \infty$ and least type $\tau < \infty$, then u is polyharmonic and thus is the restriction of an entire function v. By the last inequality

$$\limsup_{k\to\infty} \left| \frac{1}{k!^{1-1/\rho}} v^{(k)}(y) \right|^{1/k} = \tau,$$

which is the criterion for v to have exponential order ρ and type τ^ρ/ρ, by Theorem 4.1, Chapter I.

Conversely, if u is entire with exponential order ρ and type σ, if $\varepsilon > 0$, and if K is compact in $\mathbb{C}^1$, then there exists a constant C such that

$$\|u\|_{K(R)} \leq C \exp\{(\sigma+\varepsilon)R^\rho\}, \quad \text{for all } R > 0$$

(see Definition 4.4, Chapter I). By the Cauchy bounds (1.1), it follows that

$$\|\Delta^p u\|_K \leq (2p)!\, R^{-2p} C \exp\{(\sigma+\varepsilon)R^\rho\}, \quad \text{for all } R > 0, \text{ all } p.$$

For fixed p, the left-hand side is independent of R and the right-hand side is minimized at $R^\rho = 2p/\rho(\sigma+\varepsilon)$. Thus

$$\begin{aligned}\|\Delta^p u\|_K &\leq (2p)! \left(\frac{2p}{\rho(\sigma+\varepsilon)}\right)^{-2p/\rho} Ce^{2p/\rho} \\ &= (2p)!^{1-1/\rho} \left\{\frac{(2p)!}{(2p)^{2p}}\right\}^{1/\rho} C[e\rho(\sigma+\varepsilon)]^{2p/\rho}, \quad \text{for all } p \\ &\leq (2p)!^{1-1/\rho} C_1 \{\rho(\sigma+\varepsilon)\}^{2p/\rho}, \quad \text{for all } p,\end{aligned}$$

by Stirling's formula. Since K may be in D it follows that the restriction of u to D has Laplacian order ρ and type $\tau \leq (\rho\sigma)^{1/\rho}$. By the opposite inequality, which is proved above, $\sigma = \tau^\rho/\rho$.

By the nesting properties for orders and types, both Laplacian and exponential, it follows that the entire function u has (exponential) order ρ if and only if the restriction of u to D has least Laplacian order ρ. ‖

Thus, the terminology of Laplacian orders and types extends that for entire functions of one variable. It also extends the ordering of the Gevrey classes (see Chapter VI, §4).

Just as $\Delta^p(u_1+u_2) = \Delta^p u_1 + \Delta^p u_2$ leads to the conclusion that the finite degree of a sum is bounded by the finite degrees of the summands, so it also leads to the following inequalities.

Lemma 1.1. If u_1 and u_2 have least Laplacian orders λ_1 and λ_2 on a domain D, then their sum has least order $\lambda \leq \max\{\lambda_1, \lambda_2\}$; if u_1, u_2, and u_1+u_2 all have least Laplacian order λ on D and u_1 and u_2 have least types μ_1 and μ_2, then u_1+u_2 has least type $\mu \leq \max\{\mu_1, \mu_2\}$.

The inequalities are the best possible. It can certainly happen that $u_2 = -u_1$, for instance, so that $u_1 + u_2 = 0$.

It is clear that u and Δu must have the same Laplacian orders and types. The next result leads to Theorem 1.2, and generalizes this observation by showing that u and Au have the same properties if A is a member of a broad class of linear differential operators.

Lemma 1.2. Let $u \in \mathcal{P}^{\infty,\infty}(D)$ *and let* D^q *be any differentiation,* $D^q = \partial^{|q|}/\partial x_1^{q_1} \dots \partial x_n^{q_n}$.

(a) *If* u *is polyharmonic of finite degree* p *then* $D^q u$ *is also.*
(b) *If* u *is of Laplacian order* ρ *and type* τ, $0 < \rho \leq \infty$, $0 \leq \tau \leq \infty$, *then* $D^q u$ *is also.*
(c) *If* u *is of Laplacian order* ρ, $0 \leq \rho < \infty$, *then* D^q *is also.*

Proof. The proof of (a) is immediate since $\Delta^p D^q = D^q \Delta^p$, and that of (c) follows from (b) by taking an infimum (Proposition 1.1). It remains to show (b).

Let K be compact in D, let D_1 be a subdomain of D containing K and having compact closure in D, and let $\varepsilon > 0$. By the hypothesis,

$$\|\Delta^p u\|_{D_1} \leq (2p)!^{1-1/\rho}(\tau + \varepsilon)^{2p}$$

if p is large enough. Let x be any point of K and set $R = d(x, \partial D_1)/(|q| + 2)$, where $d(x, \partial D_1)$ is the distance from x to ∂D_1, the boundary of D_1. If $B = B_R(x)$ is the ball in $\mathbb{R}^n$ of radius R about x, then $\bar{B} \subset D_1$, and Proposition 3.2, Chapter I shows that

$$\begin{aligned}\|\Delta^p D^q u\|_B &= \|D^q \Delta^p u\|_B \\ &\leq \sum_{k=0}^{|q|} C_1(n, |q|, k, R)\,\|\Delta^{p+k} u\|_{D_1} \\ &\leq C_2(2p + 2\,|q|)!^{1-1/\rho}(\tau + \varepsilon)^{2p+2\,|q|}, \quad \text{for all large } p\end{aligned}$$

where C_1 and C_2 depend on B, D_1, and ε, but not on p or q. Since K can be covered by finitely many such balls,

$$\begin{aligned}\|\Delta^p D^q u\|_K &\leq C_3(2p + 2\,|q|)!^{1-1/\rho}(\tau + \varepsilon)^{2p+2\,|q|} \\ &\leq (2p)!^{1-1/\rho}(\tau + 2\varepsilon)^{2p}, \quad \text{for all large } p.\end{aligned}$$

Since $\varepsilon > 0$ may be arbitrarily small, this inequality suffices. ||

Theorem 1.2. Let u *be a function defined on a domain* $D \subset \mathbb{R}^n$ *and let* A *be a linear differential operator with polynomial coefficients whose degrees are bounded by* α.

(a) *If* u *is polyharmonic of degree* p, *then* Au *is polyharmonic of degree* $p + \alpha$.
(b) *If* u *has Laplacian order* ρ *and type* τ, $0 < \rho \leq \infty$, $0 \leq \tau \leq \infty$, *or just Laplacian order* ρ, $0 \leq \rho < \infty$, *then so does* Au.

Proof. Since the degree, order, and type of a sum are bounded by those of the summands, it suffices to prove the theorem for the special case in which $Au = x^\sigma D^q u$ where $\sigma = (\sigma_1, \ldots, \sigma_n)$, which is reduced by Lemma 1.2 to the case $Au = x^\sigma u$.

(a) The formula for the Laplacians of products (Lemma 1.1, Chapter I) can be applied to $x^\sigma u$ giving

$$\Delta^{p+|\sigma|}(x^\sigma u) = \sum_{k+l+|q|=p+|\sigma|} \frac{2^{|q|}(p+|\sigma|)!}{k!\,l!\,q!}(D^q \Delta^k x^\sigma)(D^q \Delta^l u).$$

Any non-vanishing terms in this formula must satisfy $k \geqslant 0$, $|q| + 2k \leqslant |\sigma|$, and $l < p$. Since these inequalities are incompatible with the index condition $k + l + |q| = p + |\sigma|$, all terms must be zero.

(b) The kind of argument just given in (a) must now be applied, for each σ, to

$$\Delta^p(x^\sigma u) = \sum_{k+l+|q|=p} \frac{2^{|q|}p!}{k!\,l!\,q!}(D^q \Delta^k x^\sigma)(D^q \Delta^l u). \tag{1.2}$$

Specifically, since each non-vanishing term must satisfy the inequality $|q| + 2k \leqslant |\sigma|$, it follows that k, l, and $|q|$ can each take no more than $|\sigma| + 1$ different values. This bound is independent of p. If $p > |\sigma|$, for each non-vanishing term we have

$$\begin{aligned} \frac{2^{|q|}p!}{k!\,l!\,q!} &\leqslant 2^{|\sigma|}p(p-1)\ldots(p-|\sigma|+1) \\ &\leqslant 2^{|\sigma|}p^{|\sigma|}. \end{aligned} \tag{1.3}$$

For any compact $K \subset D$, the finite collection of numbers $\|D^q \Delta^k x^\sigma\|_K$ has a finite maximum depending on σ, n, and K, but independent of q and k within the constraints explained above. Finally, because of the bound $|q| \leqslant |\sigma|$ it follows there are only finitely many functions $D^q u$ to consider, each of which has Laplacian order ρ and type τ, by Lemma 1.2. If τ is finite, then there exists C_1 dependent on K and any given $\varepsilon > 0$ such that

$$\|D^q \Delta^l u\|_K = \|\Delta^l D^q u\|_K \leqslant C_1 (2l)!^{1-1/\rho}(\tau+\varepsilon)^{2l} \quad \text{for all } l.$$

Combining with (1.3) shows that

$$\begin{aligned} &\frac{2^{|q|}p!}{k!\,l!\,q!}\|D^q \Delta^k x^\sigma\|_K \, \|D^q \Delta^l u\|_K \\ &\leqslant 2^{|\sigma|}p^{|\sigma|}C_2(2l)!^{1-1/\rho}(\tau+\varepsilon)^{2l} \\ &\leqslant \frac{C_3(2p)!^{1-1/\rho}(2p)^{|\sigma|}}{\{(2p)(2p-1)\ldots(2l+1)\}^{1-1/\rho}}\left(\frac{\tau+\varepsilon}{\tau+2\varepsilon}\right)^{2l}(\tau+2\varepsilon)^{2l} \\ &\leqslant C_3(2p)!^{1-1/\rho}(\tau+2\varepsilon)^{2l}(2p)^{|\sigma|}\left\{\frac{(2p)!}{(2p-2|\sigma|)!}\right\}^{1/\rho}\left(\frac{\tau+\varepsilon}{\tau+2\varepsilon}\right)^{2p-2|\sigma|} \end{aligned}$$

since $p \geq l \geq p - |\sigma|$ for each non-vanishing term. The inequality holds for all $p > |\sigma|$, and the product of the last three factors goes to zero. Since the number of non-zero terms of (1.2) is bounded, independent of p,

$$\|\Delta^p(x^\sigma u)\|_K \leq C_4(2p)!^{1-1/\rho}(\tau + 2\varepsilon)^{2p}, \quad \text{for all } p.$$

Since ε can be arbitrarily small, this proves the assertion for order ρ and finite type τ.

If τ is not finite then the argument can be applied using τ_K instead of τ (see Definition 1.2).

The assertion for order above follows from that for order and type by taking an infimum (Proposition 1.1). ‖

The theorem shows that these properties are inherited by the functions resulting from the use of these operators. However, the properties can vary somewhat between locations in D.

Example 1.5. Let $n = 3$ and let

$$u(x, y, z) = \exp\{zh(x + iy)\}$$

where h is the restriction of an entire function of one variable. By direct differentiation we see that $\Delta u = h^2 u$, $\Delta^2 u = h^4 u$, and in general $\Delta^p u = h^{2p} u$. Thus, for any compact K

$$\|\Delta^p u\|_K \leq \|u\|_K \, \|h\|_K^{2p}, \quad \text{for all } p,$$

and so, barring trivial cases, on a bounded domain D the least Laplacian order is 1 and the least type is $\|h\|_D$.

This example, like Ex. 1.3, shows that type is to some extent a local property. Results in Chapter IV and examples in Chapter V, §3, show that the same is true of order.

Definition 1.5. For a function $u \in \mathcal{P}^{\infty,\infty}(D)$, the *local Laplacian order* function λ is defined on D, its value $\lambda(x)$ at $x \in D$ being the infimum of all orders of u on all neighbourhoods of x. If x has a neighbourhood U on which the least Laplacian order is actually $\lambda(x)$, then the *local type* function μ is defined there by $\mu(x) = \infty$, unless x has a neighbourhood in U on which u has order $\lambda(x)$ and finite type, that is, the least order is attainable. In that case $\mu(x)$ is the infimum of all types of u on all neighbourhoods of x in U.

Proposition 1.2. Let $u \in \mathcal{P}^{\infty,\infty}(D)$.

(a) *If $x \in D$ then $\lambda(x) \geq \limsup_{y \to x} \lambda(y)$, so that λ is upper semi-continuous. The same is true of local type wherever it is defined.*

(b) *$\sup\{\lambda(x); x \in D\}$ is the least Laplacian order of u on D. The same is true of local type if it is defined on D.*

The proof of the proposition follows directly from the definitions. In Ex. 1.2, $\lambda(x)\equiv 1$ and $\mu(x)\equiv|C|^{\frac{1}{2}}$ in the first case, $\mu(x)\equiv|C|^{\frac{1}{2m}}$ in the second. In Ex. 1.4, λ and μ are constant because an entire function can be expanded in power series about any point in $\mathbb{C}^1$. In Exx. 1.3 and 1.5 it is equally clear how to evaluate the local order and type functions.

The behaviour of λ and μ under taking sums of functions is determined readily from Lemma 1.1 and the definitions.

Proposition 1.3. If u_1 and u_2 have local Laplacian order functions λ_1 and λ_2 on a domain D then their sum has local Laplacian order λ satisfying the constraints

$$\lambda(x)=\max\{\lambda_1(x),\lambda_2(x)\} \quad \textit{if} \quad \lambda_1(x)\neq\lambda_2(x)$$
$$\lambda(x)\leq\lambda_1(x) \quad \textit{if} \quad \lambda_1(x)=\lambda_2(x).$$

If λ_1, λ_2, and λ are all constant and equal on D, then the local type functions, if defined, also satisfy these constraints.

The study of local Laplacian order and type is resumed in Chapter IV.

Notes. In their general setting, the concepts of Laplacian order and type appear here for the first time. The class of polyharmonic functions, that is, functions of Laplacian order ∞ and type 0, was identified by Aronszajn [6] in 1935. The functions were called 'harmonic of infinite order'. A superfluous condition in the definition was removed in Aronszajn [7], and Lelong [59, 60] gave an alternative definition of the class by using integrals over compact sets of successive Laplacians. Lelong's definition led to a definition of distributions 'harmonic of infinite order' by Avanissian and Fernique [15], who proved that these distributions were actually functions (for further results along this line, see Avanissian [14]).

2. Serio-integral representation

It is well known that repeated applications of the second Green formula lead to equations representing a $\mathscr{C}^{2p}$ function on a bounded domain $D\subset\mathbb{R}^n$ with sufficiently smooth boundary. From formula (2.11), Chapter I,

$$\text{(2.1)}\quad \Omega_n\sum_{k=0}^{p-1}\int_{\partial D}\left\{\Delta^k u(x)\frac{\partial \imath_{k+1}(x-y)}{\partial\nu_x}-\imath_{k+1}(x-y)\frac{\partial\Delta^k u}{\partial\nu}(x)\right\}d\sigma(x)$$
$$-\Omega_n\int_D \imath_p(x-y)\Delta^p u(x)\,dx=\begin{cases}u(y), & y\in D\\ 0, & y\notin\bar{D}.\end{cases}$$

For this formula, $\{\imath_k\}_{k=1}^{p}$ must be a sequence of functions satisfying

$\Delta \imath_{k+1} = \imath_k$, $\Delta \imath_1 = 0$, and

$$\Omega_n \int_S \frac{\partial \imath_1}{\partial \nu}(x)\, d\sigma(x) = 1$$

where S is the unit $(n-1)$-sphere in $\mathbb{R}^n$. If u is polyharmonic of finite degree p then the volume integral will vanish.

The choice of $\imath_p$ determines each of the auxiliary functions $\imath_k = \Delta^{p-k}\imath_p$ as well. However, if the sequence is generated inductively beginning with a harmonic function $\imath_1$, then the auxiliary function $\imath_k$ is independent of the value of p at which the series (2.1) may terminate. A single infinite sequence $\{\imath_k\}_{k=1}^{\infty}$ is generated in this way in §2, Chapter I. In this section we consider the possibility of an infinite series analogous to (2.1). It appeared first in Aronszajn [6] following the realization that a condition such as polyharmonicity is sufficient to ensure that the sequence of volume integrals goes to zero uniformly on compact subsets of D.

In Definition 1.2 it can be observed that a function with Laplacian order ρ and type ∞ on a given domain G' has the same order but some finite type on each subdomain G with compact closure in G'. The following result begins a further development in this direction.

Theorem 2.1 (Local representation). Let u have Laplacian order ∞ and finite type $\tau > 0$ on some domain $G \subset \mathbb{R}^n$, $n \geq 1$. If D is any subdomain with compact closure in G, diameter $\operatorname{diam}(D) < 1/\tau$, *and piecewise smooth boundary, then*

$$(2.2)\quad \Omega_n \sum_{k=0}^{\infty} \int_{\partial D} \left\{ \Delta^k u(x) \frac{\partial \imath_{k+1}(x-y)}{\partial \nu_x} - \imath_{k+1}(x-y) \frac{\partial \Delta^k u}{\partial \nu}(x) \right\} d\sigma(x)$$

$$= \begin{cases} u(y), & y \in D \\ 0, & y \in E_\tau \end{cases}$$

where $\nu = \nu_x$ is the interior normal for ∂D at x,

$$E_\tau = \bigcap_{z \subset \bar{D}} B_{1/\tau}(z) \backslash \bar{D},$$

and $B_R(z)$ is the ball in $\mathbb{R}^n$ with radius R and centre z. E_τ is open and the series is uniformly convergent on compact subsets of $D \cup E_\tau$. (E_τ may be empty.)

Proof. Let K be any compact in $D \cup E_\tau$ and let $\varepsilon > 0$ be arbitrary, subject to the restriction that

$$(\tau + \varepsilon) R < 1$$

where

$$R = \sup\{|y - x| : y \in K, x \in \bar{D}\}$$

is the maximum distance from the points of K to points of $\bar{D}$.

Since u satisfies the inequalities for Laplacian order $\rho=\infty$ and type τ, by virtue of the estimates on $\imath_p$ (Proposition 3.3, Chapter I), the integrand of the volume integral in (2.1) satisfies

$$\begin{aligned}(2.3)\quad |\imath_p(x-y)\Delta^p u(x)| &\leqslant C\frac{(p+1)^{n/2}}{(2p-2)!}|x-y|^{2p-n}\\ &\quad\times(|\log|x-y||+1+\log p)(2p)!\,(\tau+\varepsilon)^{2p}\\ &\leqslant C(2p)^{n/2+3}\{R(\tau+\varepsilon)\}^{2p-n}\,|\log|x-y||\end{aligned}$$

for $y\in K$, $x\in D\backslash\{y\}$, and $p\geqslant n/2$. The logarithmic term on the right is integrable and the integral has a bound C' which is independent of y, that is

$$\left\|\Omega_n\int_D \imath_p(x-y)\Delta^p u(x)\,\mathrm{d}\sigma(x)\right\|_K \leqslant |\Omega_n|\,CC'(2p)^{n/2+3}\{R(\tau+\varepsilon)\}^{2p-n},\qquad p\geqslant n/2.$$

Since $R(\tau+\varepsilon)<1$ the right-hand side converges to zero. ‖

If u is polyharmonic, then it has Laplacian order $\rho=\infty$ and every type $\tau>0$. In this case the domain D may have arbitrarily large diameter, and the series will converge uniformly on compact subsets of

$$D\cup\left(\bigcup_{\tau>0}E_\tau\right)=\mathbb{R}^n\backslash\partial D.$$

In general, however, if u is simply a function having Laplacian order ∞ and type ∞ on a domain G_1, then by the definition of type ∞ each point has a neighbourhood on which u has finite type τ. It has a sub-neighbourhood which is a ball with diameter less than $1/\tau$. Thus each point has a neighbourhood on which u can be represented by (2.2). For this reason the representation is 'local'.

In Ex. 1.1 it was shown that every analytic function necessarily has Laplacian order ∞ and type ∞. The converse is possibly the most important single consequence of the Local Representation.

Theorem 2.2. For a function u to be analytic on a domain $G\subset\mathbb{R}^n$ it is necessary and sufficient that u have Laplacian order ∞ and type ∞ on G.

Proof. In view of Ex. 1.1, it remains to prove the sufficiency. If $z\in G$, then z has a neighbourhood U on which u has Laplacian order ∞ and finite type $\tau>0$, and another neighbourhood D with $\bar{D}\subset U$, whose boundary is piecewise smooth and whose diameter is less than $1/\tau$. Let R be one-third of the (shortest) distance from z to ∂D, a compact set. Let B and $\mathcal{B}$ be the balls in $\mathbb{R}^n$ and $\mathbb{C}^n$ respectively, with centre z and radius R.

For each $x \in \partial D$, both $\mathfrak{r}_k(x-y)$ and

$$\frac{\partial \mathfrak{r}_k(x-y)}{\partial \nu_x}$$

are analytic on B as functions of y. Therefore each term of the local representation (2.2) is also analytic on B, and the series converges uniformly on $\bar{B}$. It will be shown first that each term has analytic continuation to a holomorphic function on the simply connected domain $\mathscr{B} \subset \mathbb{C}^n$, and finally that the resulting series converges uniformly.

If $y \in \mathbb{C}^n$ then $y = \xi + i\eta$ where ξ and η are points of $\mathbb{R}^n$. If $x \in \mathbb{R}^n$, then

$$\begin{aligned} r^2(x-y) &= \sum_{j=1}^{n} \{x_j - (\xi_j + i\eta_j)\}^2 \\ &= \sum_{j=1}^{n} \{(x_j - \xi_j)^2 - \eta_j^2 - 2i(x_j - \xi_j)\eta_j\} \\ &= |x-\xi|^2 - |\eta|^2 - 2i(x-\xi, \eta) \end{aligned} \tag{2.4}$$

where $(x-\xi, \eta)$ is the inner product in $\mathbb{R}^n$. If $y = \xi + i\eta$ lies in $\mathscr{B}$, whose centre is z and whose radius is R, then

$$R^2 > \sum_{j=1}^{n} |y_j - z_j|^2 = |\xi - z|^2 + |\eta|^2.$$

If $x \in \partial D$ then $|x-z| \geqslant 3R$. Combining these two inequalities with (2.4) shows that

$$\begin{aligned} \mathscr{Re}(r^2(x-y)) &= |x-\xi|^2 - |\eta|^2 \\ &\geqslant (|x-z| - |z-\xi|)^2 - |\eta|^2 \\ &> (3R-R)^2 - R^2 \\ &> R^2, \end{aligned} \tag{2.5}$$

that is, if $x \in \partial D$, then $r^2(x-y)$ has positive real part for all $y \in \mathscr{B}$. Therefore its square root with positive real part defines on $\mathscr{B}$ a holomorphic function whose restriction to $B = \mathscr{B} \cap \mathbb{R}^n$ is $|x-y| = r(x-y)$. All the powers of the latter also have continuation as holomorphic functions on $\mathscr{B}$, so does $\log r(x-y)$, and so do the auxiliary functions $\mathfrak{r}_k(x-y)$ and their directional derivatives $\partial \mathfrak{r}_k(x-y)/\partial \nu_x$ for each $x \in \partial D$. Integration with respect to x over ∂D shows that each term of (2.2) has continuation as a function of y which is holomorphic on $\mathscr{B}$.

There remains the uniform convergence on $\mathscr{B}$. To show this we make bounds on $|r(x-y)|$ and then on the auxiliary functions for $x \in \partial D$ and

$y \in \mathcal{B}$. From (2.5) and (2.4),

$$(2.6)\qquad \begin{aligned} R < |r(x-y)| &= |r^2(x-y)|^{\frac{1}{2}} = ||x-\xi|^2 - |\eta|^2 - 2i(x-\xi, \eta)|^{\frac{1}{2}} \\ &\leq (|x-\xi|^2 + |\eta|^2 + 2|x-\xi|\,|\eta|)^{\frac{1}{2}} \\ &= |x-\xi| + |\eta| \\ &< 4R + R \\ &\leq 5/6\tau \end{aligned}$$

by the definition of R. From the Laplacian order and type of u, formula (2.6), and the bound (3.13), Chapter I, for any $\varepsilon > 0$ there exists C_ε such that

$$(2.7)\qquad \begin{aligned} &\left|\Delta^k u(x) \frac{\partial \imath_{k+1}}{\partial \nu_x}(x-y)\right| \\ &\leq C_\varepsilon (2k)!\,(\tau+\varepsilon)^{2k} C \frac{(k+1)^{1+n/2}}{(2k)!} |r(x-y)|^{2k-n+1} \\ &\quad \times \{1 + |\log r(x-y)|\}\{1+\log(k+1)\} \\ &\leq C'_\varepsilon (k+1)^{2+n/2} \left\{(\tau+\varepsilon)\frac{5}{6\tau}\right\}^{2k-n}, \qquad k \geq n/2 \\ &\leq C'_\varepsilon (k+1)^{2+n/2} \left(\frac{10}{11}\right)^{2k-n} \end{aligned}$$

where the last inequality holds for $0 < \varepsilon \leq \tau/11$.

Since $\bar{D} \subset U$, there exists a compact set $K \subset U$ with $\bar{D}$ in its interior, and for every $\varepsilon > 0$

$$\|\Delta^k u\|_K \leq C_\varepsilon (2k)!\,(\tau+\varepsilon)^{2k}, \quad \text{for all } k.$$

By Proposition 3.1, Chapter I,

$$\begin{aligned} \left\|\frac{\partial \Delta^k u}{\partial \nu}\right\|_{\partial D} &\leq C_1 C_\varepsilon (2k)!\,(\tau+\varepsilon)^{2k} + C_2 C_\varepsilon (2k+2)!\,(\tau+\varepsilon)^{2k+2} \\ &\leq C_3 C_\varepsilon (2k+2)!\,(\tau+\varepsilon)^{2k+2}, \quad \text{for all } k, \end{aligned}$$

where C_1 and C_2 are determined only by the dimension n and the minimum distance between ∂D and ∂K, which is positive. Combining this with (2.6) and the bound for $\imath_{k+1}$ in formula (3.13), Chapter I, gives bounds where $x \in \partial D$ and $y \in \mathcal{B}$ as follows:

$$\begin{aligned} &\left|\imath_{k+1}(x-y) \frac{\partial \Delta^k u}{\partial \nu}(x)\right| \\ &\leq C \frac{(k+1)^{n/2}}{(2k)!} |r(x-y)|^{2k-n+1} \{1 + |\log r(x-y)|\}\{1+\log(k+1)\} \\ &\quad \times C_\varepsilon (2k+2)!\,(\tau+\varepsilon)^{2k+2} \end{aligned}$$

$$\leqslant C''_\varepsilon (k+1)^{2+n/2} \left\{ (\tau+\varepsilon)\frac{5}{6\tau} \right\}^{2k-n}, \qquad k \geqslant n/2$$
$$\leqslant C''_\varepsilon (k+1)^{2+n/2} \left(\frac{10}{11}\right)^{2k-n}$$

where the last inequality holds for $0<\varepsilon\leqslant\tau/11$.

Using this inequality and (2.7) shows that

$$\left\| \int_{\partial D} \left\{ \Delta^k u(x) \frac{\partial \imath_{k+1}(x-y)}{\partial \nu_x} - \imath_{k+1}(x-y) \frac{\partial \Delta^k u}{\partial \nu}(x) \right\} \mathrm{d}\sigma(x) \right\|_{\mathcal{B}}$$
$$\leqslant C'''_\varepsilon \int_{\partial D} \mathrm{d}\sigma(x) \left(\frac{10}{11}\right)^{2k} (k+1)^{2+n/2}, \qquad k \geqslant n/2.$$

It follows that the series (2.2) is uniformly convergent on $\mathcal{B}$ by the Weierstrass M-test. Therefore the sum must be holomorphic on $\mathcal{B}$, and consequently u, its restriction to $\mathbb{R}^n$, must be analytic on $B = \mathcal{B} \cap \mathbb{R}^n$, a neighbourhood of the arbitrary point $z \in G$. ||

Corollary 2.1. A function u on a domain $G \subset \mathbb{R}^n$ is analytic if and only if each point $z \in G$ has a neighbourhood D on which u can be represented by the serio-integral representation (2.2). *If u is polyharmonic on G, then D can be any subdomain containing z, provided that D has compact closure in G and ∂D is piecewise smooth.*

The neighbourhood D is, of course, that appearing in Theorem 2.1 and in the proof of Theorem 2.2. That proof consisted of showing that u has analytic continuation from $D \subset \mathbb{R}^n$ to a domain in $\mathbb{C}^n$. It is reasonable to ask how far this continuation can be extended. Is there a largest domain in $\mathbb{C}^n$ to which all functions analytic on G have analytic continuation? There is not, but there is a domain, the 'harmonicity hull' of G, to which all functions polyharmonic on G have analytic continuation. This domain is determined by the continuation of the terms of the series and depends exclusively on the properties of the auxiliary functions. For polyharmonic functions, the resulting series converges uniformly on compacts. The results are summarized in Theorem 2.4.

In $\mathbb{R}^n$ let S be a compact piecewise smooth surface (of dimension $n-1$) with normal ν. If u is analytic on a neighbourhood G of S and k is a non-negative integer, the formula

$$I_k(S, u; y) = \Omega_n \int_S \left(\Delta^k u(x) \frac{\partial \imath_{k+1}(x-y)}{\partial \nu_x} - \imath_{k+1}(x-y) \frac{\partial \Delta^k u}{\partial \nu}(x) \right) \mathrm{d}\sigma(x)$$

defines on $\mathbb{R}^n \backslash S$ a function of y which is polyharmonic of finite degree $k+1$, since $\imath_{k+1}$ has that property. It is clear that I_k can have analytic continuation into $\mathbb{C}^n$ only along arcs on which $\imath_{k+1}(x-y)$ has continuation for every $x \in S$. By its definition (equations (2.3) and (2.3′), Chapter

I), $\imath_{k+1}$ is of the form

$$\text{either}\quad \frac{r^{2k+2-n}}{\gamma_k}\quad \text{or}\quad \frac{r^{2k+2-n}}{\gamma_k'}(c_{k+1-n/2}-\log r).$$

Thus I_k can have analytic continuation into $\mathbb{C}^n$ only on arcs beginning at points of $\mathbb{R}^n\backslash S$ and such that for each $x\in S$ the continuation of $r(x-y)$ does not vanish at any y on the arc. This restriction will become the defining property for the 'harmonicity hull' of $\mathbb{R}^n\backslash S$ (Definition 2.2) where I_k will be seen to have unlimited continuation.

However, the continuations of

$$r(x-y)=\{r^2(x-y)\}^{\frac{1}{2}}=\left\{\sum_{j=1}^{n}(x_j-y_j)^2\right\}^{\frac{1}{2}}$$

may branch. If we restrict continuation to arcs on which the values of $r^2(x-y)$ fall in the simply connected domain which results from cutting the complex plane along the closed negative real axis $\mathbb{R}^-=\{z: z\leqslant 0\}$, then $r(x-y)$ will have a continuation with positive real part for every $x\in S$. The domain constructed from such arcs will be called the 'reduced harmonicity hull' of $\mathbb{R}^n\backslash S$. On it both $r(x-y)$ and $\imath_{k+1}(x-y)$ have continuations which define holomorphic functions of y, for each $x\in S$. It will be shown that I_k defines a holomorphic function there.

For $n=1$ no extended construction is necessary. By virtue of Theorem 1.1, every function polyharmonic on an interval D in $\mathbb{R}^1$ has analytic continuation to $\mathbb{C}^1$. Thus, the harmonicity hull of an interval is $\mathbb{C}^1$, but otherwise the harmonicity hull of $\mathbb{R}\backslash S$ is $\mathbb{C}\backslash S$. The reduced harmonicity hull can also be given at once. For $n\geqslant 2$, however, a longer construction is required.

Definition 2.1. If x is any point of $\mathbb{C}^n$, $n\geqslant 2$, the *isotropic cone through* x is

$$V(x)=\{y\in\mathbb{C}^n\colon r^2(x-y)=0\}$$

and the *full cone through* x is

$$V'(x)=\{y\in\mathbb{C}^n\colon r^2(x-y)\leqslant 0\}.$$

The intersections of these cones with $\mathbb{R}^n$ are denoted $\sum(x)$ and $\sum'(x)$ respectively.

Remark 2.1. It should be observed that $V(x)$ and $V'(x)$ are translates of $V(0)$ and $V'(0)$, and that

$$y\in V'(0)\quad \text{if and only if}\quad ty\in V'(0)\quad \text{for all}\quad t>0$$
$$y\in V(0)\quad \text{if and only if}\quad \zeta y\in V(0)\quad \text{for all}\quad \zeta\in\mathbb{C}.$$

Thus, 'cone' is the correct term.

Furthermore, V and V' have a symmetry property, in that

$$x \in V(y) \quad \text{if and only if} \quad y \in V(x)$$
$$x \in V'(y) \quad \text{if and only if} \quad y \in V'(x).$$

This symmetry yields at once the following elementary calculational tool, which is used very frequently in the construction of harmonicity hulls.

Remark 2.2. If $x \in \mathbb{R}^n$, then

$$y \in V(x) \quad \text{if and only if} \quad x \in \sum(y)$$
$$y \in V'(x) \quad \text{if and only if} \quad x \in \sum{}'(y).$$

In fact it will follow that if D is a domain in $\mathbb{R}^n$, $n \geq 2$, then the harmonicity hull of D will be a subset of $\{z \in \mathbb{C}^n : \sum(z) \cap \partial D = \phi\}$ and the reduced harmonicity hull will be a subset of $\{z \in \mathbb{C}^n : \sum'(z) \cap \partial D = \phi\}$.

Proposition 2.1. Let $z \in \mathbb{C}^n$ with $n \geq 2$ and let $z = \xi + i\eta$ with ξ and η in $\mathbb{R}^n$. If $\eta \neq 0$ and if $\prod(z)$ is the hyperplane in $\mathbb{R}^n$ through ξ and orthogonal to η then:

(a) $\sum(z)$ is the $(n-2)$-sphere in $\prod(z)$ with centre ξ and radius $|\eta|$, whereas $\sum'(z)$ is the closed $(n-1)$-ball $\bar{B}_{|\eta|}(\xi) \cap \prod(z)$, which is the convex hull of $\sum(z)$;

(b) $\prod(z)$, $\sum(z)$, and $\sum'(z)$ can all be provided with orientations consistent with the standard orientation in $\mathbb{R}^n$.

Thus Σ (or Σ') becomes a one-to-one representation of $\mathbb{C}^n \backslash \mathbb{R}^n$ in the collection of oriented $(n-2)$-spheres (or the collection of oriented closed $(n-1)$-balls) in $\mathbb{R}^n$. The representation extends to points z in $\mathbb{R}^n$, which are represented by themselves as spheres (or balls) with zero radius.

Proof. The hyperplane $\prod(\xi + i\eta)$ consists of the points x for which the inner product $(x-\xi, \eta)$ vanishes. The ball $\bar{B}_{|\eta|}(\xi) \cap \prod(z)$ consists of those points for which, in addition,

$$|x-\xi|^2 - |\eta|^2 \leq 0.$$

Since

$$\begin{aligned} r^2(x-z) &= r^2(x-(\xi+i\eta)) \\ &= \sum_{j=1}^{n} \{(x_j-\xi_j)^2 - \eta_j^2 - 2i(x_j-\xi_j)\eta_j\} \\ &= |x-\xi|^2 - |\eta|^2 - 2i(x-\xi, \eta), \end{aligned} \tag{2.8}$$

it follows that this set is exactly that on which $r^2(x-z)$ is real and not positive in $\mathbb{R}^n$, that is

$$\bar{B}_{|\eta|}(\xi) \cap \prod(\xi+i\eta) = V'(\xi+i\eta) \cap \mathbb{R}^n = \sum{}'(\xi+i\eta).$$

$\sum(\xi+i\eta)$ is its boundary in the hyperplane $\prod(\xi+i\eta)$.

The orientation for $\prod(z)$ and $\sum'(z)$ is that which together with $\eta \in \mathbb{R}^n$ generates the standard orientation for $\mathbb{R}^n$. Since $z = \xi + i\eta$ and $\bar{z} = \xi - i\eta$ produce the same hyperplane except for orientation, the closed $(n-1)$-balls $\sum'(z)$ and $\sum'(\bar{z})$ can be distinguished only by their orientation but the correspondence Σ' is one-to-one.

The same is true for Σ, for which the orientation is that which together with the inward normal generates the orientation assigned to $\prod(z)$. ‖

The plane $\prod(z)$ varies continuously with z on $\mathbb{C}^n \backslash \mathbb{R}^n$, but spheres and balls can be said to vary continuously on all of $\mathbb{C}^n$, since for them the difficulty with $z \in \mathbb{R}^n$ is inessential. There the radius vanishes. It follows that if F is any closed set in $\mathbb{R}^n$, $n \geq 2$, then

$$\bigcup_{x \in F} V(x) \quad \text{and} \quad \bigcup_{x \in F} V'(x)$$

are both closed in $\mathbb{C}^n$. If G is an open set in $\mathbb{R}^n$, then

$$\mathbb{C}^n \backslash \bigcup_{x \notin G} V(x) \quad \text{and} \quad \mathbb{C}^n \backslash \bigcup_{x \notin G} V'(x)$$

will be open sets in $\mathbb{C}^n$.

Definition 2.2. Let G be an open set in $\mathbb{R}^n$. If $n \geq 2$, the *harmonicity hull* $\tilde{G}$ and the *reduced harmonicity hull* $\hat{G}$ respectively are the unions of those components of the open sets

$$\mathbb{C}^n \backslash \bigcup_{x \notin G} V(x) \quad \text{and} \quad \mathbb{C}^n \backslash \bigcup_{x \notin G} V'(x)$$

which intersect G. If instead $n = 1$, then

$$\hat{G} = \{z \in \mathbb{C}^1 \colon \mathcal{R}e\, z \in G\},$$

while $\tilde{G} = \mathbb{C}^1$ if G is connected. If G is not connected then $\tilde{G} = \mathbb{C}^1 \setminus \partial G$.

It is obvious that $G \subset \hat{G} \subset \tilde{G}$.

We shall see that each component of $\hat{G}$ or $\tilde{G}$ intersects $\mathbb{R}^n$ in some one component of G if $n \geq 2$.

Definition 2.3. If Γ is an arc in $\mathbb{C}^n$, $n \geq 2$, starting at $x \in \mathbb{C}^n$ and ending at $z \in \mathbb{C}^n$, then the representation Σ transforms Γ into a *tube* $\Sigma \circ \Gamma$ *in* $\mathbb{R}^n$, starting at $\sum(x)$ and ending at $\sum(z)$. Similarly, the representation Σ' transforms Γ into a *full tube* $\Sigma' \circ \Gamma$ *in* $\mathbb{R}^n$.

These tubes will become the principal tool in describing harmonicity hulls in dimension $n \geq 2$, as the following proposition and its proof suggest.

Proposition 2.2. Let G be a domain in $\mathbb{R}^n$, $n \geq 2$.

(a) *A point* $z \in \mathbb{C}^n$ *lies in* $\tilde{G}$ *if and only if there is a tube in* G *starting at a point of* G *and ending at* $\sum(z)$.

(b) *A point* $z \in \mathbb{C}^n$ *lies in* $\hat{G}$ *if and only if* $\sum'(z) \subset G$.

(c) *If* G_1 *is a domain in* $\mathbb{R}^n$ *disjoint from* G *then* $\tilde{G}$ *and* $\tilde{G}_1$ *are disjoint domains and* $(G \cup G_1)^{\sim} = \tilde{G} \cup \tilde{G}_1$. *The same is true of* $\hat{G}$ *and* $\hat{G}_1$. *Also,* $G \subset \hat{G} \subset \tilde{G}$ *and* $G = \mathbb{R}^n \cap \hat{G} = \mathbb{R}^n \cap \tilde{G}$.

Proof. Since G is a connected subset of $\tilde{G}$, each component of which intersects G, it follows that $\tilde{G}$ is connected. Hence, a point z lies in $\tilde{G}$ if and only if there exists an arc Γ in $\tilde{G}$ beginning in G and terminating at z. The same can be said of $\hat{G}$.

(a) If $y \in \tilde{G}$, then

$$y \notin \bigcup_{x \notin G} V(x),$$

so that

$$\sum(y) = \mathbb{R}^n \cap V(y) \subset G.$$

This must be so for every y on any arc Γ in $\tilde{G}$. The arc is therefore transformed by Γ into a tube lying entirely in G. Since this argument is reversible the converse is also proven.

(b) If $z = \xi + i\eta$ then on $0 \leq t \leq 1$ the function $\Gamma(t) = \xi + it\eta$ parametrizes a straight line segment from $\xi \in \mathbb{R}^n$ to z. Of course, $\Sigma' \circ \Gamma(t)$ lies entirely in $\sum'(z)$, which must lie in G if $z \in \hat{G}$, according to the definition (see Prop. 2.1(a)). Thus Γ lies in $\hat{G}$ if and only if $\sum'(z) \subset G$.

(c) This now follows at once from (a) and (b). ||

Many of the properties of $\tilde{G}$ and $\hat{G}$ are readily deduced from the representation of arcs as tubes. Among them is the following limit property, needed when we represent a function polyharmonic on G using domains a little smaller than G itself.

Corollary 2.2. If G *is the union of a nested sequence of domains* $\{D_l\}_{l=0}^{\infty}$ *in* $\mathbb{R}^n$, $n \geq 1$, *then*

$$\tilde{G} = \bigcup_l \tilde{D}_l \quad \text{and} \quad \hat{G} = \bigcup_l \hat{D}_l.$$

Proof. For $n \geq 2$ this follows at once from the fact that a (compact) arc in $\tilde{G}$ or $\hat{G}$ corresponds to a tube or full tube whose image is a compact subset of G and is therefore covered by some one of the domains D_l.

For $n = 1$ the result follows directly from Definition 2.2. ||

Proposition 2.3. Let S *be any compact piecewise smooth* $(n-1)$*-dimensional surface with normal* ν *in* $\mathbb{R}^n$, $n \geq 1$. *If* u *is any* $\mathscr{C}^\infty$ *function on some neighbourhood of* S, *then*

$$(2.9) \quad I_k(S, u; y) = \Omega_n \int_S \left\{ \Delta^k u(x) \frac{\partial \mathscr{r}_{k+1}(x-y)}{\partial \nu_x} - \mathscr{r}_{k+1}(x-y) \frac{\partial \Delta^k u(x)}{\partial \nu} \right\} d\sigma(x)$$

defines a function polyharmonic of finite degree $k+1$ on $G=\mathbb{R}^n\backslash S$ which has unlimited analytic continuation in $\tilde{G}$. Its continuation to $\hat{G}$ defines a function holomorphic on $\hat{G}$.

Proof. Let $\Gamma:[0,1]\to\mathbb{C}^n$ be an arc in $\tilde{G}$ beginning at a point $\Gamma(0)\in G\subset\mathbb{R}^n$. For each $x\in S$ we can continue $r(x-y)$ by power series development at a finite succession of points $\Gamma(t_l)$ in the usual way. Since $\tilde{G}$ is open, each $\Gamma(t_l)$ has a polydisc neighbourhood on which $r(x-y)$ has a Taylor series for each $x\in S$. The collection of such arcs, together with the polydisc neighbourhoods, defines a complex analytic manifold over $\tilde{G}$ on which $r(x-y)$ has unlimited continuation (see §4, Chapter I).

The continuation of $r(x-y)$ induces continuation of $\mathscr{r}_{k+1}(x-y)$, and therefore of $\partial\mathscr{r}_{k+1}(x-y)/\partial\nu_x$. Together, for each u, they induce an analytic continuation of $I_k(u,S;y)$.

The restriction to arcs in $\hat{G}$ makes the continuation of $r(x-y)$ a holomorphic function of y on $\hat{G}$. (For instance, if $G=(0,1)\subset\mathbb{R}^1$ and $x=1$ then

$$r(x-y)=\{(1-y)^2\}^{\frac{1}{2}}=1-y, \qquad y\in\hat{G}.$$

For general n, $r(x-y)$ will have positive real part on $\hat{G}$, although this is not an essential point.) Consequently, the auxiliary function $\mathscr{r}_{k+1}(x-y)$ and its normal derivative are holomorphic functions on $\hat{G}$, for each $x\in S$, and polyharmonic of finite degree $k+1$ as well. It follows that I_k has the same properties. ‖

Remark 2.3. If x is a point in $\mathbb{R}^n$ then the continuation of $\mathscr{r}_{k+1}(x-y)$ along a closed loop Γ in $(\mathbb{R}^n\backslash\{x\})^{\sim}$ depends in a curious way on whether the dimension n is even or odd. If $\Gamma(0)=\Gamma(1)$ is the base point of the loop, then the argument of a continuation $r(x-\Gamma(t))$ increases by πs, where s is the winding number of the curve $r^2(x-\Gamma)$ about $0\in\mathbb{C}^1$.

If n is odd, so that $\mathscr{r}_{k+1}$ is a constant multiple of r^{2k+2-n} (equation (2.3), Chapter I), it follows that

$$\begin{aligned}\mathscr{r}_{k+1}(x-\Gamma(1))&=\exp\{i\pi s(2k+2-n)\}\mathscr{r}_{k+1}(x-\Gamma(0))\\&=(-1)^s\mathscr{r}_{k+1}(x-\Gamma(0)),\end{aligned}$$

that is if n is odd then $\mathscr{r}_{k+1}(x-y)$ can have two possible values, of opposite sign. Naturally the same is true of the normal derivative.

If n is even and $k<n/2$, then $\mathscr{r}_{k+1}$ is a rational function, holomorphic on $(\mathbb{R}^n\backslash\{0\})^{\sim}$, but if $k\geq n/2$, then

$$\mathscr{r}_{k+1}(x-y)=\frac{r^{2k+2-n}(x-y)}{\gamma_k'}\{c_{k+1-n/2}-\log r(x-y)\}$$

by equation (2.3′), Chapter I. Thus,

$$\mathscr{r}_{k+1}(x-\Gamma(1))=\mathscr{r}_{k+1}(x-\Gamma(0))-\frac{\pi s i r^{2k+2-n}(x-\Gamma(0))}{\gamma_k'}$$

where s is the winding number of the curve $r^2(x-\Gamma)$ about $0 \in \mathbb{C}^1$. In this case $\imath_{k+1}(x-y)$ will have countably many different values, and the same is true of the normal derivative.

For a closed curve Γ in $(\mathbb{R}^n \backslash \{x\})^\wedge$, however, the curve $r^2(x-\Gamma)$ lies in the simply connected domain $\mathbb{C}^+ = \mathbb{C}^1 \backslash \mathbb{R}^-$ and has winding number zero about $0 \in \mathbb{C}^1$. The following statement summarizes the contrast.

Proposition 2.4. If S is a compact piecewise smooth $(n-1)$-dimensional surface with normal ν in $\mathbb{R}^n$, $n \geq 1$, if u is analytic on a neighbourhood of S, and if $x \in S$, then the analytic continuations of $\imath_{k+1}(x-y)$ (and its normal derivative) on $(\mathbb{R}^n \backslash \{x\})^\wedge$ and of $I_k(S, u; y)$ on $(\mathbb{R}^n \backslash S)^\wedge$ always define holomorphic functions there, while the continuations on $(\mathbb{R}^n \backslash \{x\})^\sim$ and $(\mathbb{R}^n \backslash S)^\sim$ may branch.

When the whole series is put together the continuations of the sum will satisfy in $\mathbb{C}^n$ conditions like those defining Laplacian order and type in $\mathbb{R}^n$. The definition which follows differs from that in $\mathbb{R}^n$ only by requiring the functions to be holomorphic as well, since an analytic function on a domain in $\mathbb{R}^n$ has anti-analytic continuation as well as analytic continuation into the complex domain.

The Laplace operator to be used on continuations into $\mathbb{C}^n$ is

$$\Delta = \sum_{j=1}^{n} \frac{\partial^2}{\partial z_j^2}$$

where $z_1, \ldots, z_n$ are the n complex co-ordinate functions.

Definition 2.4. A function u which is holomorphic on a domain $D \subset \mathbb{C}^n$ is *polyharmonic on D* if its sequence of Laplacians $\{\Delta^p u\}_{p=0}^{\infty}$ satisfies the condition that $\{(|\Delta^p u|/(2p)!)^{\frac{1}{2p}}\}_{p=0}^{\infty}$ approaches zero uniformly on every compact subset of D. Similarly, every other definition from §1 is carried over, any neighbourhoods being taken in D. Thus *Laplacian order and type, local Laplacian order and type,* and all the function classes are defined on domains in $\mathbb{C}^n$.

(Furthermore, the definitions transfer naturally to complex analytic manifolds over $\mathbb{C}^n$ if there is a globally defined Laplacian.)

(The equality of Laplacian orders and types calculated in $\mathbb{R}^n$ with those calculated in $\mathbb{C}^n$ will be shown in Proposition 2.5, but full justification of the definition of local type is not complete until Corollary 1.3, Chapter IV, shows that local Laplacian order is constant on $D \subset \mathbb{R}^n$ if and only if it is constant on $\hat{D} \subset \mathbb{C}^n$.)

Theorem 2.3 (*Local serio-integral construction*). *Let S be a compact piecewise smooth $(n-1)$-dimensional surface with normal ν in $\mathbb{R}^n$, $n \geq 1$. If u is a function analytic on a neighbourhood U of S and having there*

Laplacian order $\rho=\infty$ *and finite type* τ, *then*

$$I(S,u;y)=\sum_{k=0}^{\infty} I_k(S,u;y)$$
$$=\sum_{k=0}^{\infty}\Omega_n\int_S\left\{\Delta^k u(x)\frac{\partial \mathscr{r}_{k+1}(x-y)}{\partial \nu_x}-\mathscr{r}_{k+1}(x-y)\frac{\partial\Delta^k u}{\partial \nu}(x)\right\}d\sigma(x)$$

defines a function analytic on the open set

$$G_\tau=\bigcap_{x\in S} B_{1/\tau}(x)\backslash S\subset\mathbb{R}^n$$

where the series converges uniformly absolutely on compact subsets. (G_τ *may be empty.*)

(a) *If* $n\geq 2$, *then I has unlimited analytic continuation in* $\tilde{G}_\tau$, *where the series of continuations converges uniformly absolutely on compact polydiscs about each point. If* $n=1$ *the continuation is to*

$$\mathscr{G}_\tau=\bigcap_{x\in S}\mathscr{B}_{1/\tau}(x)\backslash S\subset\mathbb{C}^1.$$

(b) *If* $n\geq 2$ (*or* $n=1$) *and* $\mu>\tau$, *then any continuation of I which is holomorphic on some subdomain of* $\tilde{G}_\mu$ (*or* $\mathscr{G}_\mu$) *has there Laplacian order* $\rho=\infty$ *and type* $\mu\tau/(\mu-\tau)$.

(c) *If u is polyharmonic on U with Laplacian order* ρ *and type* τ' *such that* $0<\rho<\infty$ *and* $\tau'<\infty$, *then any continuation of I which is holomorphic on some subdomain of* $\tilde{G}_0=(\mathbb{R}^n\backslash S)^{\tilde{}}$ *has the same Laplacian order and type there.*

Proof. Each term of the series has analytic continuation in $\tilde{G}_\tau$ (or in $\mathscr{G}_\tau$ if $n=1$) by Proposition 2.3. The remainder of the proof depends on providing bounds on the terms I_k.

There is a neighbourhood W of S whose closure lies in U. If u is of Laplacian order ρ and type τ' on U, then for every $\varepsilon>0$ there exists a constant C_ε such that

$$\|\Delta^k u\|_S\leq\|\Delta^k u\|_W\leq C_\varepsilon(2k)!^{1-1/\rho}(\tau'+\varepsilon)^{2k},\quad \text{for all } k.$$

By Proposition 3.1, Chapter I, there exist constants C_1 and C_2 depending on S and W but not on u such that

$$\left\|\frac{\partial\Delta^k u}{\partial\nu}\right\|_S\leq C_\varepsilon C_1(2k)!^{1-1/\rho}(\tau'+\varepsilon)^{2k}+C_\varepsilon C_2(2k+2)^{1-1/\rho}(\tau'+\varepsilon)^{2k+2}$$

for all k. Since u and τ' are fixed, however, there is no difficulty in choosing C_ε so that both $\|\Delta^k u\|_S$ and $\|\partial\Delta^k u/\partial\nu\|_S$ are less than

$$C_\varepsilon(2k+2)!^{1-1/\rho}(\tau'+\varepsilon)^{2k},\quad\text{for all } k. \tag{2.10}$$

Next, bounds are obtained for $\imath_{k+1}$ and its normal derivative. Let Γ be a fixed arc in $\tilde{G}_\tau$ if $n \geqslant 2$ or in $\mathscr{G}_\tau$ if $n=1$ with initial point in G_τ and final point z. Let K be the union of the set in $\mathbb{C}^n$ corresponding to Γ with a compact polydisc P about z in $\tilde{G}_\tau$ if $n \geqslant 2$, in $\mathscr{G}_\tau$ if $n=1$.

If

$$R^2 = \max\{|r^2(x-y)|: x \in S,\ y \in K\},$$

then $R < 1/\tau$. The inequality is obvious if $n=1$. If $n \geqslant 2$, write $y = \xi + i\eta$ with $\xi, \eta \in \mathbb{R}^n$.

$$\begin{aligned}|r^4(y-x)| &= \left|\sum_{j=1}^{n} (|\xi_j - x_j|^2 - |\eta_j|^2 - 2i(\xi_j - x_j)\eta_j)\right|^2 \\ &= ||\xi - x|^2 - |\eta|^2 - 2i(\xi - x, \eta)|^2 \\ &\leqslant (|\xi - x|^2 - |\eta|^2)^2 + 4|\xi - x|^2 |\eta|^2 \\ &= (|\xi - x|^2 + |\eta|^2)^2 \leqslant (|\xi - x| + |\eta|)^4\end{aligned}$$

where $(\xi - x, \eta)$ is the usual inner product in $\mathbb{R}^n$. Since $x \in S$ and $y \in \tilde{G}_\tau$, it follows that every point of $\sum(y)$ has distance at most $1/\tau$ from x, that is $|\xi - x| + |\eta| < 1/\tau$. This combines with the previous inequality to show that $|r(x-y)| < 1/\tau$ for $x \in S$ and $y \in \tilde{G}_\tau$. Since S and K are compact, the maximum satisfies the same inequality, and so

$$R\tau < 1. \tag{2.11}$$

The compactness also assures a positive minimum:

$$d = \min\{|r^2(x-y)|^{\frac{1}{2}}: x \in S,\ y \in K\} > 0.$$

According to equation (3.13), Chapter I, if $x \in S$ and $y \in K$, then

$$\text{both} \quad |\imath_{k+1}(x-y)| \quad \text{and} \quad \left|\frac{\partial \imath_{k+1}(x-y)}{\partial \nu_x}\right|$$

are bounded by

$$\begin{aligned}&C\frac{(k+1)^{1+n/2}}{(2k)!}|r(x-y)|^{2k-n}\{|\log r(x-y)|+1\}\{1+\log(k+1)\} \\ &\leqslant C\frac{(k+1)^{2+n/2}}{(2k)!}\frac{R^{2k}}{d^n}(|\log R| + |\log d| + A + 1) \\ &\leqslant C_K\frac{(k+1)^{2+n/2}}{(2k)!}R^{2k}\end{aligned}$$

where A is the maximum of $\frac{1}{2}|\arg r^2(x-y)|$ for $x \in S$, y continued along Γ

and through P. With (2.10) applied to $\Delta^p u$ this yields the inequality

$$\text{(2.12)}\quad \sum_{k=0}^{\infty} |I_k(S, \Delta^p u; y)| \le \sum_{k=0}^{\infty} \left|\Omega_n \int_S d\sigma(x)\right| 2C_\varepsilon(2k+2p+2)^{1-1/\rho}$$

$$\times(\tau'+\varepsilon)^{2k+2p} C_K \frac{(k+1)^{2+n/2}}{(2k)!} R^{2k}$$

$$\le C'_{\varepsilon,K}(\tau'+\varepsilon)^{2p} \sum_{k=0}^{\infty} \frac{(2k+2p+2)!^{1-1/\rho}}{(2k)!}(2k+2)^{2+n/2}\{(\tau'+\varepsilon)R\}^{2k}.$$

To prove the principal assertion and (a), take $p=0$, $\rho=\infty$, and $\tau'=\tau$, and obtain

$$\sum_{k=0}^{\infty} |I_k(S, u; y)| \le C'_{\varepsilon,K} \sum_{k=0}^{\infty} (2k+2)^{4+n/2}\{(\tau+\varepsilon)R\}^{2k}.$$

Since this holds for each $\varepsilon>0$, while $\tau R<1$ by (2.11), it suffices to consider only ε so small that $(\tau+\varepsilon)R<1$ and observe that uniform absolute convergence on K follows by the Weierstrass M-test. By the arbitrariness of K the uniform absolute convergence on compacta is proved.

Since each term of the series is holomorphic on the interior of K, the series can be differentiated term by term. Because $\Delta^p I_k=0$ for $p>k$, we have

$$\Delta^p I(S, u; y) = \sum_{k=p}^{\infty} \Omega_n \int_S \left\{\Delta^k u(x) \frac{\partial \imath_{k-p+1}(x-y)}{\partial \nu_x} - \frac{\partial \Delta^k u}{\partial \nu}(x)\imath_{k-p+1}(x-y)\right\} d\sigma(x)$$

$$= \sum_{k=0}^{\infty} \Omega_n \int_S \left\{\Delta^{k+p} u(x) \frac{\partial \imath_{k+1}(x-y)}{\partial \nu_x} - \frac{\partial \Delta^{k+p} u}{\partial \nu}(x)\imath_{k+1}(x-y)\right\} d\sigma(x)$$

$$= \sum_{k=0}^{\infty} I_k(S, \Delta^p u; y).$$

To the last series we can also apply (2.12), for general p, under the assumption that u is of Laplacian order $\rho>0$ and finite type τ' on U. The kth coefficient in the power series can be bounded using the integral which defines the beta function:

$$\frac{(2p)!\,(2k+1)!}{(2k+2p+2)!} = \frac{\Gamma(2p+1)\Gamma(2k+2)}{\Gamma(2k+2p+3)}$$

$$= \int_0^1 t^{2p}(1-t)^{2k+1}\, dt$$

$$>(\beta-\alpha)\alpha^{2p}(1-\beta)^{2k+1} \quad \text{for } 0<\alpha<\beta<1$$

or

$$(2k+2p+2)! < (2p)!\,(2k+1)!\,(\beta-\alpha)^{-1}\alpha^{-2p}(1-\beta)^{-2k-1}.$$

Thus, for every choice of α and β satisfying $0<\alpha<\beta<1$,

$$(2.13)\quad |\Delta^p I(S, u; y)| \leqslant \sum_{k=0}^{\infty} |I_k(S, \Delta^p u; y)|$$

$$\leqslant C'_{\varepsilon,K}(\tau'+\varepsilon)^{2p} \sum_{k=0}^{\infty} \left\{\frac{(2p)!\,(2k+1)!}{(\beta-\alpha)\alpha^{2p}(1-\beta)^{2k+1}}\right\}^{1-1/\rho}$$

$$\times \frac{(2k+2)^{2+n/2}}{(2k)!}\{(\tau'+\varepsilon)R\}^{2k}$$

$$\leqslant C''_{\varepsilon,K}(2p)!^{1-1/\rho}\left(\frac{\tau'+\varepsilon}{\alpha^{1-1/\rho}}\right)^{2p} \sum_{k=0}^{\infty} \frac{(2k+2)^{3+n/2}}{(2k)!^{1/\rho}}$$

$$\times \left\{\frac{(\tau'+\varepsilon)R}{(1-\beta)^{1-1/\rho}}\right\}^{2k}, \quad \text{for all } p.$$

To prove part (c), for which $\rho<\infty$, choose α so near unity that

$$\frac{\tau'+\varepsilon}{\alpha^{1-1/\rho}}<\tau'+2\varepsilon$$

and observe that if $1>\beta>\alpha$ then the series itself is convergent to a number which is independent of p, that is for each $\varepsilon>0$ and $y\in K$

$$|\Delta^p I(S, u; y)| \leqslant C'''_{\varepsilon,K}(2p)!^{1-1/\rho}(\tau'+2\varepsilon)^{2p}, \quad \text{for all } p.$$

Since each point of $\tilde{G}_\tau$ (or $\mathcal{G}_\tau$) lies in the interior of some such K, the order and type assertion follows.

For part (b) we have $\rho=\infty$, $\tau'=\tau$, and $R<1/\mu$, so that (2.13) becomes

$$(2.14)\quad |\Delta^p I(S, u; y)| \leqslant C''_{\varepsilon,K}(2p)!\left(\frac{\tau+\varepsilon}{\alpha}\right)^{2p} \sum_{k=0}^{\infty} (2k+2)^{3+n/2}\left\{\frac{(\tau+\varepsilon)/\mu}{1-\beta}\right\}^{2k},$$

for all p.

To obtain convergence it suffices to consider $\varepsilon<\mu-\tau$ and

$$1-\beta>\frac{\tau+\varepsilon}{\mu}, \quad \text{so that} \quad \alpha<\beta<\frac{\mu-\tau-\varepsilon}{\mu}.$$

Taking reciprocals shows that it is sufficient to have

$$\frac{\tau+\varepsilon}{\alpha}>\frac{\mu(\tau+\varepsilon)}{\mu-\tau-\varepsilon} \quad \text{or} \quad \frac{\tau+\varepsilon}{\alpha}\geqslant\frac{\mu(\tau+2\varepsilon)}{\mu-\tau-\varepsilon},$$

that is

$$|\Delta^p I(S, u; y)| \leqslant C'''_{\varepsilon,K}(2p)!\left(\frac{\mu(\tau+2\varepsilon)}{\mu-\tau-\varepsilon}\right)^{2p},$$

for all p, all $y\in K$. The conclusion follows. ‖

Corollary 2.3. Under the hypotheses of Theorem 2.3, a continuation made from $G_\tau \subset \mathbb{R}^n$, $n \geq 2$, to its reduced harmonicity hull $\hat{G}_\tau$ rather than $\tilde{G}_\tau$, defines on $\hat{G}_\tau$ a holomorphic function. For $n = 1$, a continuation made from G_τ to $\mathcal{G}_\tau \cap \hat{G}_\tau$ rather than $\mathcal{G}_\tau$ defines a function holomorphic on $\mathcal{G}_\tau \cap \hat{G}_\tau$.

This is because each term defines a holomorphic function on $\hat{G}_\tau$, although the series converges on the larger set.

Corollary 2.4. Let S be a compact piecewise smooth $(n-1)$-dimensional surface in $\mathbb{R}^n$, $n \geq 1$, and let u have Laplacian order ∞ and type τ on a neighbourhood U of S. Let $x \in S$ be a point with a neighbourhood where S is flat, and let R be positive and such that the intersection of S with the closed n-ball $\bar{B}_R(x)$ is a closed $(n-1)$-ball lying in $U \cap G_\tau$. Designate the two components of $B_R(x)\backslash S$ as B^+ and B^-, accordingly as ν, the normal of S, points in or out respectively. If $S^+ = \partial B^+ \backslash S$ and $S^- = \partial B^- \backslash S$, the inward normal for $B_R(x)$ being chosen, and if $S_C = S \backslash B_R(x)$, then

$$I(S, u; y) = \begin{cases} I(S_C, u; y) - I(S^+, u; y), & y \in B^- \\ I(S_C, u; y) - I(S^+, u; y) + u(y), & y \in B^+, \end{cases}$$

each term on the right-hand side being polyharmonic on all of $B_R(x)$ and hence on $\hat{B}_R(x)$.

Proof. Since $\partial B^+ = S^+ \cup (S \backslash S_C)$, and since the normals for S and S^+ point into B^+,

$$u(y) = I(S^+, u; y) + I(S, u; y) - I(S_C, u; y), \qquad y \in B^+,$$

by Theorem 2.1. Similarly, for points $y \in B^-$,

$$\begin{aligned} I(S^+, u; y) + I(S^-, u; y) &= u(y) \\ &= -I(S, u; y) + I(S_C, u; y) + I(S^-, u; y) \end{aligned}$$

since the normal for S points out of B^-. Solving for $I(S, u; y)$ in these equations gives the formula.

By Theorem 2.3, the integrals on S^+, S^-, and S_C all continue into $\hat{B}_R(x)$, since it is a subset of the reduced harmonicity hulls of $\mathbb{R}^n \backslash S^+$, $\mathbb{R}^n \backslash S^-$, and $\mathbb{R}^n \backslash S_C$. ||

It is clear that 'flat' could be replaced by 'smooth' in the preceding result, provided that the balls are replaced by suitably smooth images of balls.

The theorem which follows collects in one place the statements scattered throughout the section on the application of serio-integral representation to functions which are polyharmonic. The results are taken in turn from Theorem 2.3, Proposition 2.3, Corollary 2.4, Theorem 2.1, Theorem 2.2, and Corollary 2.3.

Theorem 2.4 (Serio-integral representation of polyharmonic functions). Let S be any compact piecewise smooth $(n-1)$-dimensional surface in $\mathbb{R}^n$, $n \geqslant 1$, and let u be polyharmonic on a neighbourhood of S.

(a) *The formula*

$$\text{(2.15)} \quad I(S, u; y) = \sum_{k=0}^{\infty} \Omega_n \int_S \left\{ \Delta^k u(x) \frac{\partial \mathscr{r}_{k+1}(x-y)}{\partial \nu_x} - \mathscr{r}_{k+1}(x-y) \frac{\partial \Delta^k u(x)}{\partial \nu} \right\} \mathrm{d}\sigma(x)$$

defines a function both analytic and polyharmonic on each component of $\mathbb{R}^n \backslash S$. The kth *term has degree $k+1$.*

(b) *Across any smooth portion of S, I has analytic continuation. If I^- is the continuation in the direction of the normal and I^+ is the continuation in the opposite direction, then $I^+ - I^- = u$ on a sufficiently small connected neighbourhood.*

(c) *If in addition $S = \partial D$ for some bounded domain D such that $\partial/\partial\nu$ is the inward normal derivative, and if u is polyharmonic on a neighbourhood of $\bar{D}$, then $I \equiv u$ on D and $I \equiv 0$ on $\mathbb{R}^n \backslash \bar{D}$, that is*

$$\sum_{k=0}^{\infty} \Omega_n \int_S \left\{ \Delta^k u(x) \frac{\partial \mathscr{r}_{k+1}(x-y)}{\partial \nu_x} - \mathscr{r}_{k+1}(x-y) \frac{\partial \Delta^k u(x)}{\partial \nu} \right\} \mathrm{d}\sigma(x) = \begin{cases} u(y), & y \in D \\ 0, & y \in \mathbb{R}^n \backslash \bar{D}. \end{cases}$$

Thus, every polyharmonic function is necessarily analytic.

(d) *The function defined by I on $\mathbb{R}^n \backslash S$ has unlimited analytic continuation along all arcs in $(\mathbb{R}^n \backslash S)^{\sim}$, where the series formed from the continuations of the terms is uniformly convergent on compacta and differentiable term by term. These continuations will have every Laplacian order and type that u has on a neighbourhood of S.*

(e) *The continuations to $(\mathbb{R}^n \backslash S)^{\wedge}$ determine a holomorphic function there.*

What follows is the last of the main results, the characterization of $\tilde{G}$ as a domain deserving the name 'harmonicity hull'.

Theorem 2.5. For any domain $G \subset \mathbb{R}^n$, $n \geqslant 1$, $\tilde{G}$ is the unique largest domain in $\mathbb{C}^n$ to which every function u polyharmonic on G has analytic continuation. On $\tilde{G}$, any given continuation will have every Laplacian order and type which u has on G.

Proof. If K is any compact subset of $\tilde{G}$ the union of all $(n-2)$-spheres representing points of K is a compact subset of G. Thus there exists a bounded domain $D \subset G$ such that $K \subset \tilde{D}$, and clearly the boundary of D can be taken piecewise smooth with normal interior for D. Since any function u polyharmonic on G is represented on D by the serio-integral representation and continuable to all of $\tilde{D}$, it follows that u has continuation to $\tilde{G}$ by taking unions over all such subdomains (Corollary 2.2). Order and type follow from Theorem 2.3.

Conversely, as is well known, there exist examples of functions u harmonic on G but having a natural boundary at $\partial\tilde{G}$. To construct one, take a sequence of points $\{x_k\}$ lying in ∂G and dense there, and let

$$u(y)=\sum_{k=1}^{\infty} c_k \imath_1(x_k-y).$$

If the sequence $\{c_k\}$ converges to zero sufficiently rapidly, the series will converge uniformly absolutely on compacta in $\tilde{G}$. The sum will be harmonic and have unlimited continuation on $\tilde{G}$ but will fail to have analytic continuation at each point of $V(x_k)$, for each k. Since such points are dense in $\partial\tilde{G}$ this set will be a natural boundary. ‖

There do exist domains G such that all continuations of all functions polyharmonic on G are single valued on $\tilde{G}$ even though $\hat{G}$ is a proper subset of $\tilde{G}$. This point will be discussed and examples given in §4.

Remark 2.4. An analytic function which is not polyharmonic has a local representation in $\mathbb{R}^n$ whose extent depends on considerations of Laplacian order and type (Theorem 2.1). The local representation gives local continuation into $\mathbb{C}^n$ where the type may increase (Theorem 2.3). These facts are generally referred to as the *local serio-integral representation of analytic functions.*

Example 2.1. Let $n=1$, define $u(z)=(z-i)^{-1}$ for $z\in\mathbb{C}\backslash\{i\}$, and note that

$$\Delta^p u(z)=\frac{(2p)!}{(z-i)^{2p+1}},\qquad z\in\mathbb{C}\backslash\{i\}.$$

Thus u has Laplacian order ∞ and type τ on $\mathbb{C}\backslash\mathcal{B}_{1/\tau}(i)$. In fact it has local Laplacian order $\lambda(z)\equiv\infty$ and local type $\mu(z)=1/|z-i|$. Let G be the real interval

$$G=(-1/\sqrt{3},1/\sqrt{3})\subset\mathbb{R}.$$

If S consists of the end points of G with ν the inward normal for G, then the serio-integral representation cannot converge at i or on any set

$$\mathcal{G}_\tau=\mathcal{B}_{1/\tau}(-1/\sqrt{3})\cap\mathcal{B}_{1/\tau}(1/\sqrt{3})$$

unless $\tau\geqslant\sqrt{3}/2$, to avoid including i in $\mathcal{G}_\tau$. However, the local type at points of S is exactly $\sqrt{3}/2$. Thus the convergence region $\mathcal{G}_{\sqrt{3}/2}$ given in Theorem 2.3 is the best possible.

It should be noted that the local type on this set takes the value 1 at $z=0$, a local maximum on $G\subset\mathbb{R}$, decreases with limit $\frac{1}{2}$ at $-i\in\partial\mathcal{G}_{\sqrt{3}/2}$, and increases unboundedly near $z=i$.

Example 2.2. Let $n\geqslant 2$ and let $u(z)=\{r(z)\}^{-n}$, which is analytic on $\mathbb{R}^n\backslash\{0\}$ but not polyharmonic. It is singular only on $V(0)$, the isotropic cone through the origin. The local Laplacian order is $\lambda(z)\equiv\infty$ and the local type is $\mu(z)=1/|r(z)|$.

If y is a point of $\mathbb{R}^n\setminus\{0\}$ and $D=B_{|y|/3}(y)$, then u has Laplacian order ∞ and type $3/(2|y|)$ on D since the point of $\bar{D}$ nearest the origin is at a distance $2|y|/3$. It is evident from the tube construction of $\tilde{D}$ that the closure of $\tilde{D}$ does not intersect $V(0)$, and so $\mu(z)$ is bounded on $\tilde{D}$.

Further, $v(z)=u(z)+\imath_1(y-z)$ has singularities on $V(y)$ which are quite unrelated to its local Laplacian order and type since $\imath_1$ is harmonic.

Corollary 2.5. The definition of local Laplacian order and type in $\mathbb{C}^n$ *is consistent with that in* $\mathbb{R}^n$.

Proof. Let D be a domain in $\mathbb{C}^n$ and u be a function holomorphic on it. Let λ and μ be local Laplacian order and type as calculated using neighbourhoods in $\mathbb{C}^n$, and let λ_r and μ_r be the corresponding functions calculated using neighbourhoods in $\mathbb{R}^n$. It is clear that if $x\in D\cap\mathbb{R}^n$, then $\lambda(x)\geqslant\lambda_r(x)$ and $\mu(x)\geqslant\mu_r(x)$.

Suppose u is polyharmonic. For any $\varepsilon>0$, x has a neighbourhood $U\subset D\cap\mathbb{R}^n$ on which u has Laplacian order at most $\lambda_r(x)+\varepsilon$, and Theorem 2.5 shows that on $\hat{U}$, λ is bounded by the same number. Hence $\lambda(x)\leqslant\lambda_r(x)$, and the two functions coincide. However, if $\lambda_r(x)=\infty$, then $\lambda\geqslant\lambda_r$ implies $\lambda(x)=\infty=\lambda_r(x)$.

Now consider type. If λ is constant and finite on a neighbourhood of x in D, then u is polyharmonic and the argument based on Theorem 2.5 can be repeated for type. For the general case, though, if u has type $\mu_r(x)+\varepsilon$ on $B_R(x)$ with $2R<1/\{\mu_r(x)+\varepsilon\}$, then Theorem 2.3 shows that x has a ball neighbourhood in $\mathbb{R}^n$ with radius $\varepsilon/2\{\mu_r(x)^2+\varepsilon\mu_r(x)\}$ on whose reduced harmonicity hull $\mu<\mu_r(x)+2\varepsilon$. Thus $\mu(x)\leqslant\mu_r(x)$. ‖

In Chapter IV it will be shown that the constancy of λ in $\mathbb{R}^n$, the necessary condition for defining local type there, is enough to ensure the constancy of λ on a neighbourhood in $\mathbb{C}^n$.

Remark 2.5. The proof of the preceding corollary also shows that

$$\lambda(x)\geqslant\limsup_{z\to x}\lambda(z)\quad\text{and}\quad\mu(x)\geqslant\limsup_{z\to x}\mu(z).$$

The next result is a weak form of the Maximum Principle, to be given later (Theorem 1.3, Chapter IV).

Proposition 2.5. If u is polyharmonic on a domain $G\subset\mathbb{R}^n$, $n\geqslant 1$, *then the local Laplacian order of any continuation of u in* $\tilde{G}$ *satisfies*

$$\|\lambda\|_{\tilde{G}}=\|\lambda\|_G.$$

If in addition G is bounded and u is polyharmonic on a neighbourhood of $\bar{G}$ *in* $\mathbb{R}^n$, *then*

$$\|\lambda\|_G\leqslant\|\lambda\|_{\partial G},$$

and if $n \geqslant 2$, *then*†

$$\|\lambda\|_{\partial\tilde{G}} = \|\lambda\|_{\partial G}.$$

The same results hold for local type if λ *is constant on* $\tilde{G}$ *in the first case, on a neighbourhood of* $(\tilde{G})^-$ *in the second (on* $\mathbb{C}$ *for* $n = 1$).

Proof. The equality $\|\lambda\|_{\tilde{G}} = \|\lambda\|_G$ follows immediately from Theorem 2.5 and Corollary 2.5.

If G is bounded, if U is a neighbourhood of $\bar{G}$ in $\mathbb{R}^n$ on which u is polyharmonic, and if K is any compact sub-domain of G, then there exists a surface S in $U\backslash K$ which is piecewise smooth and the surface of a bounded sub-domain D of U containing K, such that

$$u(y) = I(S, u; y), \qquad y \in D.$$

By Theorem 2.3,

$$\|\lambda\|_K \leqslant \|\lambda\|_D \leqslant \|\lambda\|_{U\backslash K}.$$

Taking the supremum over such K and the infimum over such U shows that

$$\|\lambda\|_G \leqslant \|\lambda\|_{\partial G} \quad \text{and} \quad \|\lambda\|_{\partial G} = \|\lambda\|_{\bar{G}}.$$

If $n \geqslant 2$ then $(\tilde{G})^-$ lies in $\tilde{U}$, and so

$$\|\lambda\|_{\partial G} \leqslant \|\lambda\|_{\partial\tilde{G}} \leqslant \|\lambda\|_{(\tilde{G})^-} \leqslant \|\lambda\|_{\tilde{U}} = \|\lambda\|_U.$$

Taking the infimum over such U shows that

$$\|\lambda\|_{\partial G} = \|\lambda\|_{\partial\tilde{G}} = \|\lambda\|_{\bar{G}}. \quad \|$$

Notes

Formula (2.1) is a variant of a formula of Boggio [20]. He used a different sequence of auxiliary functions (not the functions $\imath_p$), and as a consequence his formula does not extend naturally to infinite series. Boggio's formula, as well as preliminary work by Gutzner and Mathieu, is in the book by Nicolescu [68]. The serio-integral representation for polyharmonic functions first appears in Aronszajn [6].

The groundwork for the construction of harmonicity hulls was laid by Aronszajn [6]. In a series of papers by Lelong [61–63] the construction of the hulls and the continuation of harmonic functions is made explicit, and it is shown that there is a harmonic function which cannot be continued outside (see Theorem 2.5), so that a hull is a domain of holomorphy. Kiselman [56] discusses hulls of analytic continuation for a general class of elliptic operators with constant coefficients (which includes the Laplace operator) and gives the harmonicity hull as a example of his construction.

† The exception occurs because $\tilde{G} = \mathbb{C}$ if $n = 1$, and so $\partial\tilde{G}$ is empty.

The harmonicity hulls for special, but important, domains have been studied by Siciak [81] and Jarnicki [53, 54]. Most of their work concerns the harmonicity hull of a ball, a topic treated in detail in the next section (for an application to a special case, see Teissier du Cros [84]).

It is important to point out that Theorem 2.5 applies to the *whole class* of polyharmonic functions, not just to the solutions of a fixed elliptic equation (harmonic functions, in particular), and also gives preservation of Laplacian order and type of the continuations.

3. Harmonicity hulls of domains in $\mathbb{R}^n$, $n \geqslant 2$

The harmonicity hull and reduced harmonicity hull of an open set $G \subset \mathbb{R}^n$ were introduced in Definition 2.2. For $n = 1$ they were described completely:

If G is an open subset of $\mathbb{R}^1$, *then* $\tilde{G} = \mathbb{C}^1$ if G is connected, $\mathbb{C}^1 \backslash \partial G$ otherwise, and $\hat{G} = \{z \in \mathbb{C}^1 : \mathscr{R}e\ z \in G\}$.

It was shown in Proposition 2.2 that for $n \geqslant 2$ the hulls of disjoint open sets are disjoint. Therefore, *in this section we shall restrict attention to domains* (*connected open sets*) *in* $\mathbb{R}^n$ *for* $n \geqslant 2$, giving general properties and examples, both for later use and to distinguish concepts. Convexity will be an important theme. The most important example, the harmonicity hull of a ball, will be elaborated at considerable length because it will be used extensively in later chapters.

The main tool, the tube, was introduced in Definition 2.3, and the technique of using it was illustrated in Proposition 2.2, where it was shown that

$$\mathbb{R}^n \cap \tilde{G} = \mathbb{R}^n \cap \hat{G} = G \subset \hat{G} \subset \tilde{G}. \tag{3.1}$$

Remark 3.1. Since $z \in \mathbb{C}^n$ lies in $\tilde{G}$ if and only if there is a tube in G starting at a point of G and ending at $\sum(z)$, by Proposition 2.2, and since a point y lies in $\hat{G}$ if and only if $\sum'(y) \subset G$, it follows that the points z of $\tilde{G}$ are those such that $\sum(z)$ can be reached by a tube in G beginning at $\sum(y)$ where y is a point of G itself, or of $\hat{G}$, or of $\tilde{G}$.

Remark 3.2. It is also clear that an isometry carries tubes to tubes. Thus, any isometry T of $\mathbb{R}^n$ extends to an isometry of $\mathbb{C}^n$ such that

$$T[\tilde{G}] = (TG)^{\tilde{}} \quad \text{and} \quad T[\hat{G}] = (TG)^{\hat{}}$$

(Of course, complex conjugation is an isometry on $\mathbb{C}^n$ which carries $\tilde{G}$ or $\hat{G}$ to itself.) For this reason, a distinguished point in G will usually be taken to be the origin. For instance, a *star domain* in $\mathbb{C}^n$ or $\mathbb{R}^n$ can be taken to be an open set with the property that if z is a point of it then so is tz, for each real t such that $0 \leqslant t \leqslant 1$. A *star domain with centre* z_0 is obtained by the translation $T(z) = z + z_0$.

Proposition 3.1. If G is a star domain in $\mathbb{R}^n$, or the exterior of the closure of a bounded star domain in $\mathbb{R}^n$, then $z \in \tilde{G}$ if and only if $\sum(z) \subset G$.

The proof consists only of shrinking $\sum(z)$ to the origin or expanding it to $\sum(z')$ where $\sum'(z')$ lies in G, by Remark 3.1. Similar arguments prove the following.

Proposition 3.2. If G is a domain in $\mathbb{R}^n$ the following statements are equivalent:
(a) *G is a star domain;*
(b) *$\hat{G}$ is a star domain;*
(c) *$\tilde{G}$ is a star domain.*

Proposition 3.3. If G is a convex domain in $\mathbb{R}^n$ then $\tilde{G} = \hat{G}$.

Proof. By (3.1), in all cases $\hat{G} \subset \tilde{G}$. If G is convex then $\sum(z) \subset G$ implies that $\sum'(z) \subset G$, so that $z \in \hat{G}$. Thus $\tilde{G} \subset \hat{G} \subset \tilde{G}$. ∥

The property $\tilde{G} = \hat{G}$ does not characterize convex domains, a point to be expanded upon beginning at Remark 3.7. However, domains having this property have harmonicity hulls which are easier to describe than those lacking it, for which the details of tubes are important (see Exx. 3.5–3.7). Convex domains are particularly useful because they have this property.

Proposition 3.4. If G is a domain in $\mathbb{R}^n$ the following statements are equivalent:
(a) *G is convex;*
(b) *$\hat{G}$ is convex;*
(c) *$\tilde{G}$ is convex.*

Proof. Either (b) or (c) implies (a) since G is the intersection of $\hat{G}$ or $\tilde{G}$ with $\mathbb{R}^n$ by (3.1). Since (a) implies that $\tilde{G} = \hat{G}$, by Proposition 3.3, it remains to show that (a) implies (b).

Let z^0 and z^1 lie in $\hat{G}$, so that $\sum'(z^0)$ and $\sum'(z^1)$ lie in G, and identify the convex combinations

$$z^t = (1-t)z^0 + tz^1, \qquad 0 \leqslant t \leqslant 1.$$

As usual, these points can be written

$$z^t = \xi^t + i\eta^t = \{(1-t)\xi^0 + t\xi^1\} + i\{(1-t)\eta^0 + t\eta^1\}. \tag{3.2}$$

Consider first the case in which η^0 and η^1 are linearly independent, generating a two-dimensional subspace $P \subset \mathbb{R}^n$. In P, let T be one of the two rotations through $\pi/2$ around the origin. For each t, define in P a unit vector

$$v^t = T(\eta^t)/|\eta^t|$$

and observe that

$$|\eta^t|\,v^t = T(\eta^t) = (1-t)T(\eta^0) + tT(\eta^1) = (1-t)\,|\eta^0|\,v^0 + t\,|\eta^1|\,v^1. \tag{3.3}$$

Let $\prod(\eta^0)$ and $\prod(\eta^1)$ be the $(n-1)$-dimensional subspaces of $\mathbb{R}^n$ orthogonal to η^0 and to η^1 respectively. Since η^0 and η^1 are linearly independent, the intersection is an $(n-2)$-dimensional subspace

$$\prod = \prod(\eta^0) \cap \prod(\eta^1)$$

orthogonal to η^0, η^1, and P. Together $\prod$ and v^t span the $(n-1)$-dimensional sub-space $\prod(\eta^t)$ orthogonal to η^t.

Now we are ready to show that each $\sum(z^t)$ lies in G. Fix t. Since $\sum(z^t)$ has centre ξ^t, radius $|\eta^t|$, and lies in the $(n-1)$-plane through ξ^t orthogonal to η^t, by Proposition 2.1, it follows that any of its points σ can be represented as

$$\sigma = \xi^t + |\eta^t|\,(\alpha v^t + \beta u) \tag{3.4}$$

where u is some unit vector in $\prod$ and $\alpha^2 + \beta^2 = 1$, that is $\alpha v^t + \beta u$ is a unit vector in $\prod(\eta^t)$ determined by t and σ. Substituting from (3.2) and (3.3) into (3.4) shows that

$$\begin{aligned}\sigma &= \{(1-t)\xi^0 + t\xi^1\} + \alpha\{(1-t)\,|\eta^0|\,v^0 + t\,|\eta^1|\,v^1\} + \beta\,|\eta^t|\,u \\ &= \{(1-t)\xi^0 + t\xi^1\} + \alpha\{(1-t)\,|\eta^0|\,v^0 + t\,|\eta^1|\,v^1\} + \frac{\beta\{(1-t)\,|\eta^0| + t\,|\eta^1|\}\,|\eta^t|\,u}{\{(1-t)\,|\eta^0| + t\,|\eta^1|\}} \\ &= (1-t)\sigma^0 + t\sigma^1\end{aligned}$$

where

$$\sigma^k = \xi^k + |\eta^k| \left\{\alpha v^k + \frac{|\eta^t|\,\beta u}{(1-t)\,|\eta^0| + t\,|\eta^1|}\right\}, \qquad k = 0, 1.$$

The vector $\sigma^k - \xi^k$ is orthogonal to η^k since both v^k and u are. Since

$$|\eta^t| = |(1-t)\eta^0 + t\eta^1| \leq (1-t)\,|\eta^0| + t\,|\eta^1|,$$

the magnitude of $\sigma^k - \xi^k$ is at most

$$|\eta^k|\,(\alpha^2 + \beta^2)^{\frac{1}{2}} = |\eta^k|,$$

and so $\sigma^k \in \sum'(\xi^k + i\eta^k) = \sum'(z^k)$, that is every point of $\sum(z^t)$ is a convex combination of points of $\sum'(z^0)$ and $\sum'(z^1)$. By the convexity of G it follows that $\sum'(z^t) \subset G$ and so z^t is a point of $\hat{G}$. Thus $\hat{G}$ is convex.

There remains the case in which η^0 and η^1 are not linearly independent. In this case any one of the points z^t is a convex combination of two new points $\hat{z}^0$ and $\hat{z}^1$ which are close enough to z^0 and to z^1 respectively to lie in $\hat{G}$, and yet linearly independent. By the preceding case, $z' \in \hat{G}$. ‖

Example 3.1. The harmonicity hull of a half-space. First consider the co-ordinate half-space in $\mathbb{R}^n$, $X_n = \{(x_1, \ldots, x_n) : x_n > 0\}$. Since X_n is convex it follows that $\tilde{X}_n = \hat{X}_n$. Thus, a point z lies in $\tilde{X}_n$ if and only if $\sum'(z) \subset X_n$, and conversely. If $z = \xi + i\eta$ with $\eta \neq 0$, then $\sum'(z)$ lies in X_n if and only if $\xi \in X_n$ and the $(n-1)$-plane $\prod(z)$, orthogonal to η through ξ, intersects $\mathbb{R}^{n-1}$ in a set which is either empty or an $n-2$ plane at a distance greater than $|\eta|$ from ξ, that is

$$\tilde{X}_n = \hat{X}_n = \{\xi + i\eta : \xi \in X_n, \xi_n^2 + \eta_n^2 > |\eta|^2\}.$$

If $d(x, X)$ represents the (minimum) distance from a point x to a set X, then $|x_n| = d(x, \partial X_n)$. The result of substituting this into the previous equation is invariant under isometry,

$$\tilde{G} = \hat{G} = \{\xi + i\eta : \xi \in G, d^2(\xi, \partial G) + d^2(\eta, \partial G) > |\eta|^2\}$$

which is valid for any half-space in $\mathbb{R}^n$ whatsoever.

Lemma 3.1. If G_t is a domain in $\mathbb{R}^n$ for each t in some index set T, finite or infinite, and if $\bigcap_{t \in T} G_t$ is open then

$$\left(\bigcap_{t \in T} G_t\right)^{\wedge} = \bigcap_{t \in T} \hat{G}_t.$$

Since the proof follows immediately from Remark 3.1, the interest lies in the fact that intersection cannot be replaced by union (Ex. 3.5) nor reduced harmonicity hull by harmonicity hull (Ex. 3.6), and that the lemma leads to a description of the hulls of convex domains.

Example 3.2. The harmonicity hull of a convex domain. If G is a convex domain in $\mathbb{R}^n$ then G is an intersection of half-spaces,

$$G = \bigcap_t G_t.$$

It follows from Proposition 3.3, Ex. 3.1, and Lemma 3.1 that

$$\tilde{G} = \hat{G} = \{\xi + i\eta \in \mathbb{C}^n : \xi \in G, \inf\{d^2(\xi, \partial G_t) + d^2(\eta, \partial G_t)\} > |\eta|^2\}.$$

The next example is the most important in the whole subject.

Example 3.3. The harmonicity hull of the ball $D = B_R(0) \subset \mathbb{R}^n$, $n \geq 2$. (The harmonicity hull of any ball can be derived from this one by translation.)

Since D is convex, for any point $z \in \mathbb{C}^n$ it is the case that $\sum(z) \subset D$ if and only if $\sum'(z) \subset D$, that is $\tilde{D} = \hat{D}$. Further, if $z = \xi + i\eta$ then $\sum'(z)$ lies in the intersection of $D = B_R(0)$ with the hyperplane $\prod(z)$ through ξ orthogonal to η. If z lies in $\mathbb{R}^n$, that intersection is z itself; otherwise it is an $(n-1)$-ball with centre $\eta_1 = (\xi, \eta)\eta/|\eta|^2$ and radius $\sqrt{(R^2 - |\eta_1|^2)}$ where (ξ, η) is the usual inner product in $\mathbb{R}^n$.

The $(n-1)$-ball $\sum'(\xi + i\eta)$ lies in that intersection if and only if

$$|\xi - \eta_1| + |\eta| < \sqrt{(R^2 - |\eta_1|^2)} \quad \text{or} \quad R^2 > (|\xi - \eta_1| + |\eta|)^2 + |\eta_1|^2 \tag{3.5}$$

because the point of $\sum'(\xi+i\eta)$ most distant from η_1 has distance $|\xi-\eta_1|+|\eta|$ from η_1. (That point is also the point most distant from the origin.) Since

$$|\xi-\eta_1|^2=|\xi|^2-2(\xi,\eta_1)+|\eta_1|^2=|\xi|^2-\frac{(\xi,\eta)^2}{|\eta|^2},$$

it follows that

$$|\xi-\eta_1|\,|\eta|=(|\xi|^2\,|\eta|^2-(\xi,\eta)^2)^{\frac{1}{2}} \quad\text{and}\quad |\xi-\eta_1|^2=|\xi|^2-|\eta_1|^2.$$

Thus, (3.5) implies that $\xi+i\eta$ lies in $\tilde{D}$ if and only if

$$\begin{aligned}R^2&>|\xi-\eta_1|^2+2\,|\xi-\eta_1|\,|\eta|+|\eta|^2+|\eta_1|^2\\&=|\xi|^2+|\eta|^2+2\{|\xi|^2\,|\eta|^2-(\xi,\eta)^2\}^{\frac{1}{2}},\end{aligned}$$

the inequality also being valid if $z=\xi\in\mathbb{R}^n$, for which $\eta=0$, that is

$$(3.6)\qquad \begin{cases}\tilde{B}_R(0)=\hat{B}_R(0)=\{\xi+i\eta\in\mathbb{C}^n:\xi,\eta\in\mathbb{R}^n,\,q(\xi+i\eta)<R\}\\ \text{where}\quad q(\xi+i\eta)=[|\xi|^2+|\eta|^2+2\sqrt{\{|\xi|^2\,|\eta|^2-(\xi,\eta)^2\}}]^{\frac{1}{2}}\end{cases}$$

The set $\hat{B}_1(0)$ is the same as the classical domain $\mathscr{R}_{IV}$ of E. Cartan and is sometimes referred to as the 'Lie ball' (see Hua [51]). The sets $\hat{B}_R(x)$ will be referred to as *Lie balls*. For applications in later chapters, various additional properties of the set will be presented.

Proposition 3.5 (*Homogeneity of q*). *If $z\in\mathbb{C}^n$ and $\zeta\in\mathbb{C}$ then*

$$q(\zeta z)=|\zeta|\,q(z).$$

Thus q is a norm on $\mathbb{C}^n$.

Proof. The calculation is straightforward but tedious. It consists of applying q^2 to

$$\begin{aligned}&|\zeta|\,(\cos\theta+i\sin\theta)(\xi+i\eta)\\&\quad=|\zeta|\,\{(\xi\cos\theta-\eta\sin\theta)+i(\xi\sin\theta+\eta\cos\theta)\}\end{aligned}$$

and simplifying. Because the unit ball for q, $\tilde{B}_1(0)=\{z:q(z)<1\}$, is a convex set, the homogeneity property makes q a norm on $\mathbb{C}^n$. ||

Remark 3.3. It follows that $z\in\hat{B}_R(0)$ if and only if $\zeta z\in\hat{B}_R$ for all $\zeta\in\mathbb{C}$ with $|\zeta|\leq 1$. Thus the harmonicity hull of a ball about the origin is a *complete circular domain* in the terminology of Bochner and Martin [19, p. 78].

Proposition 3.6. For each $R>0$, if $\mathscr{B}_R$ is the complex ball of radius R about 0 in $\mathbb{C}^n$, then it is the smallest ball containing $\tilde{B}_R(0)$. $\mathscr{B}_{R/\sqrt{2}}$ is the largest ball contained in $\tilde{B}_R(0)$, and $\bar{B}_{R/\sqrt{2}}$ intersects the isotropic cone $V(x)$ for each $x\in\partial B_R(0)$.

Proof. Since

$$\mathscr{B}_R=\{\xi+i\eta\in\mathbb{C}^n:|\xi|^2+|\eta|^2<R^2\},$$

it follows from the inequalities

$$|\xi|^2+|\eta|^2\leq q^2(\xi+i\eta)\leq|\xi|^2+|\eta|^2+2\sqrt{(|\xi|^2\,|\eta|^2)}$$
$$=(|\xi|+|\eta|)^2\leq 2(|\xi|^2+|\eta|^2)$$

that

$$\mathcal{B}_R\supset\tilde{B}_R(0)\supset\mathcal{B}_{R/\sqrt{2}}.$$

However, $\mathcal{B}_R$ and $\tilde{B}_R(0)$ have common boundary points, exactly those $\xi+i\eta$ such that

$$|(\xi,\eta)|=|\xi|\,|\eta|\quad\text{and}\quad|\xi|^2+|\eta|^2=R^2. \tag{3.7}$$

Further, $\tilde{B}_R(0)$ and $\mathcal{B}_{R/\sqrt{2}}$ have common boundary points where

$$(\xi,\eta)=0\quad\text{and}\quad|\xi|^2=|\eta|^2=(R/2)^2. \tag{3.8}$$

Thus, if $x\in\partial B_R(0)$, then

$$\{\xi+i\eta\colon\ \xi=x/2,\ \eta\in\textstyle\sum(i\xi)\}\subset V(x)\cap\partial\mathcal{B}_{R/\sqrt{2}}.\quad\|$$

In fact, if $x\in\partial B_R(0)$ then

$$\{\xi+i\eta\colon\ \xi=x/2,\ \eta\in\textstyle\sum(i\xi)\}=V(x)\cap\partial\mathcal{B}_{R/\sqrt{2}}.$$

Remark 3.4. The points $\xi+i\eta$ for which (3.7) holds play a special role. For them, ξ and η are linearly dependent so that $\xi=t\eta$ for some real t unless $\eta=0$. For such points,

$$\xi+i\eta=(t+i)\eta=\zeta\theta$$

where

$$\theta=R\frac{\eta}{|\eta|}\in\partial B_R(0)\quad\text{and}\quad\zeta=(t+i)\frac{|\eta|}{R}\in\mathbb{C}$$

with

$$|\zeta|^2=(t^2+1)\frac{|\eta|^2}{R^2}=\frac{|\xi|^2+|\eta|^2}{R^2}=1.$$

Proposition 3.7. In $\mathbb{C}^n$, $n\geq 2$, *if*

$$\Lambda_R=\{\zeta\theta\colon\ \theta\in\partial B_R(0),\ \zeta\in\mathbb{C},\quad\text{and}\quad|\zeta|=1\}$$

then

$$\Lambda_R=\partial\tilde{B}_R(0)\cap\partial\mathcal{B}_R$$

and consists of those points $z\in\partial\tilde{B}_R(0)$ *for which* $\sum(z)\subset\partial B_R(0)$. *These are the extreme points of the convex set* $\tilde{B}_R(0)$.

Proof. The central point is that $\xi+i\eta\in\Lambda_R$ if and only if ξ and η are linearly dependent (Remark 3.4) for then $\xi+i\eta=\zeta\theta$ implies that ξ and η are both multiples of $\theta\in\mathbb{R}^n$.

Linear dependence is equivalent to the property

$$q(\xi+i\eta)=(|\xi|^2+|\eta|^2)^{\frac{1}{2}}=|\xi+i\eta|,$$

from the definition of q, equation (3.6). Since $q(\xi+i\eta)=R$ defines $\partial\tilde{B}_R(0)$ and $|\xi+i\eta|=R$ defines $\partial\mathscr{B}_R$, it follows that

$$\Lambda_R=\partial\tilde{B}_R(0)\cap\partial\mathscr{B}_R.$$

Linear dependence is also equivalent to the orthogonality of ξ to $\prod(z)$, the $(n-1)$-plane through ξ, orthogonal to η, provided that $\eta\neq 0$. (If $\eta=0$ then $\{\xi\}=\sum(\xi+i\eta)$ lies in $\partial B_R(0)$.) This situation occurs if and only if every point of $\sum(\xi+i\eta)$ has distance $(|\xi|^2+|\eta|^2)^{\frac{1}{2}}=R$ from the origin, and so $\sum(\xi+i\eta)\subset\partial B_R(0)$.

It remains to exhibit the extreme points.

Since $\tilde{B}_R(0)$ is a subset of $\mathscr{B}_R$, a set for which every boundary point is an extreme point, it follows that every point of $\Lambda_R=\partial\tilde{B}_R(0)\cap\mathscr{B}_R$ is an extreme point.

Now consider a point $z=\xi+i\eta$ in $\partial\tilde{B}_R(0)\backslash\Lambda_R$. Observe that $\xi\neq 0$ and $\eta\neq 0$ in this case. We shall show that there is a line segment

$$\{z^t\colon z^t=(1-t)z^0+tz^1,\ 0\leq t\leq 1\}$$

lying in $\partial\tilde{B}_R(0)$ joining two points of Λ_R such that z is a point of the segment.

The point z^1 is $\xi^1+i\eta^1$, where ξ^1 is the nearest point of $\prod(z)$ to the origin and η^1 is that positive multiple of η for which z^1 falls in Λ_R. Specifically,

$$\xi^1=\frac{\eta}{|\eta|^2}(\xi,\eta)\quad\text{and}\quad\eta^1=\frac{|\eta^1|}{|\eta|}\eta$$

where $|\eta^1|=(R^2-|\xi^1|^2)^{\frac{1}{2}}$ is the distance from ξ^1 to every point of $\sum(z^1)=\prod(z)\cap\partial B_R(0)$.

The point $z^0=\xi^0$ is the intersection of $\partial B_R(0)$ with the ray from ξ^1 through ξ. In other words, z^0 is the single point which is the intersection of the $(n-1)$-sphere $\partial B_R(0)$ and the $(n-2)$-sphere $\sum(z)$. Specifically, since $|\eta^1|$ is the distance from ξ^1 to $\partial B_R(0)$ and $|\eta|$ is the distance from ξ to $\partial B_R(0)$, it is readily calculated that

$$\xi^0=\frac{|\eta^1|\,\xi-|\eta|\,\xi^1}{|\eta^1|-|\eta|},\qquad\eta^0=0$$

where $|\eta^1|-|\eta|$ is the distance from ξ^1 to ξ. Now let

$$\begin{aligned}z^t&=(1-t)(\xi^0+i\eta^0)+t(\xi^1+i\eta^1)\\&=\{(1-t)\xi^0+t\xi^1\}+it\eta^1\\&=\xi^t+i\eta^t.\end{aligned}$$

Since the distance from $\xi^t=(1-t)\xi^0+t\xi^1$ to ξ^0 is $t|\xi^1-\xi^0|=t|\eta^1|=|\eta^t|$ it follows that $\sum(\xi^t+i\eta^t)$ intersects $\partial B_R(0)$ at the single point ξ^0, for $0\leq t<1$.

Thus, each $z^t=\xi^t+i\eta^t$ is a boundary point of $\tilde{B}_R(0)$, and the given point z corresponds to $t=|\eta|/|\eta^1|$. ‖

The set Λ_R is actually the Šilov boundary of $\hat{B}_R(0)$.

Proposition 3.8. If $\theta\in\partial B_R(0)$ and if $\zeta\in\mathbb{C}$ with $|\zeta|=1$, then $r^2(x-\zeta\theta)$ is a non-vanishing function of x on $\tilde{B}_R(0)$. It has two square roots which are holomorphic on $\tilde{B}_R(0)$. If $\zeta=1$, one of these roots is the analytic continuation of $|x-\theta|$.

Proof. By its definition (Definition 2.1), the isotropic cone $V(\theta)$ is the locus of zeros of $r^2(x-\theta)$ in $\mathbb{C}^n$, and by the definition of a harmonicity hull (Definition 2.2) $r^2(x-\theta)$ does not vanish on $\tilde{B}_R(0)$. In fact, the continuation of $r(x-\theta)=|x-\theta|$ into $(\mathbb{R}^n\backslash\{\theta\})^\wedge$ is unbranched. The same is true of the continuation into the sub-domain $\hat{B}_R(0)=\tilde{B}_R(0)$.

Since $\tilde{B}_R(0)$ is a complete circular domain (Remark 3.3), if $|\zeta|=1$ then $\zeta x\in\tilde{B}_R(0)$ exactly when $x\in\tilde{B}_R(0)$. Thus $r^2(\zeta^{-1}x-\theta)$ is also holomorphic and non-zero for x in $\tilde{B}_R(0)$, a simply connected set, where it has two holomorphic square roots. The same is true of $r^2(x-\zeta\theta)=\zeta^2r^2(\zeta^{-1}x-\theta)$. ‖

Proposition 3.9. Let f be holomorphic on $\hat{B}=\hat{B}_R(0)$ and let

$$f(x)=\sum_{k=0}^{\infty}u_k(x) \tag{3.9}$$

be its (unique) representation as a series of homogeneous polynomials, u_k being of degree k. The series is uniformly absolutely convergent on compact subsets of $\hat{B}_R(0)$.

Proof. The series (3.9) is an arrangement of the Taylor series for f, and therefore is uniformly absolutely convergent on every compact polydisc about the origin lying in $\hat{B}$ where f is holomorphic. Since $\mathcal{B}_{R/\sqrt{2}}\subset\hat{B}$ and is a union of such polydiscs, the series is absolutely convergent on $\mathcal{B}_{R/\sqrt{2}}$.

Now let $x\in\hat{B}=\hat{B}_R(0)=\tilde{B}_R(0)$. It must be shown that (3.9) is uniformly absolutely convergent on some neighbourhood of x. Since $\hat{B}$ is a complete circular domain (Remark 3.3), there exists $\rho>1$ such that $\zeta x\in\hat{B}$ for each $\zeta\in\mathbb{C}$ with $|\zeta|\leq\rho$. Of course

$$f(x)=\frac{1}{2\pi i}\int_{|\zeta|=\rho}f(\zeta x)(\zeta-1)^{-1}\,d\zeta$$

$$=\sum_{k=0}^{\infty}\frac{1}{2\pi i}\int_{|\zeta|=\rho}f(\zeta x)\zeta^{-k-1}\,d\zeta \tag{3.10}$$

$$=\sum_{k=0}^{\infty}f_k(x) \tag{3.11}$$

where

$$f_k(z)=\frac{1}{2\pi i}\int_{|\zeta|=\rho} f(\zeta z)\zeta^{-k-1}\,d\zeta \quad \text{for} \quad z\in \hat{B}_{R/\rho}(0) \tag{3.12}$$

since the series

$$\sum_{k=0}^{\infty} \zeta^{-k-1}=(\zeta-1)^{-1}$$

is uniformly convergent on the circle in $\mathbb{C}$ where $|\zeta|=\rho$. Since f is bounded on $\partial\hat{B}_{R/\rho}(0)$, the series (3.11) will be uniformly absolutely convergent on $\hat{B}_{R/\rho}(0)$, which will be the required neighbourhood of x. It remains to show that (3.11) and (3.9) are the same because $f_k=u_k$.

If z is a point of $\mathscr{B}_{R/\rho\sqrt{2}}\subset\hat{B}_{R/\rho}(0)\subset\hat{B}$ and if $|\zeta|\leq\rho$, then $|\zeta z|<R/\sqrt{2}$ and the series (3.9) is uniformly absolutely convergent on $|\zeta|=\rho$. Thus the series (3.9) for $f(\zeta z)$ can be substituted into the integrand and the limiting processes interchanged:

$$\begin{aligned} f_k(z)&=\frac{1}{2\pi i}\int_{|\zeta|=\rho}\sum_{l=0}^{\infty} u_l(\zeta z)\zeta^{-k-1}\,d\zeta\\ &=\sum_{l=0}^{\infty}\frac{1}{2\pi i}\int_{|\zeta|=\rho} u_l(z)\zeta^{l}\zeta^{-k-1}\,d\zeta\\ &=u_k(z), \qquad z\in\mathscr{B}_{R/\rho\sqrt{2}}, \end{aligned}$$

that is f_k is the analytic continuation of the polynomial u_k to $\hat{B}$, so that (3.9) and (3.11) coincide. Since (3.11) has the required convergence property, (3.9) does also. ‖

See Siciak [81] for a similar result.

Remark 3.5. Unions of open polydiscs about the origin, the domains on which power series converge, are *complete Rheinhardt domains* in the terminology of Bochner and Martin [19, p. 81]. For $n=2$ it is known that the largest complete Rheinhardt domain in $\hat{B}_R(0)$ is the set $\{(z_1,z_2)\in\mathbb{C}^n : |z_1|+|z_2|<R\}$. For higher dimensions the set is not known, but it contains $\mathscr{B}_{R/\sqrt{2}}$ as a proper subset.

Our next example is the most important of the non-convex domains.

Definition 3.1. A *comball* is the complement in $\mathbb{R}^n$ of the closure of an open ball:

$$C_R=C_R(0)=\mathbb{R}^n\backslash\bar{B}_R(0), \qquad R>0$$

is the comball with *centre* at the origin and *radius* R. A comball with zero radius is the complement of a point.

Example 3.4. The harmonicity hull of the comball $C_R(0)$. If u is a function harmonic on $B_R = B_R(0)$, then

$$v(x) = u\left(\frac{x}{r^2(x)}\right) \tag{3.13}$$

is harmonic on $C_R = C_R(0)$ by the Kelvin transformation (Proposition 1.4, Chapter I). If the continuation $\tilde{u}$ from B_R to $\tilde{B}_R$ has natural boundary on $\partial\tilde{B}_R$, then the continuation $\tilde{v}$ of v from C_R to $\tilde{C}_R$ will also have natural boundary. It will be singular at points of the form

$$x^* = \frac{x}{r^2(x)}, \qquad x \in \partial\tilde{B}_R \setminus V(0) \tag{3.14}$$

where $V(0)$ is the isotropic cone though 0, on which r^2 vanishes.

Define

$$s(\xi + i\eta) = \begin{cases} 0, & \xi + i\eta = 0 \\ \dfrac{|r^2(\xi + i\eta)|}{q(\xi + i\eta)}, & \xi + i\eta \neq 0 \end{cases} \tag{3.15}$$
$$= [|\xi|^2 + |\eta|^2 - 2\sqrt{\{|\xi|^2\,|\eta|^2 - (\xi, \eta)^2\}}]^{\frac{1}{2}}$$

and observe that the homogeneity of q (Proposition 3.5) and of $|r|$ implies that s is also homogeneous:

$$s(\zeta x) = |\zeta|\, s(x) \quad \text{for all } \zeta \in \mathbb{C}. \tag{3.16}$$

Not only is it true that

$$qs = |r^2| \tag{3.17}$$

but also that

$$q(x)s(x^*) = 1, \qquad r^2(x) \neq 0. \tag{3.18}$$

Thus $q(x) < R$ if and only if $s(x^*) > 1/R$ provided that $x \notin V(0)$. It now follows that

$$\begin{cases} \tilde{C}_R = \{\xi + i\eta\colon \xi, \eta \in \mathbb{R}^n, s(\xi + i\eta) > R\} \\ \text{where } s(\xi + i\eta) = [|\xi|^2 + |\eta|^2 - 2\sqrt{\{|\xi|^2\,|\eta|^2 - (\xi, \eta)^2\}}]^{\frac{1}{2}}. \end{cases} \tag{3.19}$$

In fact, if v is harmonic on $C_R(0) = \mathbb{R}^n \setminus \bar{B}_R(0)$, then its Kelvin transform

$$v^*(x) = v\left(\frac{x}{r^2(x)}\right) = v(x^*)$$

is harmonic on $B_{1/R}(0)\setminus\{0\}$, whose harmonicity hull is

$$\begin{aligned} \tilde{B}_{1/R}(0) \setminus V(0) &= \{x^*\colon r^2(x^*) \neq 0,\, q(x^*) < 1/R\} \\ &= \{x^* : s(x) > R\} \end{aligned}$$

by (3.18), that is v^* has unlimited analytic continuation through $(B_{1/R}(0)\setminus V(0))^\sim$ if and only if v has unlimited analytic continuation throughout $\tilde{C}_R(0)$. ‖

Remark 3.6. Not only does $\tilde{C}_R(0)$ fail to be convex, it is closed under multiplication by complex numbers ζ with $|\zeta| \geq 1$. Its boundary intersects $\partial\tilde{B}_R(0)$ exactly in the set Λ_R of those points z lying also in $\mathscr{B}_R$, for which $\sum(z)$ is a subset of $\partial B_R(0)$.

In spite of the fact that the boundaries of $\tilde{B}_R(0)$ and $\tilde{C}_R(0)$ intersect in Λ_R, the following separation property holds.

Proposition 3.10. No point of $\tilde{B}_R(0)$ lies in the isotropic cone through any point of $(\tilde{C}_R(0))^{-}$, and conversely.

Proof. Let $x \in (\tilde{C}_R(0))^{-}$. The isotropic cone through x intersects $\mathbb{R}^n$ in $\sum(x)$, a subset of $\bar{C}_R(0)$, that is $r^2(x-z)$ does not vanish at any point of $\mathbb{R}^n \setminus \bar{C}_R(0)$ and in particular it is zero nowhere in $B_R(0)$. A square root can be extracted which is continuous on $B_R(0)$. With it we can define $\imath_1(x-y)$ for y in $B_R(0)$, where it is harmonic, and continue it to $\tilde{B}_R(0)$, the harmonicity hull.

Since $\imath_1(x-y)$ does not have continuation anywhere on the isotropic cone of x, it follows that $V(x)$ does not intersect $\tilde{B}_R(0)$.

The converse holds since $y \in V(x)$ if and only if $x \in V(y)$. ||

The examples given so far all have the property that $z \in \tilde{G}$ if and only if $\sum(z) \subset G$. A solid torus in $\mathbb{R}^3$ does not have this property, since it contains circular loops which are not even null-homotopic. The two elaborations which follow also provide counter-examples to the possibility that Lemma 3.1,

$$\left(\bigcap_t G_t\right)^{\wedge} = \bigcap_t \hat{G}_t,$$

might have analogues using either $\tilde{G}_t$ or unions.

Example 3.5. Domains G_1 and G_2 for which $\hat{G}_1 \cup \hat{G}_2 \neq (G_1 \cup G_2)^{\wedge}$, $\tilde{G}_1 \cup \tilde{G}_2 \neq (G_1 \cup G_2)^{\sim}$. Let G_1 be a solid torus about the x_3-axis in $\mathbb{R}^3$, let G be its convex hull, a domain consisting of all points within a given distance of a 2-disc, and let G_2 be an open neighbourhood of $G \setminus G_1$ which is a proper subdomain of G. Not only does it happen that $\hat{G}_1 = \tilde{G}_1$ and $\hat{G} = \tilde{G}$, but G_2 may be taken to have the same property. Nevertheless, there are circular arcs $\sum(z)$ about the x_3-axis in $G_1 \setminus G_2$ which are not even null-homotopic in G_1 but whose convex hulls are subsets of G. Such points are not in $\hat{G}_1$ or in $\hat{G}_2$ but are in $\hat{G}$. (Ex. 3.8 also illustrates this point.)

Example 3.6. Domains G_1 and G_2 for which $(G_1 \cap G_2)^{\sim} \neq \tilde{G}_1 \cap \tilde{G}_2$. Let G_1 be the cup-shaped domain in $\mathbb{R}^3$

$$G_1 = \{x : 1 < |x| < 2,\ x_3 < \tfrac{1}{2}\},$$

and let

$$G_2 = \{x : 1 < |x| < 2,\ x_3 > -\tfrac{1}{2}\}.$$

Each has the property that $\sum(z) \subset G_k$ if and only if $z \in \tilde{G}_k$, but

$$G = G_1 \cap G_2 = \{x : 1 < |x| < 2,\ |x_3| < \tfrac{1}{2}\}$$

is homeomorphic to a solid torus. It contains circles $\sum(z)$ such that $z \in \tilde{G}_1 \cap \tilde{G}_2$ but $z \notin \tilde{G}$.

Example 3.7. A null homotopic domain G for which some $z \notin \tilde{G}$ has $\sum(z) \subset G$. In the previous example, the cup with circular sections allowed the passage of tubes up the sides. That passage can be interrupted.

Let Q_R be the cube $\{(x_2, x_2, x_3) : |x_j| < R, j = 1, 2, 3\} \subset \mathbb{R}^3$, and consider $Q_{1+\varepsilon} \backslash \bar{Q}_1$, a thin-walled box in $\mathbb{R}^3$. Let G be the result of puncturing the top wall:

$$G = (Q_{1+\varepsilon} \backslash \bar{Q}_1) \backslash \{(0, 0, x_3) : 1 < x_3 < 1 + \varepsilon\}.$$

The result is homotopic to a punctured 2-sphere, and hence to a point. It is homeomorphic to a ball. However, the 1-spheres having centre $(0, 0, x_3)$ and lying in the 'top' of the box, in the set

$$\{x : 0 < x_1^2 + x_2^2 < 1 + \varepsilon, 1 < x_3 < 1 + \varepsilon\},$$

are not the ends of tubes beginning at a point. They can neither be slid 'down the sides' if $\varepsilon < \sqrt{2} - 1$, nor be shrunk to a point where they are, because of the puncture at $x_1^2 + x_2^2 = 0$.

Remark 3.7. Exx. 3.5–3.7 can be mimicked in $\mathbb{R}^n$ if $n \geqslant 3$, but not in $\mathbb{R}^2$, which is an extreme case. If z is a point of $\mathbb{C}^2 \backslash \mathbb{R}^2$ then $\sum(z)$ is an (ordered) pair of points in $\mathbb{R}^2$. Tubes are pairs of arcs in $\mathbb{R}^2$. Thus, if G is any domain in $\mathbb{R}^2$, then z lies in $\tilde{G}$ if (and only if) $\sum(z)$ lies in G. Since z is in $\hat{G}$ if and only if $\sum'(z) \subset G$, it follows that $\tilde{G}$ coincides with $\hat{G}$ if and only if G is convex, if $n = 2$. The same is not true for $n \geqslant 3$, as will be shown by Ex. 3.8.

Example 3.8. A non-convex domain G with $\tilde{G} = \hat{G}$. In $\mathbb{R}^3$, if

$$G = \{(x_1, x_2, x_3) : x_1 > 0 \text{ or } x_2 > 0\},$$

that is if G is $\mathbb{R}^3$ less a quadrant, then every circle in G surrounds a disc which also lies in G, so that $\tilde{G} = \hat{G}$. However, G is not convex, since the segment joining $(1, -2, 0)$ and $(-2, 1, 0)$ contains $(-\frac{1}{2}, -\frac{1}{2}, 0)$, which is not in G.

There are analogues of Ex. 3.8 in higher dimensions, but for $n > 3$ there exists a sequence of properties of which convexity and having $\tilde{G} = \hat{G}$ are only two cases. These are now introduced, to be illustrated in Ex. 3.9 and Remark 3.9.

Definition 3.2. A domain G in $\mathbb{R}^n$ will be said to be *convex in dimension* k if the convex hull of every k-sphere in G also lies in G. The property is satisfied vacuously for $k \geqslant n$, but for $k = 0, 1, \ldots, n-1$, the property may not be held universally. It will be referred to as Co(k).

Remark 3.8. Co(0) is simply convexity, since a 0-sphere is a pair of points. For a domain G in $\mathbb{R}^n$, Co($n-2$) is exactly the property that $\tilde{G}$ and $\hat{G}$ should coincide. Thus Co(0) and Co($n-2$) are the same for $n = 2$, as in Remark 3.7. The next result shows that Co(0), Co(1),... is a sequence of successively weaker properties. Any convex set has all of them. Ex. 3.9 will show that in $\mathbb{R}^n$ property Co($k+1$) is strictly weaker than Co(k) if $k < n$, although every domain has Co(n), vacuously.

Proposition 3.11. If a domain G has $\mathrm{Co}(k_0)$ *then it has* $\mathrm{Co}(k)$ *for* $k>k_0$ *as well.*

Proof. If S is a k-sphere in G with positive radius, let Π be the $(k+1)$-plane determined by S. If x lies in the $(k+1)$-ball $\mathrm{co}S$ which is the convex hull of S, then there exists a (k_0+1)-plane Π_0 through x and lying in Π. Since $\Pi_0 \cap S$ is a k_0-sphere whose convex hull contains x and lies in G by $\mathrm{Co}(k_0)$, it follows that $x \in G$. ‖

Proposition 3.12. If $D \subset \mathbb{R}^k$ *is a domain whose complement has no bounded components and if* $n>k>0$ *then* $D\times\mathbb{R}^{n-k}$ *is a domain in* $\mathbb{R}^n$ *having property* $\mathrm{Co}(k-1)$.

Proof. If $k=1$ then D is an interval. Since $D\times\mathbb{R}^{n-k}$ will be convex it has $\mathrm{Co}(0)$ *a fortiori.*

Let $k \geq 2$, let $G = D\times\mathbb{R}^{n-k}$, let Π be a k-plane in $\mathbb{R}^n$, let S be a $(k-1)$-sphere in $\Pi\cap G$, and let $P: G\to D$ be the orthogonal projection map onto $\mathbb{R}^k \cap G$.

Consider first the case in which Π contains a line L parallel to $\mathbb{R}^{n-k}$. If x is any point of $\mathrm{co}S$, then the line L' through x parallel to L either lies wholly in $G = D\times\mathbb{R}^{n-k}$, or never intersects G. Since L' contains two points of $S\subset G$, it follows that $x \in L' \subset G$. Thus, $\mathrm{co}S\subset G$.

In the alternative case, no line in Π is parallel to $\mathbb{R}^{n-k}$ and so the projection P may be restricted to Π making it a homeomorphism of $\Pi\cap G$ to D. Since $\mathrm{co}S$ disconnects Π, its interior can only intersect bounded components of $\Pi\backslash G$, which correspond by P to bounded components of $\mathbb{R}^k\backslash D$, of which there are none by hypothesis. Therefore $\mathrm{co}S$ lies in $\Pi\cap G$. ‖

Example 3.9. Domains in $\mathbb{R}^n$ *showing that* $\mathrm{Co}(k)$ *is strictly stronger than* $\mathrm{Co}(k+1)$ *if* $0\leq k$ *and* $k+1\leq n$. For $k=0$, let D be a bounded domain in $\mathbb{R}^2$ which is simply connected but not convex. For the case $n=2$ it has $\mathrm{Co}(1)$ by the simple connectivity, but lacks $\mathrm{Co}(0)$, which is convexity. For $n\geq 2$, take $G=D\times\mathbb{R}^{n-2}$, which has property $\mathrm{Co}(1)$ by Proposition 3.12 although it lacks $\mathrm{Co}(0)$, convexity, because D lacks it.

For $k=n-1$, take $G=\mathbb{R}^n\backslash\{0\}$, which has the vacuous property $\mathrm{Co}(n)$ but lacks $\mathrm{Co}(n-1)$.

For the remaining cases, $n\geq 3$.

If $k=n-2$, let $Q_R=\{x\in\mathbb{R}^n : |x_j|<R, j=1,\ldots,n\}$ and let $G=Q_R\backslash(Q_{3R/4})^-$. The $(n-1)$-sphere circumscribed about $Q_{3R/4}$, which has radius $3R\sqrt{n}/4$, intersects the exterior of $\bar{Q}_R$. Therefore there are no $(n-1)$-spheres in G whose convex hulls contain points of $(Q_{3R/4})^-$, so that G has $\mathrm{Co}(n-1)$. However, if x is a vertex of $Q_{3R/4}$ and Π is the $(n-1)$-plane orthogonal to the line through x and the origin then Π contains $(n-2)$-spheres in G with centre x. Their convex hulls contain x and do not lie in G. Hence G lacks $\mathrm{Co}(n-2)$.

For $1\leq k\leq n-3$ let $D_1=\mathbb{R}^{k+1}\backslash\{0\}$, a domain lacking $\mathrm{Co}(k)$. Define $D=D_1\times\mathbb{R}^1$, a domain in $\mathbb{R}^{k+2}$ whose complement has no bounded components, and let

$$G = D \times \mathbb{R}^{n-k-2} = D_1 \times \mathbb{R}^{n-k-1}.$$

By Proposition 3.12, G has Co$(k+1)$. However, it lacks Co(k) since $G \cap \mathbb{R}^{k+1} = D_1$ lacks it.

Remark 3.9. The construction given in the preceding example shows that for $n \geq 3$ Co(0) and Co(1) are distinct, so that Co(0) and Co$(n-2)$ are also distinct, that is there exist domains G lacking convexity but having the property $\hat{G} = \tilde{G}$. If D is any non-convex set in $\mathbb{R}^2$ then $D \times \mathbb{R}^{n-2}$ will be of this sort. For instance, see Ex. 3.8.

Example 3.10. A class of complements of closed convex sets. Let D be a domain in $\mathbb{R}^2$ whose complement is convex. For $n \geq 3$, let $G = D \times \mathbb{R}^{n-2}$ and let P be the orthogonal projection onto $\mathbb{R}^2$. If z is any point of $\hat{G}$, then $\sum'(z)$ is a convex set in G and $P(\sum'(z))$ is a convex set in D, separated from $\mathbb{R}^2 \backslash D$ by a supporting line L_t for the latter. In fact, $L_t \times \mathbb{R}^{n-2}$ separates $\mathbb{R}^n$ into two half-spaces, $\sum'(z)$ falling in the one, X_t, which does not intersect $\mathbb{R}^n \backslash G$. By Ex. 3.1,

$$\tilde{X}_t = \hat{X}_t = \{\xi + i\eta : \xi \in X_t, d^2(\xi, \partial X_t) + d^2(\eta, \partial X_t) > |\eta|^2\}$$

and $z \in \hat{X}_t$. It follows that

$$(D \times \mathbb{R}^{n-2})^{\wedge} = \{\xi + i\eta : \xi \in D \times \mathbb{R}^{n-2}, \sup\{d^2(\xi, L_t \times \mathbb{R}^{n-2}) + d^2(\eta, L_t \times \mathbb{R}^{n-2})\} > |\eta|^2\}.$$

If $n = 3$ and $\mathbb{R}^2 \backslash D$ is unbounded, or if $n \geq 4$, then Proposition 3.12 can be applied to show that $\tilde{G} = \hat{G}$, so that the formula holds for $\tilde{G}$ as well.

Proposition 3.13. If D is a domain in $\mathbb{R}^{n-1}$ and $Q = (-R, R)$ with $0 < R \leq \infty$, then

$$(D \times Q)^{\sim} \cap \mathbb{C}^{n-1} = \hat{D} \cap \{\xi + i\eta \in \mathbb{C}^{n-1} : |\eta| < R\}.$$

Proof. If $\xi + i\eta \in (D \times Q)^{\sim} \cap \mathbb{C}^n$, then it must happen that ξ lies in $D \times \{0\}$, with $\xi_n = 0$, and η in $\mathbb{R}^{n-1} \times \{0\}$, with $\eta_n = 0$. In particular, $\sum(\xi + i\eta)$ lies in a plane containing a normal to $D \times \{0\}$.

If $n = 2$ then $\sum(\xi + i\eta)$ is a 0-sphere, a pair of points:

$$\sum(\xi + i\eta) = \{(\xi_1, \pm\eta_1)\} \subset D \times Q \subset \mathbb{R}^2.$$

Since $D \times Q$ is a cartesian product, such a pair lies in $D \times Q$ if and only if $\xi_1 \in D$ and $|\eta_1| < R$. Because $D \subset \mathbb{R}^1$, this happens if and only if $\xi_1 + i\eta_1$ lies in $\hat{D}$ and $|\eta_1| < R$, by Definition 2.2.

If $n > 2$ then $\sum(\xi + i\eta)$, an $(n-2)$-sphere, intersects $D \times \{0\}$ in an $(n-3)$-sphere whose convex hull must also lie in $D \times \{0\}$ because it is the projection of $\sum(\xi + i\eta)$ itself. Thus $\hat{D} \supset (D \times Q)^{\sim} \cap \mathbb{C}^{n-1}$. Since $\sum(\xi + i\eta)$ lies in a plane containing a normal to $D \times \{0\}$, it can have a radius no greater than R, that is $|\eta| < R$, or

$$(D \times Q)^{\sim} \cap \mathbb{C}^{n-1} \subset \hat{D} \cap \{\xi + i\eta \in \mathbb{C}^{n-1} : |\eta| < R\}.$$

To prove the opposite inclusion one has only to observe that if $\sum'(x+iy)$ is an $(n-2)$-ball in D with radius less than R, then $\xi=(x,0)$ and $\eta=(y,0)$ determine a point $\xi+i\eta\in\mathbb{C}^n$ such that $\sum'(\xi+i\eta)\in D\times Q$. ‖

Notes

See the Notes in §2 for references concerning harmonicity hulls.

4. Monodromy groups of harmonicity hulls

Let D and U be domains in $\mathbb{R}^n$ and $\mathbb{C}^n$ respectively, $D\subset U$ and suppose that the analytic function u on D has unlimited analytic continuation in U (see §4, Chapter I). We shall define a monodromy group of u in U, a quotient group of the first fundamental group. After some terminology and results are recalled from algebraic topology, we shall show that the harmonicity hull $\tilde{D}$ has a natural monodromy group of its own which is constructed as a quotient group of a subgroup of the singular (homology) $n-1$ cycles. The monodromy group of a function u polyharmonic on D will then be identified with a subgroup of the monodromy group of the harmonicity hull, and finally the monodromy group of u will be realized in $\mathbb{R}^n$ rather than $\mathbb{C}^n$.

Let D be a domain in $\mathbb{R}^n$ or $\mathbb{C}^n$ and let $D\subset U$ where U is a domain in $\mathbb{C}^n$. Assume that u is analytic (or holomorphic) on D and has unlimited analytic continuation in U. For a fixed $x^0\in D$, among the closed paths in U starting at x^0 we distinguish those along which u returns to its original Taylor development at x^0. This class of paths is preserved under homotopy and comprises a normal subgroup $\Pi^0_{x^0}$ of the first fundamental group Π_{x^0} of U. The quotient group $\Pi_{x^0}/\Pi^0_{x^0}$ is called the *monodromy group of u in U*. By the connectedness of D, this group is independent of the base point x^0. The monodromy group is either finite or enumerable, the cardinality being equal to the multiplicity of the function u continued in U. The continuation of u along paths in U leads to a region $\check{U}$ on the Weierstrassian manifold of u (see §4, Chapter I). For every point in U the number of points in $\check{U}$ over this point is equal to the cardinality of the monodromy group and is therefore constant.

We digress to recall some facts from algebraic topology. Let D be a domain in $\mathbb{R}^n$ and let $F=\mathbb{R}^n\backslash D$. The homology and cohomology groups which we shall consider will have coefficients in a discrete abelian group $\mathcal{G}$.

The cohomology group of interest to us is formed by the set of all continuous functions $\varphi: F\to\mathcal{G}$ vanishing outside a compact set. These functions form an additive group which can be identified with $\check{H}^0_c(F;\mathcal{G})$, the Cech cohomology group with compact supports of dimension 0 of F relative to $\mathcal{G}$.†

† See, for example, Spanier [82].

Next we consider the singular $(n-1)$-dimensional homology group of D relative to $\mathcal{G}$. This is the quotient group

$$H_{n-1}(D;\mathcal{G}) = Z_{n-1}(D;\mathcal{G})/B_{n-1}(D;\mathcal{G})$$

where the elements of $Z_{n-1}(D;\mathcal{G})$ are the singular $n-1$ cycles in D and the elements of $B_{n-1}(D;\mathcal{G})$ are the singular $n-1$ boundaries in D. If $C \in Z_{n-1}(D;\mathcal{G})$, there is a function $w(p;C)$ defined for $p \in \mathbb{R}^n \backslash \text{support } C$, with values in $\mathcal{G}$, called the *index† of p with respect to C*, which (intuitively) counts the number of times that C winds around p. (In $\mathbb{R}^2$, it is the winding number of C about p.) In the simplicial case, considering C as a cycle in $\mathbb{R}^n$, it is the coefficient of a simplex to which p belongs in a simplicial subdivision of K, where $\partial K = C$. The index carries over to the singular case.

Theorem (*Alexander Duality*). *There is a canonical isomorphism between* $H_{n-1}(D;\mathcal{G})$ *and* $\check{H}^0_c(F;\mathcal{G})$.

The isomorphism can be described in the simplicial case as follows. If C is an $n-1$ cycle in D, define $\varphi(p) = w(p, C)$ for $p \in F$, and then $\varphi \in \check{H}^0_c(F;\mathcal{G})$. Conversely, if $\varphi \in \check{H}^0_c(F;\mathcal{G})$ define $F_q = \{x \in F : \varphi(x) = q\}$, $q \in \mathcal{G}$. The sets F_q are finite in number, compact for $q \neq 0$, mutually disjoint, and have union F. There are mutually disjoint open sets U_q such that $F_q \subset U_q$ and for $q \neq 0$ each $\bar{U}_q$ is formed by a finite number of simplices. If each $\bar{U}_q$, $q \neq 0$, is oriented consistently with $\mathbb{R}^n$ and provided with coefficient 1, we obtain a chain K_q with support equal to $\bar{U}_q$. The boundary of the chain $\sum_{q\neq 0} qK_q$, the $(n-1)$-dimensional cycle $\sum_{q\neq 0} q\,\partial K_q$, is the cycle corresponding to the given φ.

Now let u be polyharmonic on a domain $G \subset \mathbb{R}^n$. Since u has unlimited analytic continuation in the harmonicity hull $\tilde{G}$, it has a well-determined monodromy group as described before. $\tilde{G}$ also has a monodromy group which we now construct.

In §2, particularly Remark 2.3 and Theorems 2.3 and 2.4, we saw that the behaviour of the continuation of u along a closed path Γ in $\tilde{G}$, starting and ending at $x^0 \in G$, depended essentially on the winding number $s(x)$ of the path $r(x-\Gamma)$ in $\mathbb{C}^1$ relative to 0 where x is some fixed point of $\mathbb{R}^n \backslash G$, $V(x) \cap \Gamma = \phi$, and r is the function defined in §2. We shall express this winding number by a quantity defined in $\mathbb{R}^n$. We can replace Γ by an arbitrarily close homotopic closed path, starting and ending at x^0, formed by a finite number of segments. From now on we shall consider only such paths. We have

$$\Gamma = \bigcup_{k=0}^{N-1} [z^k; z^{k+1}], \qquad z^0 = z^N = x^0, \qquad [z^k; z^{k+1}] \cap V(x) = \phi.$$

† For a development of this topic, see Aleksandrov [2]; for connections with cup products, see Spanier [82].

For our present purposes we shall also assume that $z^k = \xi^k + i\eta^k$, ξ^k and η^k in $\mathbb{R}^n$, $\xi^k = \xi^{k+1}$ for k even and $\eta^k = \eta^{k+1}$ for k odd. Furthermore, we shall choose ξ^k and η^k (by submitting them to small changes if necessary) so that x does not belong to the $(n-1)$-plane of $\sum(z^k)$ for $k = 0,\ldots, N$. Thus $[\xi^k; z^k] \cap V(x) = \phi$.

Each segment $[z^k; z^{k+1}]$ can be replaced in Γ by a closed path

$$\Gamma_k + [\xi^k; z^k] + [z^k; z^{k+1}] - [\xi^{k+1}; z^{k+1}] - \Gamma_{k+1} = \Gamma'_k \tag{4.1}$$

where the Γ_k are any paths in $\mathbb{R}^n \backslash \{x\}$ joining x^0 to ξ^k. Note that the paths Γ'_k avoid $V(x)$. We now obtain the representation

$$\Gamma = \sum_{k=0}^{N-1} \Gamma'_k \tag{4.2}$$

in which all the terms are closed paths starting and ending at x^0. To calculate the winding number $s(x)$ of $r(x-\Gamma)$, it is enough to calculate and add the winding numbers $s_k(x)$ of $r(x-\Gamma'_k)$.

Consider the representation $\sum(\Gamma'_k)$. There are two cases: k odd and k even. When k is odd, the points of $[z^k; z^{k+1}]$ are

$$z^k(t) = (1-t)\xi^k + t\xi^{k+1} + i\eta^k$$

for $0 \leqslant t \leqslant 1$. The planes of $\sum(z^k(t))$ are mutually parallel and thus at most one of them contains x.

When k is even, the points of $[z^k; z^{k+1}]$ are

$$z^k(t) = \xi^k + i\{(1-t)\eta^k + t\eta^{k+1}\}.$$

It is immediately seen that in all cases the planes of $\sum(z^k(t))$ avoid x except perhaps for one such plane. In both cases all the planes corresponding to points in $[\xi^k; z^k]$ coincide and do not contain x.

It follows that whether k is odd or even, either all $(n-1)$-balls $\sum'(z)$ corresponding to the points of Γ'_k do not contain x or at most one $\sum'(z)$ corresponding to $z = z^k(t)$, $0 < t < 1$, contains x. In the first case Γ'_k is clearly homotopic in $\mathbb{C}^n \backslash V(x)$ to $\Gamma_k + [\xi_k; \xi_{k+1}] - \Gamma_{k+1}$, which means that the winding number is zero.

In the second case it is easy to see (by considering the homotopy of tubes) that we can choose an η small enough so that Γ'_k is homotopic in $\mathbb{C}^n \backslash V(x)$ to a nearby path $\check{\Gamma}_k$ such that $\check{\Gamma}_k$ is equal to $\Gamma''_k + \hat{\Gamma}_k - \Gamma'''_k$ where Γ''_k and Γ'''_k are paths in $\mathbb{R}^n \backslash \{x\}$ joining x^0 with $x-\eta$ and $x+\eta$ respectively and $\hat{\Gamma}_k$ is a path joining $x-\eta$ with $x+\eta$, whose points correspond by the transformation Σ to intersections of the $(n-1)$-sphere whose centre is x and radius $|\eta|$ with planes orthogonal to η, the orientation together with the direction of η giving the orientation of $\mathbb{R}^n$.

Since the winding number s_k is unchanged by homotopy and depends only on changes in the argument of $r(x-y)$, the paths Γ''_k and Γ'''_k, which lie in $\mathbb{R}^n \backslash (x)$, will not influence the winding number. Therefore the winding number is determined completely by the change in the argument of $r(x-y)$ on the path $\hat{\Gamma}_k$. An easy calculation shows that the argument changes from 0 to $\pm 2\pi$, the minus sign corresponding to the case when the orientation of the spheres $\sum(z)$ for $z \in \hat{\Gamma}_k$ gives to the whole $(n-1)$-sphere $|z-x|=|\eta|$ the positive orientation of $\mathbb{R}^n$, the plus sign corresponding to the other case. Therefore the winding number s_k equals the index $w(x;\{z:|z-x|=|\eta|\})$, the sphere being taken with the suitable orientation. It follows that the winding number about the origin of the path $r(x-\Gamma)$ in $\mathbb{C}^1$ is equal to the index $w(x;\Sigma\circ\Gamma)$, where $\Sigma\circ\Gamma$ is considered as an $(n-1)$-cycle in $\mathbb{R}^n$ with orientation determined by the orientation of the spheres $\sum(z)$ for $z \in \Gamma$.

In §2 we considered the serio-integral representation of a polyharmonic function u, initially defined in a domain $G \subset \mathbb{R}^n$, along a path Γ in $\tilde{G}$ starting from a point x^0 in G. We used a bounded region $D \subset \bar{D} \subset G$ with piecewise smooth boundary and in the serio-integral representation the integration was over ∂D, the path $\Gamma \subset \tilde{D}$. In the formulae in Remark 2.3 we used the winding number relative to the origin in $\mathbb{C}^1$ of the path $r(x-y)$ for y varying on Γ, $x \in \partial D$, which we can now replace by the index $w(x;\Sigma\circ\Gamma)$. Using this representation we obtain when n is odd

$$(4.3)\quad u(\hat{x}^0)=\Omega_n \sum_{l=0}^{\infty}\int_{\partial D}(-1)^{w(x;\Sigma\circ\Gamma)} \times\left(\Delta^l u(x)\frac{\partial \mathscr{r}_{l+1}(x-x^0)}{\partial \nu_x}-\mathscr{r}_{l+1}(x-x^0)\frac{\partial \Delta^l u(x)}{\partial \nu_x}\right)d\sigma(x),$$

and when $n=2m$ is even

$$(4.3')\quad u(\hat{x}^0)=\Omega_n \sum_{l=0}^{\infty}\int_{\partial D}\left(\Delta^l u(x)\frac{\partial \mathscr{r}_{l+1}(x-x^0)}{\partial \nu_x}-\mathscr{r}_{l+1}(x-x^0)\frac{\partial \Delta^l u(x)}{\partial \nu_x}\right)d\sigma(x)$$
$$-\pi i w(x,\Sigma\circ\Gamma)\Omega_n \sum_{l=m-1}^{\infty}\int_{\partial D}\left(\Delta^l u(x)\frac{\partial r^2(x-x^0)}{\partial \nu_x}\right.$$
$$\left.\times\frac{(l+1-m)r^{2l-n}(x-x^0)}{\gamma_l'}-\frac{\partial \Delta^l u(x)}{\partial \nu_x}r^{2l+2-n}(x-x^0)\right)d\sigma(x).$$

Here $\hat{x}^0$ is the point over x^0 (on the Weierstrassian manifold) corresponding to continuation from x^0 to x^0 along Γ.

From these considerations it is clear that the equality of the winding number about the origin of the path $r(x-\Gamma)$ with the index $w(x;\Sigma\circ\Gamma)$ is valid for any x belonging to $\mathbb{R}^n\backslash D$. Outside of the support of $\Sigma\circ\Gamma$ the index $w(x;\Sigma\circ\Gamma)$ is a continuous function of x which in each component of $\mathbb{R}^n\backslash$support $(\Sigma\circ\Gamma)$ takes a constant value. If we denote the union of the components where $w(x;\Sigma\circ\Gamma)=j$ by G_j (some G_j may be empty), then clearly $\partial D \subset \bigcup G_j$. Each set $\partial D \cap G_j$ is open and closed in ∂D and is a union of a finite number of components of ∂D, each of these components bounding a region outside of D. We can now write (4.3) and (4.3′) as

$$(4.4)\quad u(\hat{x}^0) = \Omega_n \sum_{l=0}^{\infty} \sum_j \int_{\partial D \cap G_j} (-1)^j \times \left(\Delta^l u(x) \frac{\partial \mathscr{r}_{l+1}(x-x^0)}{\partial \nu_x} - \mathscr{r}_{l+1}(x-x^0) \frac{\partial \Delta^l u(x)}{\partial \nu_x} \right) d\sigma(x)$$

$$(4.4')\quad u(\hat{x}^0) = \Omega_n \sum_{l=0}^{\infty} \int_{\partial D} \left(\Delta^l u(x) \frac{\partial \mathscr{r}_{l+1}(x-x^0)}{\partial \nu_x} - \mathscr{r}_{l+1}(x-x^0) \frac{\partial \Delta^l u(x)}{\partial \nu_x} \right) d\sigma(x) + \Omega_n \sum_{l=m-1}^{\infty} \sum_j \int_{\partial D \cap G_j} -\pi i j \left(\Delta^l u(x) \frac{r^2(x-x^0)}{\partial \nu_x} \times \frac{(l+1-m) r^{2l-n}(x-x^0)}{\gamma_l'} - \frac{\partial \Delta^l u(x)}{\partial \nu_x} \frac{r^{2l-2-n}(x-x^0)}{\gamma_l'} \right) d\sigma(x).$$

If D' is another bounded region with piecewise smooth Lipschitzian boundary such that $\bar{D} \subset D' \subset \bar{D}' \subset G$, we can write formulae (4.4) and (4.4′), replacing D by D'. The exterior regions bounded by components of $\partial D \cap G_j$ contain all the components of $\partial D' \cap G_j$ and the regions between the components of $\partial D \cap G_j$ and the components of $\partial D' \cap G_j$ lie in G. The integration along the boundaries of the latter regions in (4.4) and (4.4′) give zero. This means that for n odd

$$(4.5)\quad \sum_{l=0}^{\infty} \int_{\partial D \cap G_j} \left(\Delta^l u(x) \frac{\partial \mathscr{r}_{l+1}(x-x^0)}{\partial \nu_x} - \mathscr{r}_{l+1}(x-x^0) \frac{\partial \Delta^l u(x)}{\partial \nu_x} \right) d\sigma(x) = \sum_{l=0}^{\infty} \int_{\partial D' \cap G_j} \left(\Delta^l u(x) \frac{\partial \mathscr{r}_{l+1}(x-x^0)}{\partial \nu_x} - \mathscr{r}_{l+1}(x-x^0) \frac{\partial \Delta^l u(x)}{\partial \nu_x} \right) d\sigma(x).$$

The same formula is valid for $n = 2m$ even, but in addition we have

$$(4.5')\quad \sum_{l=m-1}^{\infty} \int_{\partial D \cap G_j} -\pi i j \left(\Delta^l u(x) \frac{\partial r^2(x-x^0)}{\partial \nu_x} \frac{(l+1-m) r^{2l-n}(x-x^0)}{\gamma_l'} - \frac{\partial \Delta^l u(x)}{\partial \nu_x} \frac{r^{2l+2-n}(x-x^0)}{\gamma_l'} \right) d\sigma(x) = \sum_{l=m-1}^{\infty} \int_{\partial D' \cap G_j} -\pi i j \left(\Delta^l u(x) \frac{\partial r^2(x-x^0)}{\partial \nu_x} \frac{(l+1-m) r^{2l-n}(x-x^0)}{\gamma_l'} - \frac{\partial \Delta^l u(x)}{\partial \nu_x} \frac{r^{2l+2-n}(x-x^0)}{\gamma_l'} \right) d\sigma(x).$$

Equations (4.5) and (4.5′) follow from Theorem 2.4(c). The above considerations lead us to the following definition.

Definition 4.1. In a region $G \subset \mathbb{R}^n$ consider the class of all $(n-1)$-cycles $\Sigma \circ \Gamma$ where Γ is a closed path in $\tilde{G}$ starting and ending at $x^0 \in G$. We consider this class as a subgroup of the group $Z_{n-1}(G)$ of singular $(n-1)$-cycles in G relative to the group of integers mod 2 for n odd and the group of all integers for n even.† This subgroup will be denoted by $\check{Z}(G)$. The subgroup of $\check{Z}(G)$ of cycles homologous to zero in G will be denoted $\check{Z}^0(G)$. The quotient group $\check{Z}(G)/\check{Z}^0(G)$ is the *monodromy group* of $\tilde{G}$.

To justify this definition we note that for any such closed path Γ in $\tilde{G}$ we can take a bounded region D with piecewise smooth Lipschitzian boundary such that $x^0 \in \text{support}(\Sigma \circ \Gamma) \subset D \subset \bar{D} \subset G$. If u is any polyharmonic function in G, its determination when contined along Γ does not change if and only if the index $w(x; \Sigma \circ \Gamma)$ is zero for all $x \in \partial D$. This means that the cycle $\Sigma \circ \Gamma$ is homologous to zero in D and therefore in G.

Theorem 4.1. Every function u polyharmonic in G has a monodromy group which is the quotient of $\check{Z}(G)$ by a group which contains $\check{Z}^0(G)$.

Corollary 4.1. The monodromy group of $\tilde{G}$ and the monodromy groups of all polyharmonic functions are abelian.

By definition, the monodromy group of $\tilde{G}$ is a subgroup of the homology group $H_{n-1}(G)$ (with a suitable group of coefficients). We can apply the Alexander Duality Theorem which gives a canonical representation of the above homology group on the zeroth cohomology group $\check{H}^0_c(\mathbb{R}^n \backslash G)$, and the monodromy group of $\tilde{G}$ becomes a subgroup of $\check{H}^0_c(\mathbb{R}^n \backslash G)$. The simplest case occurs when the monodromy group is equal to $\check{H}^0_c(\mathbb{R}^n \backslash G)$ since the determination of $\check{H}^0_c(\mathbb{R}^n \backslash G)$ is very direct: it is the additive group of all continuous functions on $\mathbb{R}^n \backslash G$ vanishing outside of a compact set and taking only integral values for n even and only integral values mod 2 for n odd. In particular, if the monodromy group of $\tilde{G}$ equals $\check{H}^0_c(\mathbb{R}^n \backslash G)$, all polyharmonic functions on G have single-valued continuations in $\tilde{G}$ if and only if $\check{H}^0_c(\mathbb{R}^n \backslash G) = (0)$. Hence, even though the reduced harmonicity hull $\hat{G}$ may be a proper subset of $\tilde{G}$, all the polyharmonic functions on G continue as single-valued functions on $\tilde{G}$. (Consider, for instance, a dented ball in $\mathbb{R}^n$.) This last phenomenon can also occur if the monodromy group of $\tilde{G}$ is a proper subgroup of $\check{H}^0_c(\mathbb{R}^n \backslash G)$, that is the monodromy group of $\tilde{G}$ can be (0). Example 4.1 will demonstrate this point.

Theorem 4.2. Let the monodromy group of $\tilde{G}$ equal $\check{H}^0_c(\mathbb{R}^n \backslash G)$. All polyharmonic functions on G have single-valued continuations to $\tilde{G}$ if and only if $F = \mathbb{R}^n \backslash G$ has no bounded connected component.

† Recall that the cycles $\Sigma \circ \Gamma$ are special in that, for each $z \in \Gamma$, $\Sigma(z)$ is an $(n-1)$-sphere rather than simply a continuous image of an $(n-1)$-sphere.

In view of the preceding discussion, this result is equivalent to the following.

Proposition 4.1. For any closed set $F \subset \mathbb{R}^n$, $\check{H}^0_c(F) = (0)$ *if and only if F has no bounded connected component.*

Proof. If every component of F is unbounded then $\check{H}^0_c(F) = (0)$ by definition. For the converse it will suffice to prove that if F has a bounded component B then there exists an open set $\mathcal{U}$ such that $B = \bar{B} \subset \mathcal{U}$ and $\mathcal{U}$ does not intersect any unbounded component of F.

Let $\mathcal{B}$ be an open ball in $\mathbb{R}^n$ such that $B \subset \mathcal{B}$. The space $F \cap \bar{\mathcal{B}}$ is a compact space. We shall show that it has a subset U, closed–open in $F \cap \bar{\mathcal{B}}$, such that $B \subset U$ and $U \cap \partial\mathcal{B} = \phi$. Supposing for the moment that this has been done, we see that U cannot intersect any unbounded component of F because any such component which intersects $\bar{\mathcal{B}}$ must intersect $\partial\mathcal{B}$ and U would provide a disconnection. Having such a U, we find a set $\mathcal{U}$ open in $\mathbb{R}^n$, lying in $\mathcal{B}$, such that $U = \mathcal{U} \cap F$ and $\mathcal{U}$ intersects no unbounded component of F.

Now for the construction. In a compact space every component is the intersection of all closed–open sets containing it (see Gillman and Jerison [44, p. 246]). Thus $B = \bigcap \{C : C$ closed–open in $F \cap \bar{\mathcal{B}}, C \supset B\}$. Let $K = F \cap \partial\mathcal{B}$. Now K is compact in $F \cap \bar{\mathcal{B}}$ and $K \cap B = \phi$, so $\bigcap \{K \cap C : C$ closed–open in $F \cap \bar{\mathcal{B}}, C \supset B\} = \phi$. Since each $K \cap C$ is a compact subset of a compact space, by the finite intersection property there exists a finite collection $C_1, \ldots, C_m$ such that

$$\bigcap_{j=1}^{m} (K \cap C_j) = \phi.$$

However,

$$U = \bigcap_{j=1}^{m} C_j$$

is closed–open in $F \cap \bar{\mathcal{B}}$, $U \supset B$, and $U \cap \partial\mathcal{B} = \phi$. The construction is complete and the proposition is proved. ||

Example 4.1. We show that there are domains G such that the monodromy group of $\tilde{G}$ is (0) and $\check{H}^0_c(\mathbb{R}^n \setminus G) \neq (0)$. Let $n = 3$ and let Q_R be the cube $\{(x_1, x_2, x_3) : |x_j| < R, j = 1, 2, 3\}$. Take $\varepsilon < 0{\cdot}01$ and let $G = Q_{1+\varepsilon} \setminus \bar{Q}_1$. The set $\mathbb{R}^3 \setminus G$ has only one bounded component, namely $\bar{Q}_1$, and therefore the group $\check{H}^0_c(\mathbb{R}^3 \setminus G)$ has one generator. Now let Γ be a closed path in $\tilde{G}$, starting and ending at a point in G. The path Γ can be taken piecewise linear, and the support of the corresponding tube $\sum(\Gamma)$ in G separates G into a finite number of components. If p is any point in $\mathbb{R}^3 \setminus Q_{1+\varepsilon}$, the index $w(p; \Sigma(\Gamma))$ is clearly zero. If p is in $\bar{Q}_1$, the index must also be zero. This follows because if the index were not zero, one of the components of G determined by the support of $\Sigma(\Gamma)$ would have to contain all of ∂Q_1 in its closure. This would imply that there is a circle in $\Sigma(\Gamma)$, and thus in G, whose convex hull contains the origin. By the choice of ε, this is impossible. Thus all cycles determined by tubes are homologous to zero, so the monodromy group of $\tilde{G}$ is (0).

Note that $\hat{G}$ is a proper subset of $\tilde{G}$, but that all polyharmonic functions on G have single-valued continuations to $\tilde{G}$. If ε is chosen large enough, however, the new domain G is homeomorphic to the G chosen above; $\check{H}_c^0(\mathbb{R}^3 \backslash G)$ still has one generator, but the monodromy group of $\tilde{G}$ equals $\check{H}_c^0(\mathbb{R}^3 \backslash G)$ and there are polyharmonic functions which branch in $\tilde{G}$.

III

SINGULARITIES IN $\mathbb{R}^n$

1. Principal parts

Definition 1.1. Let u be a function which is analytic on an open set $D \subset \mathbb{R}^n$. If x lies in ∂D and D is dense in some neighbourhood of x, then x is a *singular point* of u. If there exists a neighbourhood U of x and a function v analytic on U such that $u \equiv v$ on $D \cap U$, then x is a *removable singular point* of u, for u has analytic continuation to $D \cup U$, a neighbourhood of x.

This section concerns the behaviour of polyharmonic functions near singular points. Holomorphic capacity will be used briefly (see Appendix). The first theorem is certainly the most fundamental. It will be very much generalized in §1, Chapter IV.

Theorem 1.1. A holomorphic function which results from analytic continuation of a polyharmonic function from a domain D to a larger domain D′ in $\mathbb{R}^n$ or $\mathbb{C}^n$ is necessarily polyharmonic.

Proof. It suffices to let D' be a domain in $\mathbb{C}^n$ with u holomorphic on D' and polyharmonic on some subset $D \subset \mathbb{R}^n \cap D'$ which is open in $\mathbb{R}^n$. Let A be a subdomain of D which is relatively compact, and let G be a subdomain of D', relatively compact in D' and containing A. Thus, G can be taken to contain any given compact subset K of D'.

Since u has Laplacian order ∞ and type ∞, there exists $M < \infty$ such that

$$\|\Delta^p u\|_G \leq (2p)!\, M^{2p} \quad \text{for all large } p.$$

Also, for any $\varepsilon > 0$

$$\|\Delta^p u\|_A \leq (2p)!\, \varepsilon^{2p}$$

for all large p. If α is the infimum of the holomorphic capacity $\alpha(A, G; z)$ on K, then

$$\begin{aligned}\log\|\Delta^p u\|_K &\leq \alpha \log\{(2p)!\, \varepsilon^{2p}\} + (1-\alpha)\log\{(2p)!\, M^{2p}\} \\ &= \log(2p)! + 2p \log(\varepsilon^\alpha M^{1-\alpha}).\end{aligned}$$

Since A is open in $\mathbb{R}^n$, α is actually positive (Corollary A3.1 and Theorem A3.2, Appendix), and so $\varepsilon^\alpha M^{1-\alpha}$ can be made arbitrarily small by choosing ε small, that is for any $\eta > 0$

$$\|\Delta^p u\|_K \leq (2p)!\, \eta^{2p} \quad \text{if } p \text{ is large}.$$

Since K was an arbitrary compact set, it follows that u is polyharmonic on D'. ‖

Thus, the removal of a removable singularity of a polyharmonic function yields a polyharmonic function.

In $\mathbb{R}^n$ it is possible for an analytic function u to have isolated singular points. It will be shown that, if u is polyharmonic with isolated singular point x, then u can be written as a sum

$$u = R_u + A_u$$

where A_u is polyharmonic on a neighbourhood, a ball about x, and R_u is polyharmonic on $\mathbb{R}^n\backslash\{x\}$ and exhibits all the singular behaviour of u near x. The functions A_u and R_u will be called the 'analytic part' and the 'principal part' of u near x, because they are special cases of a more general construction to be given now.

Theorem 1.2. Let K be a compact set in $\mathbb{R}^n$, and let u be any function polyharmonic on some open set of the form $D\backslash K$, where D is open and contains K. There exists a function R_u polyharmonic on $\mathbb{R}^n\backslash K$ such that for each $y \notin K$ and each open set $G \subset D$ containing K but not y

$$\begin{aligned} R_u(y) &= -I(\partial G, u; y) \\ &= -\sum_{k=0}^{\infty} \Omega_n \int_{\partial G} \left\{ \Delta^k u(x) \frac{\partial \mathscr{r}_{k+1}(x-y)}{\partial \nu_x} - \mathscr{r}_{k+1}(x-y) \frac{\partial \Delta^k u}{\partial \nu}(x) \right\} d\sigma(x) \end{aligned}$$

provided that G is bounded, that $\partial G = \partial \bar{G}$ is an oriented piecewise smooth surface in D, and that ν is the interior normal for G. There also exists a function A_u, polyharmonic on D, such that if z lies in such a set G then $A_u(z) = I(\partial G, u; z)$. Together these functions satisfy

$$R_u(x) + A_u(x) = u(x), \qquad x \in D\backslash K.$$

Proof. Let G_1 and G_2 be any sets having the properties required for G. It suffices to consider the case $\bar{G}_1 \subset G_2$, since there exist appropriate sets G such that $\bar{G}_1 \cup \bar{G}_2 \subset G$.

Let E be any component of $G_2\backslash\bar{G}_1$. Since $\partial G_1 \cap \partial G_2 = \phi$, it follows that E has an oriented piecewise smooth boundary whose interior normal agrees with ν on ∂G_1 and with $-\nu$ on ∂G_2. Of course, u is polyharmonic on E and $y \notin \bar{E}$. Thus

$$(1.1) \qquad I(\partial G_2, u; y) - I(\partial G_1, u; y) = \begin{cases} u(y), & y \in G_2\backslash\bar{G}_1 \\ 0, & y \in \mathbb{R}^n\backslash(\bar{G}_2\backslash G_1) \end{cases}$$

by the serio-integral representation (Theorem 2.4, Chapter II), and so

$$-I(\partial G_2, u; y) = -I(\partial G_1, u; y)$$

for $y \notin \bar{G}_2$. This common value is $R_u(y)$.

Now

$$A_u(z) = I(\partial G_2, u; z)$$

defines a function polyharmonic on G_2. On G_1 it agrees with $I(\partial G_1, u; z)$, by (1.1), so that careful selection of G allows it to be continued to every point of D, where it is polyharmonic. Finally, if $z \in G_2 \backslash \bar{G}_1$ then

$$A_u(z) + R_u(z) = u(z)$$

by (1.1). ||

Example 1.1. If H is a bounded domain in $\mathbb{R}^2$ and v is a function which is polyharmonic on all of $\mathbb{R}^2$, then the function

$$u(y) = \begin{cases} v(y), & y \notin \bar{H} \\ 0, & y \in H \end{cases}$$

has $A_u \equiv v$ on $\mathbb{R}^2$ and

$$R_u(y) = \begin{cases} 0, & y \notin \bar{H} \\ -v(y), & y \in H. \end{cases}$$

For a function on an open subset of $\mathbb{R}^1$ let

$$u(y) = \begin{cases} v_1(y), & y < 0 \\ v_2(y), & y > 0 \end{cases}$$

where v_1 and v_2 are polyharmonic. By Theorem 1.1, Chapter II, v_1 and v_2 have analytic continuation to the whole real line. It is not hard to see that $A_u = (v_1 + v_2)/2$ and that

$$R_u(y) = \begin{cases} (v_1(y) - v_2(y))/2, & y > 0 \\ (v_2(y) - v_1(y))/2, & y < 0. \end{cases}$$

Remark 1.1. The usual concept of 'singularity' requires the exceptional compact set K to have no interior, but the construction is valid in the general case and the terminology will be introduced accordingly.

Definition 1.2. If K is a compact set in $\mathbb{R}^n$ and if u is polyharmonic on $D \backslash K$ where D is a domain containing K, then the functions R_u and A_u defined in the previous theorem and polyharmonic on $\mathbb{R}^n \backslash K$ and D respectively are the *principal part* and *analytic part of u near* K. Sometimes they will be designated $R(u, K; \cdot)$ and $A(u, K; \cdot)$.

Corollary 1.1. If K is a compact set in $\mathbb{R}^n$ and if u is polyharmonic on $D \backslash K$ where D is a neighbourhood of K, then $u = R_u + A_u$ on $D \backslash K$. Further, R_u has on $\mathbb{R}^n \backslash K$ every Laplacian order (and type) which u has on any neighbourhood of K, and A_u has on D every Laplacian order (and type) which u has on any neighbourhood of ∂D in D. If v is any function polyharmonic on $D \backslash K$ such that K consists only of removable singularities for $u - v$, then $R_u \equiv R_v$.

Proof. The first assertion is obvious from Theorem 1.2. The second comes from the properties of the serio-integral representation.

If $u-v$ has continuation to all of D then it is polyharmonic there and so $R_u - R_v \equiv R_{u-v} \equiv 0$. ‖

The corollary shows how the principal part summarizes the singular behaviour and justifies the terminology. There is also an analogue of the residue theorem.

Proposition 1.1. If K_1 and K_2 are disjoint compact sets in $\mathbb{R}^n$, if D is a domain in $\mathbb{R}^n$ containing K_1 and K_2, and if u is polyharmonic on $D\backslash(K_1 \cup K_2)$, then

$$R(u, K_1 \cup K_2; z) = R(u, K_1; z) + R(u, K_2; z)$$

for all z in $\mathbb{R}^n\backslash(K_1 \cup K_2)$.

The proof follows immediately from the integral representation of principal parts and the fact that K_1 and K_2 have disjoint neighbourhoods in D where suitable surfaces of integration can be constructed.

For a specific function v one can ask if it is the principal part of a function u. Of course, if v is the principal part for some function then it is also its own principal part. Here are examples of this kind.

Example 1.2. Let K consist of the origin in $\mathbb{R}^n$, and let u be the auxiliary function $\imath_p$. If $y \neq 0$ is a fixed point in $\mathbb{R}^n$ and $\varepsilon < |y|$, then

$$\begin{aligned} R_u(y) &= -I(\partial B_\varepsilon(0), \imath_p; y) \\ &= -\lim_{\varepsilon\to 0} I(\partial B_\varepsilon(0), \imath_p; y) \\ &= -\Omega_n \sum_{k=0}^{p-1} \lim_{\varepsilon\to 0} \int_{\partial B_\varepsilon(0)} \left(\imath_{p-k}(x)\frac{\partial \imath_{k+1}(x-y)}{\partial \nu_x} - \imath_{k+1}(x-y)\frac{\partial \imath_{p-k}}{\partial \nu}(x)\right) d\sigma(x). \end{aligned}$$

In the limit, every term but one has limit zero. Proposition 3.3, Chapter I, shows that

$$\begin{aligned} &\lim_{\varepsilon\to 0}\left|\int_{\partial B_\varepsilon(0)} \imath_{p-k}(x)\frac{\partial \imath_{k+1}(x-y)}{\partial \nu_x} d\sigma(x)\right| \\ &\leq \lim_{\varepsilon\to 0}\left[\max\left\{\left|\frac{\partial \imath_{k+1}(x-y)}{\partial \nu_x}\right| : x \in \partial B_\varepsilon(0)\right\} C\varepsilon^{2p-2k-n} \log \varepsilon \int_{\partial B_\varepsilon(0)} d\sigma(x)\right] \\ &\leq C' \lim_{\varepsilon\to 0}[\varepsilon^{2p-2k-1} \log \varepsilon] \leq C' \lim_{\varepsilon\to 0}[\varepsilon \log \varepsilon] = 0 \end{aligned}$$

since $k \leq p-1$. All but one of the other terms also vanishes in the limit. The exception, appearing at $k=p-1$, gives

$$\begin{aligned} R_u(y) &= -\Omega_n \lim_{\varepsilon\to 0} \int_{\partial B_\varepsilon(0)} \left\{ -\mathscr{h}_p(x-y)\frac{\partial \mathscr{h}_1}{\partial \nu}(x) \right\} d\sigma(x) \\ &= \Omega_n \mathscr{h}_p(0-y) \lim_{\varepsilon\to 0} \int_{\partial B_\varepsilon(0)} \frac{\partial \mathscr{h}_1}{\partial \nu}(x)\, d\sigma(x) \\ &= \mathscr{h}_p(y), \end{aligned}$$

by the definition of Ω_n (see formula (2.10), Chapter I), that is each of the functions $\mathscr{h}_p$ is its own principal part.

It is easy to see that if c is any constant then

$$u = \sum_{k=0}^{\infty} c^k \mathscr{h}_k$$

is also its own principal part, by observing the absolute convergence of all the series of integrals. Since the convergence is also uniform on compacts in the reduced harmonicity hull of $\mathbb{R}^n\backslash\{0\}$, it follows that $\Delta u = cu$, and so u has Laplacian order 1 and type $|c|^{\frac{1}{2}}$ (see Ex. 1.2, Chapter II).

Now consider $u = D^\alpha \mathscr{h}_p$ where α is a multi-index (§1, Chapter I) and $|\alpha|$ is large. If $y \neq 0$ is a fixed point of $\mathbb{R}^n$ and $R > |y|$, then

$$\begin{aligned} |u(y) - R_u(y)| &= |A_u(y)| \\ &= |I(\partial B_R(0), D^\alpha \mathscr{h}_p; y)| \\ &= \lim_{R\to\infty} |I(\partial B_R(0), D^\alpha \mathscr{h}_p; y)| \\ &\leq \Omega_n \sum_{k=0}^{p-1} \lim_{R\to\infty} \int_{\partial B_R(0)} \left| D^\alpha \mathscr{h}_{p-k}(x) \frac{\partial \mathscr{h}_{k+1}(x-y)}{\partial \nu_x} \right. \\ &\qquad\qquad \left. - \mathscr{h}_{k+1}(x-y) \frac{\partial D^\alpha \mathscr{h}_{p-k}(x)}{\partial \nu} \right| d\sigma(x) \\ &\leq C \lim_{R\to\infty} \left[R^{2p-2n-|\alpha|+1} (1+|\log R|)^2 \int_{\partial B_R(0)} d\sigma(x) \right] \\ &= 0, \text{ provided that } 2p-2n-|\alpha|+1 < 1-n \end{aligned}$$

where the estimate comes from Proposition 3.3, Chapter I. Thus, if $|\alpha| > 2p-n$ then $u = R_u$, that is, $D^\alpha \mathscr{h}_p$ is its own principal part if $|\alpha| > 2p-n$.

Remark 1.2. It is clear that if D' is a component of $\mathbb{R}^n\backslash K$, then the restriction of R_u to D' has continuation to $\tilde{D}'$, the restriction of A_u to D has continuation to $\tilde{D}$, and on $\tilde{D} \cap \tilde{D}'$ the property $R_u + A_u = u$ holds for the analytic continuations. If $n=1$ then $\tilde{D} = \tilde{D}' = \mathbb{C}$, so that the continuation should be done component by component, but if $n \geq 2$ then

$$R_u + A_u = u \quad \text{on} \quad \tilde{D} \cap (\mathbb{R}^n\backslash K)^{\sim}.$$

The proof of Theorem 1.2 is somewhat reminiscent of that for the Laurent series for holomorphic functions. If K is a single point $\{\xi\}$, then the analogy goes as far as the representation of R_u as a series of constant multiples of the functions $D^q \mathscr{r}_p$. Recall that $B_R(\xi)$ is a ball in $\mathbb{R}^n$ and $C_R(\xi)$ the comball $\mathbb{R}^n\backslash\overline{B_R(\xi)}$, while $\mathscr{B}_R(\xi)$ is a ball in $\mathbb{C}^n$.

Lemma 1.1. Let $n \geq 1$ and let A be any linear combination of the functions r^k and $r^k \log r$ on $\mathbb{R}^n$.

(a) *If ξ is an arbitrary expansion point in $\mathbb{R}^n$, then the Taylor series*

$$A(x-y) = \sum_{|q|\geq 0} \frac{(x-\xi)^q}{q!} D^q A(\xi - y) \tag{1.2}$$

is uniformly absolutely convergent for $(x; y)$ on compact subsets of

$$\begin{cases} \mathscr{B}_R(\xi) \times \hat{C}_{\sqrt{2}R}(\xi) & \text{if} \quad n \geq 2 \\ \mathscr{B}_R(\xi) \times \mathbb{C}\backslash\bar{\mathscr{B}}_R(\xi) & \text{if} \quad n = 1 \end{cases}$$

for any $R > 0$.

(b) *If $\xi, y \in \mathbb{R}^n$ and $y \neq \xi$, then the rearrangement of* (1.2) *in homogeneous polynomials*

$$A(x-y) = \sum_{\alpha=0}^{\infty} \sum_{|q|=\alpha} \frac{(x-\xi)^q}{q!} D^q A(\xi - y) \tag{1.3}$$

is uniformly convergent for x in compact subsets of $\hat{B}_{|y|}(\xi)$ if $n \geq 2$, of $\mathscr{B}_{|y|}(\xi)$ if $n = 1$.

Proof. For $n \geq 2$, the isotropic cone through a point of $\tilde{B}_{\sqrt{2}R}(\xi)$ does not intersect $\tilde{C}_{\sqrt{2}R}(\xi)$, the harmonicity hull of the complementary comball, and the isotropic cone through a point of $\tilde{C}_{\sqrt{2}R}(\xi)$ does not intersect $\tilde{B}_{\sqrt{2}R}(\xi)$ (see Proposition 3.10, Chapter II). $\mathscr{B}_R(\xi)$ is the largest ball about ξ contained entirely in $\hat{B}_{\sqrt{2}R}(\xi)$ (see Proposition 3.6, Chapter II). It follows that the function $u(x, y) = A(x-y)$ is holomorphic on $\mathscr{B}_R(\xi) \times \hat{C}_{\sqrt{2}R}(\xi)$. If η is any point of $C_{\sqrt{2}R}(\xi)$, then u is holomorphic on a neighbourhood of a compact polydisc about $(\xi; \eta) \in \mathbb{C}^{2n}$, and the power series for u about $(\xi; \eta)$ is uniformly absolutely convergent there. Now part (a) follows, for (1.2) is an arrangement of that series and $\mathscr{B}_R(\xi) \times \hat{C}_{\sqrt{2}R}(\xi)$ is a union of such compact polydiscs.

For $n = 1$, the disjointness of the factor sets is enough to ensure that $r(x-y) = |x-y|$ is non-zero for x in one and y in the other. Hence u is holomorphic on the product.

Part (b) is obvious for $n = 1$ since $|q| = q$. For $n \geq 2$ it follows from Proposition 3.9, Chapter II. ‖

Proposition 1.2. If D is a domain in $\mathbb{R}^n$, if ξ is a point of D, and if u is polyharmonic on $D\backslash\{\xi\}$, then the principal part of u near $K = \{\xi\}$ can be written

$$R_u(y) = \sum_{k=0}^{\infty} \sum_{|q|\geq 0} \frac{1}{q!} D^q \mathscr{r}_{k+1}(\xi - y) J_{k+1,q}(R) \tag{1.4}$$

where $R>0$ is chosen such that

$$R<\begin{cases}|y-\xi|/\sqrt{2}, & n\geqslant 2\\ |y-\xi|, & n=1\end{cases}$$

with $\bar{B}_R(\xi)\subset D$, and

$$(1.5)\qquad J_{k+1,q}(R)=\Omega_n\int_{\partial B_R(\xi)}(x-\xi)^q\left\{\frac{|q|}{R}\Delta^k u(x)+\frac{\partial\Delta^k u}{\partial\nu}(x)\right\}d\sigma(x).$$

The series (1.4) *is uniformly absolutely convergent on compact subsets of $C_{\sqrt{2}R}(\xi)$ if $n\geqslant 2$, of $C_R(\xi)$ if $n=1$. If u is polyharmonic of finite degree p, then $J_{k+1,q}=0$ for $k\geqslant p$.*

Proof. Let K be compact in $\mathbb{R}^n\backslash\{\xi\}$ and let R be small enough that K does not intersect $\bar{B}_R(\xi)$, or $\bar{B}_{\sqrt{2}R}(\xi)$ if $n\geqslant 2$. By the previous lemma, the series

$$(1.6)\qquad \imath_{k+1}(x-y)=\sum_{|q|\geqslant 0}\frac{(x-\xi)^q}{q!}D^q\imath_{k+1}(\xi-y)$$

of holomorphic functions is uniformly absolutely convergent for $(x;y)$ in a $\mathbb{C}^{2n}$ neighbourhood of $\bar{B}_R(\xi)\times K$. There it can be differentiated and integrated arbitrarily, term by term. Differentiation with respect to the x variables gives

$$\begin{aligned}\frac{\partial\imath_{k+1}(x-y)}{\partial\nu_x}&=\sum_{|q|\geqslant 0}\frac{1}{q!}\frac{\partial(x-\xi)^q}{\partial\nu_x}D^q\imath_{k+1}(\xi-y)\\&=-\sum_{|q|\geqslant 0}\frac{1}{q!}\frac{|q|}{R}(x-\xi)^qD^q\imath_{k+1}(\xi-y)\end{aligned}$$

on the surface $\partial B_R(\xi)$. Substitution of this and (1.6) into the serio-integral representation for R_u (Theorem 1.2) gives the result. ||

That the coefficients $J_{k+1,q}(R)$ do indeed vary with R is shown by the following example, which is a continuation of Ex. 1.2.

Example 1.3. Let $u(x)=D^\alpha\imath_p(x)$ where the multi-index $\alpha=(\alpha_1,\ldots,\alpha_n)$ and p are fixed, and let $\xi=0$. It was seen in the proof of Proposition 3.3, Chapter I, that $\Delta^k u$ can be written

$$\begin{aligned}\Delta^k u(x)&=D^\alpha\Delta^k\imath_p(x)=D^\alpha\imath_{p-k}(x)\\&=\{P_{\alpha,p-k}(x)\log r(x)+Q_{\alpha,p-k}(x)\}r^{2p-2k-n-2|\alpha|}(x)\end{aligned}$$

where P and Q are polynomials in the components of x, homogeneous of degree $|\alpha|$. The normal derivative for the ball is of the form

$$\frac{\partial\Delta^k u(x)}{\partial\nu}=\{P'_{\alpha,p-k}(x)\log r(x)+Q'_{\alpha,p-k}(x)\}r^{2p-2k-n-2|\alpha|-1}(x)$$

where P' and Q' are also homogeneous polynomials of degree $|\alpha|$. Substitution into formula (1.5) followed by integration of the polynomials shows that

$$J_{k+1,q}(R) = R^{2(p-k-1)+|q|-|\alpha|}(c_{\alpha,p-k,q} \log R + b_{\alpha,p-k,q}),$$

which depends on p unless both c is zero and either b or $2(p-k-1)+|q|-|\alpha|$ is zero. If, for instance, $n=2$ and $u = \mathscr{k}_1 = D^0\mathscr{k}_1 = -\log r$, then $|\alpha|=0$ and the coefficients are zero for $k \geqslant 1$. For $k=0$, however,

$$P_{0,1} \equiv -1, \; Q_{0,1} \equiv 0, \; P'_{0,1} \equiv 0, \quad \text{and} \quad Q'_{0,1} \equiv 1.$$

Thus

$$\begin{aligned} J_{1,q}(R) &= \Omega_2 \int_{\partial B_R(0)} x^q R^{-1}(-|q| \log R + 1)\, d\sigma(x) \\ &= R^{|q|}(1-|q| \log R) \frac{1}{2\pi} \int_0^{2\pi} \cos^{q_1} \theta \sin^{q_2} \theta \, d\theta. \end{aligned}$$

If either q_1 or q_2 is odd, then the integral is zero. If both q_1 and q_2 are even, then

$$\frac{1}{2\pi} \int_0^{2\pi} \cos^{q_1}\theta \sin^{q_2}\theta \, d\theta = q! \frac{(|q|/2)!}{(q_1/2)!\,(q_2/2)!} 2^{-|q|}((|q|/2)!)^{-2}$$

so that the series for R_u has coefficients which do depend on R. However, it can be rearranged into a series of terms homogeneous in R:

$$\begin{aligned} R_u(y) &= \sum_{k=0}^{\infty} R^{2k} \frac{(1-2k \log R)}{2^{2k}(k!)^2} \sum_{|q|=2k} \frac{(|q|/2)!}{(q_1/2)!\,(q_2/2)!} D^q \mathscr{k}_1(y) \\ &= \sum_{k=0}^{\infty} \frac{R^{2k}(1-2k \log R)}{2^{2k}(k!)^2} \Delta^k \mathscr{k}_1(y). \end{aligned}$$

These terms are all zero except for $k=0$.

Remark 1.3. A polyharmonic function has a principal part near any singular point isolated in $\mathbb{R}^n$. An analytic function which is not polyharmonic may or may not have such a principal part, but even if it has the most one can expect is a 'local principal part'.

Specifically, let D be a domain in $\mathbb{R}^n$, $n \geqslant 1$, and let u be analytic on $D \backslash \{x_0\}$, where $x_0 \in D$. For each $\sigma > 0$ define $D_\sigma = D \cap B_{1/\sigma}(x_0)$. If there exists a finite τ such that u is of Laplacian order ∞ and type τ on $D_\tau \backslash \{x_0\}$, then u has 'local principal and analytic parts', R_u and A_u, which can be given as follows.

If $y \in D_\tau \backslash \{x_0\}$ and $B = B_R(x_0) \subset D_\tau$ with R less than the minimum of $|y-x_0|$ and $(\tau - |y-x_0|)^{-1}$, then

$$R_u(y) = -I(\partial B, u; y), \qquad A_u(y) = u(y) - R_u(y)$$

where the inward normal for the surface ∂B is used. The continuation of

A_u to a neighbourhood of x_0 is obtained by

$$A_u(z) = I(\partial B, u; z), \qquad |z - x_0| < 1/2\tau$$

where B has radius less than $(\tau - |z - x_0|)^{-1}$.

The existence of these functions is assured by the local serio-integral construction (Theorem 2.3, Chapter II). The other properties are obtained from arguments like those for polyharmonic functions in the proof of Theorem 1.2.

2. Separation of singularities

In the preceding section a study was made of the singular behaviour of a function polyharmonic on a set $D \backslash K$ where K is a compact subset of D. In this section the singular set of a polyharmonic function u can be an arbitrary closed set F without interior. For any decomposition $F = F_1 \cup F_2$ where F_1 and F_2 are closed but not necessarily disjoint, it will be shown that u can be written $u = u_1 + u_2$, where the singular set of u_i is F_i, $i = 1, 2$. First we prove an approximation lemma, reminiscent of Runge's Theorem. The qualitative aspects of this lemma are summarized in Proposition 2.1, but the detailed estimates in the proof will be used in the proof of the main result on separation of singularities.

Lemma 2.1. Let $n \geqslant 1$, let S be a bounded piecewise smooth surface in $\mathbb{R}^n$, let $u \in \mathscr{C}^{2L-1}$ on some neighbourhood of $\bar{S}$, and let T be the polyharmonic function of finite degree L defined on $\mathbb{R}^n \backslash \bar{S}$ by

$$(2.1) \quad T(y) = \sum_{l=0}^{L-1} \Omega_n \int_S \left(\Delta^l u(x) \frac{\partial \mathscr{h}_{l+1}}{\partial \nu_x}(x-y) - \mathscr{h}_{l+1}(x-y) \frac{\partial \Delta^l u}{\partial \nu}(x) \right) d\sigma(x).$$

Let K be a compact subset of $\mathbb{R}^n \backslash \bar{S}$.

(a) *If ξ is a point of $\mathbb{R}^n$ and if there exists $R > 0$ such that $\bar{S} \subset B_R(\xi)$ and $K \subset \mathbb{R}^n \backslash \bar{B}_{\sqrt{2}R}(\xi)$, or $K \subset \mathbb{R} \backslash \bar{B}_R(\xi)$ if $n = 1$, then for every $\varepsilon > 0$ and $P \geqslant 0$ there exists a function f polyharmonic of degree L on $\mathbb{R}^n \backslash \{\xi\}$ such that*

$$(2.2A) \qquad \|D^\alpha T - D^\alpha f\|_K \leqslant \varepsilon \quad \textit{if} \quad |\alpha| \leqslant P, \quad \textit{and}$$

$$(2.2B) \qquad \|\Delta^p T - \Delta^p f\|_K \leqslant \varepsilon \quad \textit{for all } p \geqslant 0.$$

(b) *If $\xi \in \mathbb{R}^n \backslash \bar{S}$ and if there exists $R > 0$ such that $K \subset B_R(\xi)$ and $\bar{S} \subset \mathbb{R}^n \backslash B_{\sqrt{2}R}(\xi)$, or $\bar{S} \subset \mathbb{R} \backslash \bar{B}_R(\xi)$ if $n = 1$, then for every $\varepsilon > 0$ and $P \geqslant 0$ there exists a function g polyharmonic of degree L on all of $\mathbb{R}^n$ such that*

$$(2.3A) \qquad \|D^\alpha T - D^\alpha g\|_K \leqslant \varepsilon \quad \textit{if} \quad |\alpha| \leqslant P, \quad \textit{and}$$

$$(2.3B) \qquad \|\Delta^p T - \Delta^p g\|_K \leqslant \varepsilon \quad \textit{for all } p \geqslant 0.$$

Proof. Since f and g are both to be polyharmonic of finite degree L, the inequalities (2.2B) and (2.3B) are automatically satisfied for $p \geqslant L$. For

$p<L$ they follow from instances of (2.2A) and (2.3A) in which $P \geqslant 2L$ and ε is replaced by a smaller value.

(a) In (2.1), if $\mathscr{r}_{l+1}(x-y)$ is replaced by its power series in x about ξ (equation (1.2)), then the uniform convergence given by Lemma 1.1 yields a series of functions of y:

$$T(y) = \sum_{l=0}^{L-1} \Omega_n \int_S \left[\Delta^l u(x) \sum_{|q| \geqslant 0} \frac{1}{q!} D^q \mathscr{r}_{l+1}(\xi - y) \frac{\partial (x-\xi)^q}{\partial \nu_x} - \frac{\partial \Delta^l u}{\partial \nu}(x) \sum_{|q| \geqslant 0} \frac{1}{q!} \{D^q \mathscr{r}_{l+1}(\xi - y)\}(x-\xi)^q \right] d\sigma(x)$$
$$= \sum_{|q| \geqslant 0} \frac{1}{q!} \sum_{l=0}^{L-1} D^q \mathscr{r}_{l+1}(\xi - y) \times \Omega_n \int_S \left\{ \Delta^l u(x) \frac{\partial (x-\xi)^q}{\partial \nu_x} - (x-\xi)^q \frac{\partial \Delta^l u}{\partial \nu}(x) \right\} d\sigma(x). \tag{2.4}$$

The compact set K has a neighbourhood in $\mathbb{C}^n$ on which this series of holomorphic functions converges uniformly. Therefore some finite partial sum of (2.4) will have the required properties (2.2A) and (2.2B).

(b) The roles of $x \in \bar{S}$ and $y \in K$ must be interchanged. The power series for $\mathscr{r}_{l+1}$ cannot be replaced by the Laurent series, whose region of convergence is wrong if $n \geqslant 2$, so the terms of the power series will be transformed by reciprocal radii (reflected in the unit sphere about ξ) using the following identities in $\mathbb{R}^n$:

$$\begin{cases} x^* - \xi = (x-\xi)/|x-\xi|^2 \\ |x-y| = |x-\xi|\,|y-\xi|\,|x^*-y^*|. \end{cases} \tag{2.5}$$

The continuations of these identities into the complex domain are

$$\begin{aligned} x^* - \xi &= (x-\xi) r^{-2}(x-\xi) \\ r(x-y) &= r(x-\xi) r(y-\xi) r(x^*-y^*), \end{aligned} \tag{2.6}$$

and because they are less familiar than the real forms they will be written out wherever convenient.

Since S has compact closure in $\mathbb{R}^n \backslash \bar{B}_{\sqrt{2}R}(\xi)$, its transform S^* has compact closure in $B_{1/(\sqrt{2}R)}(\xi)$. The transform K^* of K lies in $\mathbb{R}^n \backslash B_{1/R}(\xi)$, and if $\xi \notin K$ then K^* is compact. We temporarily assume that $\xi \notin K$. Finally, note that without loss of generality we can assume that K contains a sphere with centre ξ. Under these conditions, the assumptions in (a) are fulfilled with S replaced by S^*, K replaced by K^*, and R replaced by $1/(\sqrt{2}R)$. The new variables are x^* and y^*. In order to make the substitution in (2.1) of the power series for $\mathscr{r}_{l+1}$, as was done in (2.4), we need to transform the series to a series in x^*. The terms r^{2l+2-n} will be treated first, and then the terms $r^{2l+2-n} \log r$.

First we obtain

$$
\begin{aligned}
(2.7)\quad r^{2l+2-n}(x-y) &= |x-y|^{2l+2-n} \\
&= (|x-\xi|\,|y-\xi|)^{2l+2-n}\,|x^*-y^*|^{2l+2-n} \\
&= (|x-\xi|\,|y-\xi|)^{2l+2-n} \sum_{|q|\geqslant 0} \frac{1}{q!}\{D^q r^{2l+2-n}(\xi-y^*)\} \\
&\quad \times (x^*-\xi)^q \\
&= (|x-\xi|\,|y-\xi|)^{2l+2-n} \sum_{|q|\geqslant 0} \frac{1}{q!} \\
&\quad \times \{Q_{q,l+1}(y^*-\xi)\,|y^*-\xi|^{2l+2-n-2|q|}\}(x^*-\xi)^q \\
&= r^{2l+2-n}(x-\xi) \sum_{|q|\geqslant 0} \frac{1}{q!} Q_{q,l+1}(y-\xi)(x^*-\xi)^q
\end{aligned}
$$

where

$$
D^q r^k(\xi-y^*) = \frac{\partial^{|q|}}{\partial t_n^{q_n} \dots \partial t_1^{q_1}} r^k(t_1,\dots,t_n)\bigg|_{t=(\xi-y^*)}
$$

and where $Q_{q,l+1}$ is a homogeneous polynomial of degree $|q|$ (as in the proof of Proposition 3.3, Chapter I). By Lemma 1.1, the continuation of the series (2.7) is uniformly absolutely convergent for x^* in compact subsets of $\mathscr{B}_{1/(\sqrt{2}R)}(\xi)$, y^* in compact subsets of $(\mathbb{R}^n \setminus \bar{B}_{1/R}(\xi))^{\wedge}$, so the series converges uniformly absolutely for x in some compact neighbourhood of $\bar{S}$ in $\mathbb{C}^n$, y in some compact neighbourhood of K in $\mathbb{C}^n$. The power series in (2.7) may be differentiated term by term with respect to the components of not only x^* but also x itself, for the factors $\partial x_j^*/\partial x_i$ which are so introduced do not affect the convergence. Finally, since r^{2l+2-n} is polyharmonic of finite degree at most $l+1$, the function

$$
Q_{q,l+1}(y-\xi) = |y-\xi|^{2l+2-n} D^q r^{2l+2-n}\left(\frac{\xi-y}{|\xi-y|^2}\right)
$$

is polyharmonic of the same degree on $\mathbb{R}^n \setminus \bar{B}_{|\xi|}(0)$ by Proposition 1.4, Chapter I. It follows that $Q_{q,l+1}$ is polyharmonic of the same degree on all of $\mathbb{R}^n$, and that is so even if $\xi=0$. Thus (2.7) is the series for $r^{2l+2-n}(x-y)$ to be substituted in (2.1). If n is odd or L sufficiently small, then the series for $\imath_{l+1}$ is determined. If not, the logarithm term appears:

$$
\begin{aligned}
(2.8)\quad & |x-y|^{2l+2-n}\log|x-y| \\
&= -|x^*-y^*|^{2l+2-n}(|x-\xi|\,|y-\xi|)^{2l+2-n}\log|x^*-\xi| \\
&\quad + (|x-\xi|\,|y-\xi|)^{2l+2-n}\left(|x^*-y^*|^{2l+2-n}\log\frac{|x^*-y^*|}{|y^*-\xi|}\right).
\end{aligned}
$$

The series for the first term of (2.8) is $-\log|x^*-\xi|$ times the series (2.7).

The series for the second term is

$$(2.9)\quad |x-\xi|^{2l+2-n}\sum_{|q|\geq 0}\frac{1}{q!}\Big(|y-\xi|^{2l+2-n} \times D^q\Big(|x^*-y^*|^{2l+2-n}\log\frac{|x^*-y|}{|y^*-\xi|}\Big)\Big|_{x^*=\xi}\Big)(x^*-\xi)^q$$

$$= r^{2l+2-n}(x-\xi)\sum_{|q|\geq 0}\frac{1}{q!}\Big(r^{2l+2-n}(y-\xi)\frac{\partial^{|q|}}{\partial t_n^{q_n}\dots\partial t_1^{q_1}}[r^{2l+2-n}(t)\{\log r(t) - \log|y^*-\xi|\}]|_{t=y^*-\xi}\Big)(x^*-\xi)^q$$

$$= r^{2l+2-n}(x-\xi)\sum_{|q|\geq 0}\frac{1}{q!}P_{q,l+1}(y-\xi)(x^*-\xi)^q$$

where $P_{q,l+1}$ is a homogeneous polynomial of degree $|q|$ (as in the proof of Proposition 3.3, Chapter I). The analysis of (2.9) is the same as that of (2.7), and yields the following information: $P_{q,l+1}(y-\xi)$ is polyharmonic on $\mathbb{R}^n$ with finite degree $l+1$, the series is uniformly absolutely convergent for x in a compact neighbourhood of $\bar{S}$ in $\mathbb{C}^n$ and for y in a compact neighbourhood of K in $\mathbb{C}^n$, and the same is true of all term-by-term differentiations.

A linear combination of (2.7) and (2.9) gives a series in powers of $(x^*-\xi)$, which (2.6) transforms into

$$\mathscr{r}_{l+1}(x-y)=\sum_{|q|\geq 0}R_{q,l+1}(y-\xi)r^{2l+2-n-2|q|}(x-\xi)(x-\xi)^q$$

where $R_{q,l+1}$ is a polynomial, polyharmonic of degree $l+1$ and homogeneous of degree $|q|$, and the series (together with its term-by-term derivatives) is uniformly absolutely convergent for (x, y) in a neighbourhood of $\bar{S}\times K$ in $\mathbb{C}^{2n}$. As in (a), this series can be substituted in (2.1):

$$T(y)=\sum_{|q|\geq 0}\sum_{l=0}^{L-1}R_{q,l+1}(y-\xi)\Omega_n\int_S\Big[\Delta^l u(x)\frac{\partial\{r^{2l+2-n-2|q|}(x-\xi)(x-\xi)^q\}}{\partial\nu_x} - r^{2l+2-n-2|q|}(x-\xi)(x-\xi)^q\frac{\partial\Delta^l u}{\partial\nu}(x)\Big]d\sigma(x).$$

Some finite partial sum g will have the required properties (2.3A) and (2.3B).

There remains the possibility that $\xi\in K$. Since the series above is a series of homogeneous polynomials uniformly absolutely convergent on a complete sphere about ξ in $\mathbb{R}^n$, the uniform convergence extends to a ball containing ξ. The same is true of all the derived series, so the result follows. ‖

Proposition 2.1. Let u be polyharmonic of finite degree L on an open set

$D \subset \mathbb{R}^n$ and let K be a compact subset of D. For each number $P>0$ and each $\varepsilon>0$ there exists a finite point set $F\subset\partial D$ and a function v, polyharmonic of degree L on $\mathbb{R}^n\backslash F$, such that

$$\|D^\alpha u - D^\alpha v\|_K \leqslant \varepsilon \quad \textit{if} \quad |\alpha|\leqslant P, \quad \textit{and}$$
$$\|\Delta^p u - \Delta^p v\|_K \leqslant \varepsilon \quad \textit{for all } p\geqslant 0.$$

Proof. Let B be a ball about some point $\xi\in K$ with radius at least as large as $\sqrt{2}\,\|r(x-\xi)\|_K$. Cover $\partial D\cap\bar{B}$ with finitely many discs, $B_{r_k}(\xi_k)$, $k=1,\dots,N$, such that $\xi_k\in\partial D$ and no point of K lies in $\bar{B}_{\sqrt{2}\,r_k}(\xi_k)$. (The points ξ_k will constitute the set F.) Choose a bounded open set $G\subset D$ such that $K\subset G$ and such that ∂G is a piecewise smooth surface lying in

$$\bigcup_{k=1}^{N} B_{r_k}(\xi_k)\cup\partial B.$$

Decompose ∂G into a finite number of piecewise smooth surfaces $S_0, S_1,\dots, S_N$ such that $S_0\subset\partial B$ and $S_k\subset B_{r_k}(\xi_k)$, $k=1,\dots,N$. For $y\in G$, formula (2.1), Chapter II, with $p=L$, and formula (2.15), Chapter II, give

$$u(y)=I(\partial G, u; y)=\sum_{k=0}^{N} I(S_k, u; y).$$

By the preceding lemma there exist functions g and $f_1,\dots,f_N$, polyharmonic of finite degree L on $\mathbb{R}^n$, $\mathbb{R}^n\backslash\{\xi_1\},\dots$, and $\mathbb{R}^n\backslash\{\xi_N\}$ respectively, such that if $|\alpha|\leqslant P$ then

$$\|D^\alpha I(S_0, u;\cdot)-D^\alpha g\|_K\leqslant\varepsilon/(N+1)$$
$$\|D^\alpha I(S_k, u;\cdot)-D^\alpha f_k\|_K\leqslant\varepsilon/(N+1), \qquad k=1,\dots,N.$$

The function

$$v=g+\sum_{k=1}^{N} f_k$$

will satisfy the first requirement. The second can also be satisfied in the usual way. ||

This brings us to the theorem on separation of singularities (first proved by Aronszajn [6]). There are two things to note. One is that if u is defined on an open set $G\subset\mathbb{R}^n$, then u can be provided with arbitrary continuation on the disjoint open set $\mathbb{R}^n\backslash\bar{G}$. The point is a minor one, but the hypothesis that u is defined on a set dense in $\mathbb{R}^n$ simplifies the proof of the theorem. The second point is that while this theorem holds for polyharmonic functions there is a weaker version for analytic functions restricted to $\mathbb{R}^n$ (§1, Chapter VI).

Theorem 2.1. (*Additive separation of singularities*). *Let* $n \geq 1$ *and let* u *be polyharmonic on a dense open set* $G \subset \mathbb{R}^n$. *If* F_1 *and* F_2 *are arbitrary closed subsets of* $F = \mathbb{R}^n \backslash G = \partial G$ *such that* $F = F_1 \cup F_2$, *then there exists a pair of functions* u_1 *and* u_2 *polyharmonic on* $G_1 = \mathbb{R}^n \backslash F_1$ *and* $G_2 = \mathbb{R}^n \backslash F_2$ *respectively, such that*

(a) $u = u_1 + u_2$ *on* G,

(b) *if* u *is of Laplacian order* ρ (*and type* $\tau < \infty$) *on* G, *then* u_1 *and* u_2 *are of Laplacian order* ρ (*and type* τ) *on* G_1 *and* G_2 *respectively, and*

(c) *if* u *is polyharmonic of finite degree* L *on* G, *then* u_1 *and* u_2 *are of degree* L *on* G_1 *and* G_2, *respectively.*

Proof. To ensure that u_1 and u_2 satisfy (b) for all ρ (and τ) we must use the least Laplacian order (and type) of u. Specifically, if u has finite Laplacian order, let ρ be the least Laplacian order of u on G (and if u has finite type as well let τ be the least type on G) (see Definition 1.4, Chapter II). Otherwise, let $\rho = \infty$ and $\tau = 0$. It must be shown that u_1 and u_2, when constructed, have Laplacian order ρ (and type τ). For the estimates, two cases and a subcase are distinguished.

Case 1: u has positive least Laplacian order ρ, $0 < \rho \leq \infty$, and specified least type $\tau < \infty$.

Case 2: u has finite least Laplacian order ρ, $0 \leq \rho < \infty$, but no specified type.

Subcase: u is polyharmonic of finite degree L.

The germ of the proof is the use of the serio-integral representation to construct a sequence $\{v_k\}$ of functions of the right Laplacian order (and type) or of the right degree which have singular points only on F_1 and the boundary surface of a region G_0 which separates F_2 from $F_1 \backslash F_2$. In fact, we shall construct a decreasing sequence $\{V_k\}$ of neighbourhoods of $F_1 \cap F_2$ such that $F_1 \cap F_2 = \cap V_k$ and all the singular points of v_k will lie in $F_1 \cup \bar{V}_k$. The sequence $\{v_k\}$ may not converge to the desired function u_1. Since

$$v_N = v_1 + \sum_{k=0}^{N-1} (v_{k+1} - v_k),$$

we shall alter v_k by finding appropriate functions f_k such that

$$v_1 + \sum_{k=0}^{\infty} (v_{k+1} - v_k - f_k)$$

converges on $\mathbb{R}^n \backslash F_1$. Lemma 2.1 is used to construct the functions f_k, and then it is shown that the above series converges to an appropriate u_1. Of course, $u_2 = u - u_1$.

Construction of the sets and definition of v_k and w_k. If $F_1=F_2$ there is nothing to prove, so it can be assumed that $F_1\backslash F_2\neq\phi$. The neighbourhoods V_k of $F_1\cap F_2$ in $\mathbb{R}^n$ will also be 'neighbourhoods of ∞'. They will be chosen inductively as follows. We choose

$$V_1=(\mathbb{R}^n\backslash\bar{Q}_{1,0})\cup\bigcup_{i=0}^{N(1)}Q_{1,i},$$

where $\bar{Q}_{1,0}$ is some (large) compact cube centred at the origin and the $Q_{1,i}$, $i=1,\ldots,N(1)$, are (small) bounded open cubes with centres $\xi_{1,i}$, $i=1,\ldots,N(1)$ respectively such that $F_1\cap F_2\subset V_1$ and $\mathbb{R}^n\backslash V_1\neq\phi$. In fact, $N(1)$ is finite and each $\xi_{1,i}$ is a point of $F_1\cap F_2$. If $F_1\cap F_2=\phi$, take $N(1)=0$. All cubes have faces parallel to the co-ordinate hyperplanes. Choose

$$V_2=(\mathbb{R}^n\backslash\bar{Q}_{2,0})\cup\bigcup_{i=0}^{N(2)}Q_{2,i}$$

where the same conditions are satisfied as in the construction of V_1, but in addition $F_1\cap F_2\subset V_2\subset\bar{V}_2\subset V_1$.

For the induction step, suppose $V_1,\ldots,V_{k-1}$ are chosen. Choose

$$V_k=(\mathbb{R}^n\backslash\bar{Q}_{k,0})\cup\bigcup_{i=1}^{N(k)}Q_{k,i}$$

where $\bar{Q}_{k,0}$ is some (large) compact cube centred at the origin, $N(k)$ is finite, $Q_{k,i}$ is a (small) bounded open cube with centre $\xi_{k.i}\in F_1\cap F_2$, $i=1,\ldots,N(k)<\infty$, $F_1\cap F_2\subset V_k\subset\bar{V}_k\subset V_{k-1}$, and if $k\geq 3$

$$(2.10)\qquad\begin{cases}d(\{0\},\partial Q_{k,0})>\sqrt{2}\operatorname{diam}Q_{k-2,0}\\ \operatorname{diam}Q_{k,i}<\dfrac{1}{\sqrt{2}}d(F_1\cap F_2,\partial V_{k-2})\quad\text{for}\quad 1\leq i\leq N(k)\end{cases}$$

where $\operatorname{diam}A=\sup\{|x-y|:x,y\in A\}$ and $d(A,B)=\inf\{|x-y|:x\in A,$ $y\in B\}$. For completeness, take $V_0=\mathbb{R}^n$ and note that

$$F_1\backslash F_2=\bigcup_{k=0}^{\infty}F_1\cap(\bar{V}_k\backslash V_{k+1}).$$

(The set $\mathbb{R}^n\backslash V_k$ is a kind of annular region between a large cube and the union of small ones.)

Each set $F_1\cap(\bar{V}_k\backslash V_{k+1})$, whether empty or not, is compact and may be covered by finitely many bounded open cubes, with sides parallel to the co-ordinate hyperplanes, each with centre in $F_1\cap(\bar{V}_k\backslash V_{k+1})$ and of diameter less than $\frac{1}{2}d(F_1\cap(\bar{V}_k\backslash V_{k+1}),F_2\cup\partial V_{k-1})$. Let G_1 be the union over k of all these (countably many) cubes. Define G_0, the set separating F_2 from $F_1\backslash F_2$, to be the interior of $\bar{G}_1$.

The properties of G_0 are

$$G_0 \cap F = F_1 \backslash F_2$$
$$(\partial G_0) \cap F \subset F_1 \cap F_2$$
$$\bar{G}_0 \cap F \subset F_1$$
$$(\partial G_0) \backslash (F_1 \cap F_2) \text{ is a (countable) piecewise flat surface.}$$

For each $k \geq 1$, let $S_k = (\partial G_0) \backslash V_k$. This surface is bounded and piecewise flat with a finite number of pieces. For the normal to S_k take the exterior normal for ∂G_0. Finally,

$$S_k \subset G = \mathbb{R}^n \backslash F \quad \text{and} \quad \cup S_k = (\partial G_0) \backslash (F_1 \cap F_2).$$

Define

$$v_k(y) = \begin{cases} I(S_k, u; y) + u(y), & y \in G_0 \backslash F_1 = G_0 \cap G \\ I(S_k, u; y), & y \in \mathbb{R}^n \backslash \bar{G}_0 \end{cases}$$

and observe that v_k has only removable singular points on the interior of S_k relative to ∂G_0. To see this, let y_0 be a point of the interior of S_k relative to ∂G_0. By the construction of S_k as a finite union of faces of cubes, the faces parallel to the co-ordinate hyperplanes, there exists a small open cube K with centre y_0 such that $D = K \cap G_0$ is a connected region, disjoint from F, with piecewise flat boundary $\partial D = (S_k \cap \bar{K}) \cup (G_0 \cap \partial K)$. Let $S = S_k \cap \bar{K}$ and $S' = G_0 \cap \partial K$, and choose the exterior normal on ∂D. Since u is polyharmonic on a neighbourhood of $\bar{D}$ we have

$$-I(S, u; y) - I(S', u; y) = \begin{Bmatrix} u(y), & y \in D \\ 0, & y \in \mathbb{R}^n \backslash \bar{D} \end{Bmatrix}$$

by Theorem 2.4(c), Chapter II. By Theorem 2.4(a), Chapter II, $I(S', u; y)$ is polyharmonic on $\mathbb{R}^n \backslash S'$, which is a neighbourhood of y_0, and of the same order and type there as u. By the preceding equation, the function

$$v(y) = \begin{cases} I(S, u; y) + u(y), & y \in D \\ I(S, u; y), & y \in \mathbb{R}^n \backslash \bar{D} \end{cases}$$

agrees with $-I(S', u; y)$ on $\mathbb{R}^n \backslash S'$. Therefore it has only removable singularities on K. The same must also be true of $v_k(y) = I(S_k \backslash S, u; y) + v(y)$, for $I(S_k \backslash S, u; y)$ has no singularities on K, a neighbourhood of y_0. Since y_0 was arbitrary on the interior of S_k, all singular points are removable.

In fact, the argument above shows that $v_k(y)$ can be continued to the complement of $F_1 \cup (\partial G_0 \cap \bar{V}_k)$ and has the same order and type (or degree) there as u. From now on v_k will be considered to be so continued.

The intersection of the sets of singular points of $\{v_k\}$ is

$$F_1 \cup \bigcap_k (\partial G_0 \cap \bar{V}_k) = F_1.$$

If $w_k = u - v_k$, then the definition of v_k and the above argument show that w_k is polyharmonic with the right properties on the complement of $F_2 \cup (\partial G_0 \cap \bar{V}_k)$, and the intersection of the sets $F_2 \cup (\partial G_0 \cap \bar{V}_k)$ is F_2. If $\{v_k\}$ were convergent (to u_1), then $\{w_k\}$ would converge (to u_2). Of course it would also be necessary to show that the limit functions had the right order and type (or degree).

Definition of the modifying functions f_k. Naturally

$$v_N = v_1 + \sum_{k=0}^{N-1} (v_{k+1} - v_k)$$

and

$$w_N = w_1 + \sum_{k=0}^{N-1} (w_{k+1} - w_k) = w_1 - \sum_{k=0}^{N-1} (v_{k+1} - v_k),$$

so the sequences $\{v_k\}$ and $\{w_k\}$ are convergent if and only if $\sum_{k=0}^{\infty} (v_{k+1} - v_k)$ is convergent, which it may not be. To gain convergence we modify $v_{k+1} - v_k$ by subtracting a function f_k, polyharmonic of finite degree on $\mathbb{R}^n \backslash (F_1 \cap F_2)$, which approximates $v_{k+1} - v_k$ and its Laplacians. From the definition of the functions v_k,

(2.11) $\quad v_{k+1}(y) - v_k(y)$

$$= \Omega_n \sum_{l=0}^{\infty} \int_{S_{k+1} \backslash S_k} \left\{ \Delta^l u(x) \frac{\partial \mathscr{v}_{l+1}}{\partial \nu_x} (x-y) - \mathscr{v}_{l+1}(x-y) \frac{\partial \Delta^l u}{\partial \nu} (x) \right\} d\sigma(x).$$

This series is finite in the subcase ($\Delta^L u \equiv 0$), but otherwise it is uniformly absolutely convergent on $\mathbb{R}^n \backslash V_{k-2}$, a compact set disjoint from the closure of $S_{k+1} \backslash S_k$ (and so is the series for each of its Laplacians, which may be taken term by term). The remainders are therefore small. They will be estimated at (2.19). A particular partial sum T_k is a function polyharmonic of finite degree, which will be approximated by f_k using Lemma 2.1.

The first step is to pick the length p_k of the sum, leaving the remainder small (see (2.19)). There exists an integer k_0 such that if $k > k_0$ then v_k is not identically zero. For $k \geqslant k_0$ define

$$|S_{k+1} \backslash S_k| = \int_{S_{k+1} \backslash S_k} d\sigma(x),$$

a positive number. For $k \geq 3$ define $U_{k-2} = \mathbb{R}^n \backslash \bar{V}_{k-2}$, a bounded non-empty open set whose closure is disjoint from that of $S_{k+1} \backslash S_k$, and for $k \geq k_0 + 3$ let

$$R_k = \sup\{|r(x-y)| : x \in S_{k+1} \backslash S_k,\ y \in \hat{U}_{k-2}\},$$

a positive number. Since $\{U_{k-2}\}$ is a nested increasing sequence whose union is $\mathbb{R}^n \backslash (F_1 \cap F_2)$, the sequence $\{R_k\}$ clearly has limit ∞. Thus there exists $\delta > 0$ such that $R_k > \delta$ for $k \geq k_0 + 3$. Let $\Pi(l)$ be the polynomial of degree at most $1 + n/2$ occurring in Proposition 3.3, Chapter I, for $|q| = 0$.

For Case 1 (Laplacian order $\rho > 0$, type $\tau < \infty$), for $k \geq k_0 + 3$, let

$$\alpha_k^{-1} = 2\,|\Omega_n|\,(1+R_k)^3 R_k^{-n} \sum_{l=0}^{\infty} \frac{\Pi(l)}{(2l)!^{1/\rho}} \left\{\left(\tau + \frac{2^{-2k}}{R_k}\right)(2^{k+1})^{1-1/\rho} R_k\right\}^{2l},$$

a positive finite number since the series is convergent for all ρ and τ in Case 1. (If $\rho = \infty$ then $\tau = 0$.) Choose p_1 to be an integer no less than $n/2$, and choose p_k inductively, $p_k > p_{k-1}$, such that if $k \geq k_0 + 3$ and $p > p_k$ then

$$\text{(2.12)}\qquad \left.\begin{matrix} \|\Delta^p u\|_{S_{k+1}\backslash S_k} \\ \left\|\dfrac{\partial \Delta^p u}{\partial \nu}\right\|_{S_{k+1}\backslash S_k} \end{matrix}\right\} \leq \alpha_k 2^{-k}\,|S_{k+1} \backslash S_k|^{-1} (2p)!^{1-1/\rho} \left(\tau + \frac{2^{-2k}}{R_k}\right)^{2p}.$$

That defines p_k and therefore also

(2.13) $T_k(y)$

$$= \Omega_n \sum_{l=0}^{p_k} \int_{S_{k+1}\backslash S_k} \left\{\Delta^l u(x) \frac{\partial \ell_{l+1}}{\partial \nu}(x-y) - \ell_{l+1}(x-y) \frac{\partial \Delta^l u}{\partial \nu}(x)\right\} \mathrm{d}\sigma(x)$$

for Case 1.

For Case 2 (Laplacian order $\rho < \infty$, without type) let $\rho(k) = \rho + 2^{-k}$, and for $k \geq k_0 + 3$ let

$$\alpha_k^{-1} = 2\,|\Omega_n|\,(1+R_k)^3 R_k^{-n} \sum_{l=0}^{\infty} \frac{\Pi(l)}{(2l)!^{1/\rho(k)}} (2^{k+1} R_k)^{2l}.$$

Choose p_1 to be an integer no less than $n/2$, and choose p_k inductively, $p_k > p_{k-1}$, such that if $k \geq k_0 + 3$ and $p > p_k$ then

$$\text{(2.14)}\qquad \left.\begin{matrix} \|\Delta^p u\|_{S_{k+1}\backslash S_k} \\ \left\|\dfrac{\partial \Delta^p u}{\partial \nu}\right\|_{S_{k+1}\backslash S_k} \end{matrix}\right\} \leq \alpha_k 2^{-k}\,|S_{k+1} \backslash S_k|^{-1} (2p)!^{1-1/\rho(k)}.$$

For the subcase (degree L) define $p_k = L + k$ for all k.

In each case or subcase, T_k is defined by (2.13). Rewriting,

$$T_k = \sum_{i=0}^{N(k)} T_{k,i}$$

where

$$T_{k,i}(y) = \Omega_n \sum_{l=0}^{p_k} \int_{S_{k,i}} \left\{ \Delta^l u(x) \frac{\partial \mathscr{E}_{l+1}}{\partial \nu}(x-y) - \mathscr{E}_{l+1}(x-y) \frac{\partial \Delta^l u}{\partial \nu}(x) \right\} \mathrm{d}\sigma(x),$$

and the surfaces $S_{k,i}$ give a disjoint decomposition of

$$S_{k+1} \backslash S_k = \bigcup_{i=0}^{N(k)} S_{k,i}$$

with $S_{k,0} \subset \mathbb{R}^n \backslash Q_{k,0}$, $S_{k,i} \subset Q_{k,i}$, $i = 1, \ldots, N(k)$. Now the special construction of the cubes (2.10) leads to the use of Lemma 2.1.

(a) If $1 \leqslant i \leqslant N(k)$, then

$$\begin{aligned} d(\{\xi_{k,i}\}, \mathbb{R}^n \backslash V_{k-2}) &\geqslant d(F_1 \cap F_2, \partial V_{k-2}) \\ &> \sqrt{2} \operatorname{diam} Q_{k,i} \geqslant \sqrt{2} \operatorname{diam}(\{\xi_{k,i}\} \cup S_{k,i}) \end{aligned}$$

so that there exists a function $f_{k,i}$, polyharmonic of degree $p_k + 1$ on $\mathbb{R}^n \backslash \{\xi_{k,i}\}$, such that if $p \geqslant 0$, $|q| \leqslant 2p_k$, and $y \in \mathbb{R}^n \backslash V_{k-2}$ then

$$|\Delta^p T_{k,i}(y) - \Delta^p f_{k,i}(y)| < \frac{(2^{-k})^{2p_k+1} (2p)!^{-2^k}}{N(k)+1}$$

$$|D^q T_{k,i}(y) - D^q f_{k,i}(y)| < \frac{2^{-k}}{N(k)+1}.$$

(b) Both the origin and $\mathbb{R}^n \backslash V_{k-2}$ lie in $Q_{k-2,0}$, and

$$d(\{0\}, S_{k,0}) \geqslant d(\{0\}, \partial Q_{k,0}) > \sqrt{2} \operatorname{diam} Q_{k-2,0}$$

so there exists a function g_k, polyharmonic of degree $p_k + 1$ on $\mathbb{R}^n$, such that if $p \geqslant 0$, $|q| \leqslant 2p_k$, and $y \in \mathbb{R}^n \backslash V_{k-2}$ then

$$|\Delta^p T_{k,0}(y) - \Delta^p g_k(y)| < \frac{(2^{-k})^{2p_k+1} (2p)!^{-2^k}}{N(k)+1}$$

$$|D^q T_{k,0}(y) - D^q g_k(y)| < \frac{2^{-k}}{N(k)+1}.$$

Define $f_k = g_k + f_{k,1} + \ldots + f_{k,N(k)}$ on $\mathbb{R}^n \backslash (F_1 \cap F_2)$. Observe that f_k is polyharmonic of finite degree $p_k + 1$ and that if $y \in \mathbb{R}^n \backslash V_{k-2}$ then

$$\begin{cases} |\Delta^p T_k(y) - \Delta^p f_k(y)| < (2^{-k})^{2p_k+1} (2p)!^{-2^k}, & \text{for all } p \\ |D^q T_k(y) - D^q f_k(y)| < 2^{-k}, & \text{if } |q| \leqslant 2p_k. \end{cases}$$

For a fixed q we have $2p_k > |q|$ for all sufficiently large k so that, for any positive integer K,

$$D^q \sum_{k=K+1}^{\infty} (T_k - f_k) = \sum_{k=K+1}^{\infty} D^q(T_k - f_k)$$

on $\mathbb{R}^n \setminus \bar{V}_{K-2}$, by the previous inequality. Thus

$$\left|\Delta^p \sum_{k=K+1}^{\infty} (T_k - f_k)\right| = \left|\sum_{k=K+1}^{\infty} \Delta^p(T_k - f_k)\right| < (2^{-K})^{2p}(2p)!^{-2K} \tag{2.15}$$

for all $K \geq 2$, all $p \geq 0$, on $U_{K-2} = \mathbb{R}^n \setminus \bar{V}_{K-2}$.

Definition and properties of u_1 and u_2. Define

$$u_1 = v_1 + \sum_{k=1}^{\infty} (v_{k+1} - v_k - f_k), \quad \text{and} \tag{2.16}$$

$$u_2 = w_1 + \sum_{k=1}^{\infty} (w_{k+1} - w_k + f_k) = w_1 - \sum_{k=1}^{\infty} (v_{k+1} - v_k - f_k), \tag{2.17}$$

let $\mathscr{K}$ be any compact set in $\mathbb{R}^n \setminus (F_1 \cap F_2)$, and let $\varepsilon > 0$. Choose $K > k_0 + 3$ such that $2^{-K} < \varepsilon$ and $\mathscr{K} \subset U_{K-2} = \mathbb{R}^n \setminus \bar{V}_{K-2}$. For Case 1, choose K large enough that in addition $2^{-K} < 1 - 1/\rho$ and

$$(\tau + \delta^{-1}2^{-2K})(1 + 2^{-K})^{1-1/\rho} < \tau + \varepsilon$$

where δ is the positive minimum of the distance estimates R_k. For Case 2, choose K large enough that in addition $-2^K < 1 - 1/(\rho + \varepsilon)$.

Suppose first that $\mathscr{K}$ does not intersect F_1, and consider the convergence near $\mathscr{K}$ of (2.16), rewritten

$$u_1 = v_{K+1} - \sum_{k=1}^{K} f_k + \sum_{k=K+1}^{\infty} (v_{k+1} - v_k - f_k).$$

Each f_k has no singular points on U_{K-2}; neither does $v_{k+1} - v_k$ if $k \geq K+1$. It will be shown that the series has all the required properties on U_{K-2}. The first term, v_{K+1}, has singular points only on $F_1 \cup \bar{V}_{K+1} \subset F_1 \cup (\mathbb{R}^n \setminus U_{K-2})$, and since these facts are for arbitrary K it will follow that u_1 has singular points only on F_1. (The analysis of (2.17) is similar.)

$$u_1 = v_{K+1} - \sum_{k=1}^{K} f_k + \sum_{k=K+1}^{\infty} (T_k - f_k) \tag{2.18}$$
$$+ \sum_{k=K+1}^{\infty} \Omega_n \sum_{l=p_k+1} \int_{S_{k+1} \setminus S_k} \left(\Delta^l u \frac{\partial \imath_{l+1}}{\partial \nu} - \imath_{l+1} \frac{\partial \Delta^l u}{\partial \nu}\right) d\sigma.$$

The first infinite series is convergent on U_{K-2} by (2.15). In the subcase (finite degree L) this finishes the proof (for the double sum is then zero, while (2.15) shows that the infinite series has finite degree L since v_k, T_k, and f_k do, for all k). In Case 1 the pth Laplacian of that series is bounded on U_{K-2} by $(2p)!^{1-1/\rho}\varepsilon^{2p}$; in Case 2, by $(2p)!^{1-1/(\rho+\varepsilon)}\varepsilon^{2p}$.

There remains the double series in (2.18):

$$(2.19)\quad \left|\sum_{l=p_k+1}^{\infty} \Omega_n \Delta^p \int_{S_{k+1}\setminus S_k} \left\{\Delta^l u(x) \frac{\partial \mathscr{r}_{l+1}}{\partial \nu_x}(x-y) - \mathscr{r}_{l+1}(x-y)\frac{\partial \Delta^l u}{\partial \nu}(x)\right\} d\sigma(x)\right|$$
$$\leqslant \sum_{\substack{l\geqslant 0\\ l+p\geqslant p_k+1}} \left|\Omega_n \int_{S_{k+1}\setminus S_k} \left\{\Delta^{l+p} u(x) \frac{\partial \mathscr{r}_{l+1}}{\partial \nu_x}(x-y) - \mathscr{r}_{l+1}(x-y)\frac{\partial \Delta^{l+p} u}{\partial \nu}(x)\right\} d\sigma(x)\right|.$$

In Case 1, (2.12) is applied, along with formulae (2.12)–(2.14), Chapter II, and the inequality

$$(1+2^{-k})^{2p+2l} \geqslant \frac{2^{-2kl}(2p+2l)!}{(2p)!\,(2l)!}$$

to show that for $y \in \hat{U}_{K-2}$, the right-hand side of (2.19) is at most

$$2^{-k}(2p)!^{1-1/\rho}(\tau+R_k^{-1}2^{-2k})^{2p}(1+2^{-k})^{2p(1-1/\rho)} \leqslant 2^{-k}(2p)!^{1-1/\rho}(\tau+\varepsilon)^{2p}$$

since $k \geqslant K$. For $p=0$ this shows that the double series is uniformly convergent on the open set $\hat{U}_{K-2}$ in $\mathbb{C}^n$ and therefore arbitrarily differentiable term by term. The pth Laplacian of the double series is therefore at most $(2p)!^{1-1/\rho}(\tau+\varepsilon)^{2p}$ for each $y \in \hat{U}_{K-2}$. The same was seen to be true of each of the other terms of (2.18), since $\mathscr{K} \cap F_1 = \phi$. Since ε was arbitrary, u_1 is of Laplacian order ρ and type τ on $\hat{U}_{K-2}$, a neighbourhood of $\mathscr{K}$. By the arbitrariness of $\mathscr{K}$, the same is true on $G_1 = \mathbb{R}^n \setminus F_1$.

For Case 2, (2.14) is used in place of (2.12) to show that if $y \in \hat{U}_{K-2}$ then the right-hand side of (2.19) is at most

$$2^{-k}(2p)!^{1-1/\rho(k)}2^{2p} \leqslant 2^{-k}(2p)!^{1-1/(\rho+\varepsilon)}2^{2p}$$

since $\rho(k) = \rho + 2^{-k} < \rho + \varepsilon$ for $k \geqslant K$. The proof of convergence and differentiability is as in Case 1, and the estimates for $\Delta^p u_1$ on $\hat{U}_{K-2}$ are made similarly. It follows that u_1 has Laplacian order ρ on $G_1 = \mathbb{R}^n \setminus F_1$.

As was pointed out earlier, the same infinite series appears in the analysis of (2.17), and the finite sums have the right Laplacian order (and type) on $\mathbb{R}^n \setminus F_2$ so that u_2 has the right Laplacian order (and type) on $\mathbb{R}^n \setminus F_2$. ∥

IV

LOCAL LAPLACIAN ORDER AND TYPE

1. Basic properties and extremal principles

One of the most important results in this section is the following fact.

Theorem 1.1. If u is analytic on a domain D in $\mathbb{R}^n$ *and if*

$$\lim_{p\to\infty}\left|\frac{\Delta^p u(x)}{(2p)!}\right|^{\frac{1}{2p}}=0 \tag{1.1}$$

for each x in a measurable subset A with positive measure, then u is polyharmonic on D, that is

$$\lim_{p\to\infty}\left\|\frac{\Delta^p u}{(2p)!}\right\|_K^{\frac{1}{2p}}=0 \tag{1.2}$$

for each compact $K\subset D$.

Since the result is a special case of the zero minimum principle, (Theorem 1.4), its proof will be deferred. However, the fact that a weak pointwise condition (1.1) can be made to imply a stronger local condition (1.2) forms the basis of the section. The difference between the two is essentially that between the pointwise and the local Laplacian type. In the proofs full use is made of the holomorphic capacity functions.

Recall from Definition 1.4, Chapter II, the concept of least Laplacian order of an analytic function u on a domain D. If U is any subdomain then the least order there, λ_U, is the lesser of $+\infty$ and the infimum of all numbers σ satisfying

$$\|\Delta^p u\|_U\leqslant(2p)!^{1-1/\sigma}\quad\text{for all large } p.$$

The local Laplacian order at $x\in D$ is the infimum $\lambda(x)$ of the least orders on neighbourhoods of x.

On a subdomain U where u has least order λ the least type is defined only if u is of order λ and type ∞ on U. In this case the least type μ_U is the lesser of $+\infty$ and the infimum of all numbers σ satisfying

$$\|\Delta^p u\|_U\leqslant(2p)!^{1-1/\lambda}\sigma^{2p}\quad\text{for all large } p.$$

For comparison with (1.2) these conditions can easily be rewritten as follows: λ is the least order on U if and only if

$$\exp(1-\lambda^{-1})=\limsup_{p\to\infty}\|\Delta^p u\|_U^{1/\log(2p)!}.$$

The local Laplacian order at x, $\lambda(x)$, satisfies

$$(1.3)\qquad F(x)=\exp\{1-\lambda(x)^{-1}\}$$
$$=\inf\left\{\limsup_{p\to\infty}\|\Delta^p u\|_K^{1/\log(2p)!}:\ K \text{ is a neighbourhood of } x\right\}.$$

If u has constant local order λ on some neighbourhood of x, then the local type is

$$(1.4)\quad \mu(x)=\inf\left\{\limsup_{p\to\infty}\left\|\frac{\Delta^p u}{(2p)!^{1-1/\lambda}}\right\|_K^{1/2p}:\ K \text{ is a neighbourhood of } x\right\}.$$

Thus, in the conclusion of Theorem 1.1, condition (1.2) is that u has no higher local Laplacian order and type than $\lambda(x)=\infty$ and $\mu(x)=0$, for each $x\in D$.

The fact that the infima in (1.3) and (1.4) are taken over neighbourhoods of x leads immediately to the conclusion that

$$\lambda(x)\geqslant\limsup_{z\to x}\lambda(z)\quad\text{and}\quad\mu(x)\geqslant\limsup_{z\to x}\mu(z).$$

In other words, *both λ and μ are upper semi-continuous.* The next task is to show that each is the upper semi-continuous envelope of one of the functions to be defined now.

Definition 1.1. If u is holomorphic on a domain G in $\mathbb{C}^n$ then the *pointwise Laplacian order* l is defined at $x\in G$ by

$$f(x)=\exp\{1-l(x)^{-1}\}=\limsup_{p\to\infty}|\Delta^p u(x)|^{1/\log(2p)!}.$$

If u has constant local Laplacian order λ on some neighbourhood of x, then the *pointwise type* at x is

$$m(x)=\limsup_{p\to\infty}\left|\frac{\Delta^p u(x)}{(2p)!^{1-1/\lambda}}\right|^{\frac{1}{2p}}$$

with the case $\lambda=\infty$ allowed by taking $(2p)!^{1-1/\lambda}=(2p)!$.

Example 1.1. If h is a function holomorphic on a domain E in the complex plane and if $G\subset\mathbb{C}^3$ is the set of points (z_1, z_2, z_3) such that $z_1+iz_2\in E$, then G is a domain on which

$$u(z_1, z_2, z_3)=\exp\{z_3h(z_1+iz_2)\}$$

is holomorphic. Since $h(z_1+iz_2)$ is a harmonic function of the pair (z_1, z_2), it follows that

$$\Delta u(z_1, z_2, z_3)=\frac{\partial^2}{\partial z_3^2}u(z_1, z_2, z_3)$$
$$=\{h(z_1+iz_2)\}^2u(z_1, z_2, z_3).$$

In fact, every derivative of h is also harmonic in (z_1, z_2) and so

$$\Delta^p u(z_1, z_2, z_3) = \frac{\partial^{2p}}{\partial z_3^{2p}} u(z_1, z_2, z_3)$$
$$= \{h(z_1 + iz_2)\}^{2p} u(z_1, z_2, z_3).$$

Thus $l(z_1, z_2, z_3) = 1$ except on the set where

$$h(z_1 + iz_2)u(z_1, z_2, z_3) = 0$$

and so $\lambda \equiv 1$ on G, except in trivial cases. Similarly, $m(z_1, z_2, z_3) = |h(z_1 + iz_2)|$ except on the same set, and so $\mu = |h|$ on G.

Theorem 1.2. If u is holomorphic on a domain $G \subset \mathbb{C}^n$ then λ, the local Laplacian order, is the upper semi-continuous envelope of l, the pointwise Laplacian order, and also of the restriction of l to $G \cap \mathbb{R}^n$. If λ is constant on G then μ, the local type, is the upper semi-continuous envelope of m, the pointwise type, and also of its restriction to $G \cap \mathbb{R}^n$, wherever μ is finite.

Proof. First consider type, in which case λ must be constant on G', a neighbourhood in G of $x \in G \cap \mathbb{R}^n$, and let μ be finite there. Let $\hat{m}$ be the upper semi-continuous envelope of m and let $\hat{m}_r$ be that for the restriction of m to $G \cap \mathbb{R}^n$. Since μ is upper semi-continuous we have

$$\mu(z) \geqslant \hat{m}(z) \geqslant \hat{m}_r(z) \geqslant m(z)$$

for all $z \in G \cap \mathbb{R}^n$. By Theorem A5.4, Appendix, applied to $L = m$, for any $\varepsilon > 0$ there exists a neighbourhood U of x in $G' \cap \mathbb{R}^n$ such that

$$\hat{m}_r(x) + \varepsilon \geqslant \limsup_{p \to \infty} \left\| \frac{\Delta^p u}{(2p)!^{1-1/\lambda}} \right\|_U^{\frac{1}{2p}} \geqslant \mu(x) \geqslant \hat{m}(x) \geqslant \hat{m}_r(x).$$

Since ε is arbitrary it must follow that

$$\mu(x) = \hat{m}(x) = \hat{m}_r(x).$$

Since this is true for each x in each translate of $\mathbb{R}^n$, it holds for all $x \in G'$.

The argument for order is similar, but it should be applied to

$$F(x) = \exp\{1 - \lambda(x)^{-1}\}$$
$$\text{and} \quad f(x) = \exp\{1 - l(x)^{-1}\} = \limsup_{p \to \infty} |\Delta^p u(x)|^{1/\log(2p)!}$$

instead of λ and l directly. The uniform bound required for the application of Theorem A5.4, Appendix, is simply the fact that for each $\varepsilon > 0$, each x, and some neighbourhood U of x, the bound

$$\varepsilon + \mathrm{e} \geqslant \varepsilon + \exp\{1 - \lambda(x)^{-1}\} \geqslant \|\Delta^p u\|_U^{1/\log(2p)!}$$

holds for all large p. ‖

Actually, the statement of Theorem 1.2 leaves unanswered the question whether the constancy of λ on a domain in $\mathbb{R}^n$ is enough to ensure constancy on some domain in $\mathbb{C}^n$. The affirmative answer (Corollary 1.3) will be given later.

Example 1.2. This is an elaboration of Ex. 1.1. For each z_1, z_2 define

$$v(z_3)=u(z_1, z_2, z_3)=\exp\{h(z_1+iz_2)g(z_3)\}$$

where $h(\zeta)=h(z_1+iz_2)$ is a holomorphic function of one variable, and hence is harmonic as a function of z_1 and z_2. The function g is to be entire, so that v is also entire. As in Ex. 1.1,

$$\Delta^k u(z_1, z_2, z_3)=\frac{d^{2k}v(z_3)}{dz_3^{2k}}=\Delta_1^k v(z) \tag{1.5}$$

where Δ_1 is the Laplace operator for functions of one variable only. The relationship between exponential order (and type) and Laplacian order (and type) for functions of one variable (Theorem 1.1, Chapter II) affords us a method for constructing examples in which direct differentiation would be impractical.

Since

$$v(z_3)=[\exp\{h(z_1+iz_2)\}]^{g(z_3)}$$

has no zeros, the Hadamard factorization of entire functions shows that either v is not of finite (exponential) order or the order of v is a positive integer k. In this case g must be a polynomial of degree k:

$$g(z_3)=a_k z_3^k+\dots$$

By Theorem 1.1, Chapter II, k is also the least Laplacian order of v on $\mathbb{C}$, and the least (Laplacian) type of v is $|ka_k h(z_1+iz_2)|^{1/k}$. By (1.5), then, u itself has

$$l(z_1, z_2, z_3)=\begin{cases} k, & h(z_1+iz_2)\neq 0 \\ 0, & h(z_1+iz_2)=0 \end{cases}.$$

Barring the trivial cases, it follows that $\lambda(z_1, z_2, z_3)\equiv k$ and

$$m(z_1, z_2, z_3)=|ka_k h(z_1+iz_2)|^{1/k}.$$

Of course if v does not have exponential order, for instance if

$$u(z_1, z_2, z_3)=\exp\{e^{z_3}h(z_1+iz_2)\},$$

then u must have $\lambda(z_1, z_2, z_3)=\infty$. If h is entire then u will be entire also. In this case it will be shown that u is necessarily polyharmonic (Theorem 2.1, Chapter VI), so that the local type will vanish identically.

Example 1.3. This is a further generalization,

$$u(z_1, z_2, z_3)=\exp\{\phi(z_1+iz_2, z_3)\}$$

in which ϕ is an arbitrary holomorphic function of two variables. For given z_1, z_2 define $v(z_3)=u(z_1, z_2, z_3)$.

For those z_1, z_2 such that ϕ is a polynomial in z_3 with degree $k(z_1+iz_2)$, the pointwise order of u can be calculated as in the previous example and found to be $k(z_1+iz_2)$, for then v extends to be an entire function. Since there is no other circumstance in which v is the restriction of an entire function of (finite) exponential order, it follows that

$$l(z_1, z_2, z_3) = \begin{cases} k(z_1+iz_2), & \text{if } \phi(z_1+iz_2, z_3) \text{ is a} \\ & \text{polynomial in } z_3 \text{ with degree } k(z_1+iz_2) \\ \infty, & \text{elsewhere.} \end{cases}$$

Thus, either ϕ is a polynomial in z_3, in which case u is entire and the situation of Ex. 1.2 is obtained, or else $\lambda(z_1, z_2, z_3) \equiv \infty$.

Example 1.4. The function

$$u(z_1, z_2, z_3) = \exp\left(\frac{z_3}{z_1+iz_2}\right)$$

has singularities in $\mathbb{R}^3$ at points of the form $(0, 0, z)$, $z \in \mathbb{R}$. From the examples above it is clear that $\lambda \equiv 1$ and $\mu(z_1, z_2, z_3) = |z_1+iz_2|^{-1}$. In this case μ is unbounded near the singularities.

By contrast, consider

$$\begin{aligned} u(z_1, z_2, z_3) &= \mathscr{z}_1(z_1, z_2, z_3) + \exp\{z_3(z_1+iz_2)\} \\ &= (z_1^2+z_2^2+z_3^2)^{-\frac{1}{2}} + \exp\{(z_1+iz_2)z_3\}. \end{aligned}$$

The first summand is harmonic, and so the local order and type of the sum are those of the second summand, $\lambda \equiv 1$ and $\mu(z_1, z_2, z_3) = |z_1+iz_2|$. The type has limit zero at the origin, the unique singularity in $\mathbb{R}^3$. (Actually, the first summand is the principal part and the second part is the analytic part since $\mathscr{z}_1$ is its own principal part.)

Example 1.5. The function

$$u(z) = \sum_{k=1}^{\infty} C^k \mathscr{z}_k(x), \quad x \in \mathbb{R}^n$$

was shown in Ex. 1.2, Chapter III, to be its own principal part. Since $\Delta^p u = C^p u$, it follows that $\lambda \equiv 1$ and $\mu \equiv |C|^{\frac{1}{2}}$.

Example 1.6. If $(x, y, z) \in \mathbb{R}^3$ and

$$u(x, y, z) = \exp\{ze^{(x+iy)}\} + \exp\{ze^{-(x+iy)}\},$$

then each summand has $\lambda \equiv 1$, but the first has $\mu_1(x, y, z) = e^x$ while the second has $\mu_2(x, y, z) = e^{-x}$. By Proposition 1.3, Chapter II, the local order of u must be $\lambda \equiv 1$, and the local type

$$\mu(x, y, z) = m(x, y, z) = \max\{e^x, e^{-x},\} = e^{|x|}.$$

This shows that the local type can be continuous without being smooth. Example 3.7, Chapter V, will show that local type need not even be continuous.

Theorem 1.3. Let u be holomorphic on a domain $G \subset \mathbb{C}^n$.

(a) (*Maximum principle*). *Either λ is constant on G or it does not assume its supremem $\|\lambda\|_G$, at even a single point of G. If λ is constant on G, then μ satisfies the maximum principle instead, provided that μ is finite at every* *point of G. In any case, if U is open with compact closure in G and* $x \in U$ *then* $\mu(x) \leq \|\mu\|_{\partial U}$.

(b) *If D is a connected open subset of* $G \cap \mathbb{R}^n$ *then* λ *satisfies the maximum principle on D. If* λ *is constant on D, then* μ *is defined on D. If, in addition, u is polyharmonic on D, then* μ *satisfies the maximum principle there, provided that* μ *is finite at every point of D.*

(c) *If there exists a set* $A \subset G \cap \mathbb{R}^n$, *with positive Lebesgue measure, such that* $\|l\|_A < \|\lambda\|_G$ *then*

$$l(x) \leq \lambda(x) < \|\lambda\|_G \quad \text{for all } x \in G.$$

If λ *is constant on G and* μ *is finite at every point of G, then the result holds for* μ *and m instead of* λ *and l.*

Proof. Part (c) will be treated first.

Let $x \in G$, let A' be a subset of A with positive measure and compact closure in G, and let D be a subdomain of G containing x and A' and having compact closure in G. Consider the functions f and F defined from the pointwise and local Laplacian orders respectively. (See Definition 1.1 and formula (1.3).)

For each $z \in D$ and each $\varepsilon > 0$ there is a neighbourhood U of z in D such that

$$\|\Delta^p u\|_U^{1/\log(2p)!} \leq \varepsilon + \exp(1 - \|\lambda\|_G^{-1}) \quad \text{if } p \text{ is large,}$$

by (1.3). Since D has compact closure in G, a finite cover argument shows that

$$\limsup_{p \to \infty} \|\Delta^p u\|_D^{1/\log(2p)!} \leq \exp(1 - \|\lambda\|_G^{-1}).$$

Theorem A5.1, Appendix, can now be applied to the sequence of logarithms, giving each compact $K \subset D$ a number d_K strictly less than $\exp(1 - \|\lambda\|_G^{-1})$ such that

$$\limsup_{p \to \infty} \|\Delta^p u\|_K^{1/\log(2p)!} \leq d_K < \exp(1 - \|\lambda\|_G^{-1}).$$

It follows that for each $x \in K$,

$$f(x) = \exp(1 - l^{-1}(x)) \leq \exp(1 - \lambda^{-1}(x)) = F(x) \leq d_K < \exp(1 - \|\lambda\|_G^{-1})$$

and for each x in the interior of K

$$l(x) \leq \lambda(x) \leq (1 - \log d_K)^{-1} < \|\lambda\|_G.$$

This proves assertion (c) for λ.

Now consider the local type μ, for which λ is taken to be constant on G. Let x, A', and D be as before, and $\varepsilon > 0$. Since μ is assumed to be finite, each z in G has a neighbourhood U such that

$$\left\| \frac{\Delta^p u}{(2p)!^{1-1/\lambda}} \right\|_U^{\frac{1}{2p}} \leq \varepsilon + \|\mu\|_G \text{ if } p \text{ is large,}$$

by (1.4). As was true for local order, a finite cover argument and an application of Theorem A5.1, Appendix, give each compact $K \subset D$ a number d_K strictly less than $\|\mu\|_G$ such that

$$\limsup_{p\to\infty} \left\| \frac{\Delta^p u}{(2p)!^{1-1/\lambda}} \right\|_K^{\frac{1}{2p}} \leq d_K < \|\mu\|_G.$$

Thus, for each x in the interior of K

$$m(x) \leq \mu(x) \leq d_K < \|\mu\|_G.$$

Now consider assertion (a), for type, λ being constant on G. If $x \in D$ and $\mu(x) \neq \|\mu\|_G$, then there exists $\varepsilon > 0$ such that $\mu(x) + \varepsilon < \|\mu\|_G$ and a neighbourhood U of x in G on which $\mu(z) \leq \mu(x) + \varepsilon$ for all $z \in U$. If A is the intersection of U with the translate of $\mathbb{R}^n$ through x, then the measure of A is positive and so assertion (c) can be applied. For the last sentence, apply Proposition 2.5, Chapter II if $\lambda < \infty$. Note that $\mu(x) < \infty$ if $\lambda \equiv \infty$. The argument for λ is similar.

There remains assertion (b). From D, u has analytic continuation into the domain G in $\mathbb{R}^n$. If u is not polyharmonic then $\|\lambda\|_G = \infty = \|\lambda\|_D$. If u is polyharmonic, then restrict attention to $\hat{D}$ where the serio-integral representation (Theorem 2.4, Chapter II) and the consistency property (Corollary 2.5, Chapter II) together show that $\|\lambda\|_{\hat{D}} = \|\lambda\|_D$. By assertion (a), λ is constant on $\hat{D}$ or G if and only if $\lambda(x) = \|\lambda\|_D$ for some $x \in D$, in which case it is constant on D *a fortiori*.

The crucial difference in the argument for type is that $\|\mu\|_{\hat{D}} \leq \|\mu\|_D$ only for polyharmonic functions, to which the serio-integral representation can be applied. With this, though, the argument runs as in the preceding paragraph. ||

That the finiteness condition for type is essential will be shown by Example 3.7, Chapter V. The function constructed there has $\mu \equiv 0$ on a ball in $\mathbb{R}^2$ and $\mu \equiv \infty$ on the complement of the ball.

We now show that the polyharmonicity condition is essential if type is to satisfy the maximum principle on domains in $\mathbb{R}^n$.

Example *1.7.* If u is analytic but not polyharmonic on a domain in $\mathbb{R}^n$, then $\lambda \equiv \infty$ but the local type may have a maximum. For instance, if $n = 1$ and $u(x) = (x-i)^{-1}$, then

$$\Delta^p u(x) = (2p)!\,(x-i)^{-1-2p}, \qquad x \neq i$$

so that $\lambda \equiv \infty$ and $\mu(x) = |x-i|^{-1}$. This local type has a maximum in $\mathbb{R}^1$ at $x = 0$.

Similar functions in n variables but constant in the last $n-1$ show that maxima can be obtained in $\mathbb{R}^n$ for $n \geqslant 1$.

Other examples of functions which are analytic but not polyharmonic are given in §1, Chapter VI.

The maximum principle reverses the inequalities in the upper semi-continuity of λ and μ, as follows.

Corollary 1.1. On any domain in $\mathbb{C}^n$ *or* $\mathbb{R}^n$,

$$\lambda(x) = \limsup_{z \to x} \lambda(z) \quad \textit{and} \quad \mu(x) = \limsup_{z \to x} \mu(z).$$

Proof. Let u be given. Since λ and μ are upper semi-continuous (Remark 2.5, Chapter II), only the opposite inequalities are needed.

On any domain G in $\mathbb{C}^n$ or $\mathbb{R}^n$, if λ is not constant then

$$\lambda(x) < \|\lambda\|_G$$

by the maximum principle (Theorem 1.3). Therefore

$$\lambda(x) \leqslant \limsup_{\substack{z \to x \\ z \in G}} \lambda(z).$$

The same holds for μ, by the same argument, if G lies in $\mathbb{C}^n$ or if u is polyharmonic.

Now consider the remaining case, in which $G \subset \mathbb{R}^n$ and u is analytic but not polyharmonic, and let $x \in G$ be given. Since u is of Laplacian order ∞ and type ∞, x has a neighbourhood D on which μ is bounded by, say, τ and which is of diameter less than $1/\tau$. Here u has local serio-integral representation.

For any $\sigma > \tau$ let $S_\sigma = \partial B_{1/\sigma}(x)$. Since $u(x) = I(S_\sigma, u; x)$, it follows from Theorem 2.3, Chapter II, that for each $\varepsilon > 0$

$$\begin{aligned} \mu(x) &\leqslant \frac{\sigma \|\mu\|_S (1+\varepsilon)}{\sigma - \|\mu\|_S} \\ &\leqslant \frac{\sigma}{\sigma - \tau} \|\mu\|_S (1+\varepsilon) \\ &\leqslant (1+\varepsilon)^2 \|\mu\|_S \end{aligned}$$

provided that σ is large enough. Taking the limit as σ increases and the radius shrinks now shows that

$$\mu(x) \leqslant \limsup_{\substack{z \to x \\ z \in G}} \mu(z) \quad \|$$

Corollary 1.2. If u is analytic on some domain $D \subset \mathbb{R}^n$ and if l is bounded on some subset having positive measure, for instance if λ is finite at even a single point of D, then l and λ are finite at every point of D and u is polyharmonic on all of D.

If l or λ is infinite at even a single point then $\lambda \equiv \infty$. In this case the local and pointwise types are defined and finite on D.

(The assertions about λ are not valid for type, as Ex. 7.3, Chapter V, will show.)

Corollary 1.2, in which a seemingly simple condition on a small set may have important global consequences, introduces a collection of related results, of which Theorem 1.4 and Theorem 1.1 are the most important in this section. Corollary 1.3 answers the question raised after Theorem 1.2.

Corollary 1.3. Let u be analytic on a domain D with analytic continuation to some domain G in $\mathbb{C}^n$ and let λ be constant on D.

(a) *If $\lambda \equiv \infty$ on D then $\lambda \equiv \infty$ on G; if $\lambda < \infty$ on D then λ assumes the same constant value on G if $G \subset \tilde{D}$.*

(b) *If in addition u is polyharmonic on D and has finite local type μ which is constant on D, then μ takes the same constant value on G if $G \subset \tilde{D}$.*

Proof

(a) If the constant value on D is infinite, then λ assumes its maximum on G and is constant. Otherwise, λ is no larger on $\tilde{D}$ than on D by the serio-integral representation. Thus, the constant value on D is the supremum over G in all cases. By Theorem 1.3, λ must be constant. The same kind of argument proves (b) as well. ||

Theorem 1.4 (Zero minimum principle). Let u be either holomorphic on a domain D in $\mathbb{C}^n$ or analytic on a domain D in $\mathbb{R}^n$.

(a) *If u has pointwise Laplacian order l equal to zero on a measurable set A in $D \cap \mathbb{R}^n$ with positive measure there, then u has Laplacian order zero on D.*

(b) *If for some ρ, u has Laplacian order ρ and type ∞ on D and if*

$$\lim_{p\to\infty} \left| \frac{\Delta^p u(x)}{(2p)!^{1-1/\rho}} \right|^{\frac{1}{2p}} = 0 \tag{1.6}$$

for each x in a measurable set A in $D \cap \mathbb{R}^n$ with positive measure there, then u has type zero on D.

Proof

(b) Even if D is not a domain in $\mathbb{C}^n$, there exists some domain D' in $\mathbb{C}^n$ which contains D and on which u or some continuation is holomorphic with constant local Laplacian order $\lambda \equiv \rho$ and type ∞, by Corollary 1.3. Therefore local type μ is defined. Let A' be a subset of A with positive measure and with compact closure in D', and let G be a subdomain of D' containing A'. Of course G may be taken to contain an arbitrary compact subset K of D and to have compact closure in D'.

Since u has Laplacian order ρ and finite type on G, it follows that

$$\limsup_{p\to\infty}\left(\sup_G \log\left|\frac{\Delta^p u}{(2p)!^{1-1/\rho}}\right|^{\frac{1}{2p}}\right)$$

is not $+\infty$. Further, (1.6) shows that

$$\limsup_{p\to\infty} \log\left|\frac{\Delta^p u(x)}{(2p)!^{1-1/\rho}}\right|^{\frac{1}{2p}}=-\infty$$

for each $x\in A'$. By Theorem A5.1, Appendix, it follows that

$$\limsup_{p\to\infty}\left(\sup_K \log\left|\frac{\Delta^p u}{(2p)!^{1-1/\rho}}\right|^{\frac{1}{2p}}\right)=-\infty,$$

that is for each K compact in D,

$$\limsup_{p\to\infty}\left\|\frac{\Delta^p u}{(2p)!^{1-1/\rho}}\right\|_K^{\frac{1}{2p}}=0.$$

For each x in the interior of K this limit supremum is an upper bound for $\mu(x)$, and so

$$\mu(x)=0 \quad \text{for each } x\in D.$$

Thus u has type zero.

The proof of (a) is similar. ‖

This result generalizes Theorem 1.1, which is the special case for which $\lambda\equiv\infty$. It also generalizes Corollary 1.2.

The principle of permanence of functional equations suffices to show that polyharmonicity of finite degree is a global property. The zero minimum principle shows that in general polyharmonicity is a global property.

Corollary 1.4. If u is analytic on a domain $D\subset\mathbb{R}^n$ and if u is polyharmonic on some open subset no matter how small, then u is polyharmonic on D itself.

Having Laplacian order zero is also a global property, but having zero type for some order is not, unless having type ∞ is known already.

These global constancy properties will be expanded upon in the next section, but the following interesting special case is a consequence of the maximum principle.

Proposition 1.1. If u is an entire function of one variable then λ and μ are constant.

Proof. If $\lambda(x)=\infty$ at some x then $\lambda\equiv\infty$ by the maximum principle, and $\mu\equiv 0$ since u is necessarily polyharmonic (see Theorem 1.1, Chapter II).

If not, then the supremum of λ on $\mathbb{C}$ is the same as the supremum of λ on any bounded interval, and so λ must be constant by the maximum principle. The same argument can now be applied to μ. ‖

Proposition 1.2. If u is an analytic function of one variable but not polyharmonic, then $\lambda(x)\equiv\infty$ and at each x the local type $\mu(x)$ is the reciprocal of the radius of convergence of the Taylor series for u about x.

Proof. The Taylor series can be written

$$u(z)=\sum_{k=0}^{\infty}\frac{u^{(2k)}(x)}{(2k)!}(z-x)^{2k}+\sum_{k=0}^{\infty}\frac{u^{(2k+1)}(x)}{(2k+1)!}(z-x)^{2k+1}.$$

The radius of convergence of the first series is

$$R_0(x)=\liminf_{k\to\infty}\left|\frac{u^{(2k)}(x)}{(2k)!}\right|^{-1/2k}=\liminf_{k\to\infty}\left|\frac{\Delta^k u(x)}{(2k)!}\right|^{-1/2k}=\frac{1}{m(x)},$$

and that of the second series is

$$\begin{aligned}R_1(x)&=\liminf_{k\to\infty}\left|\frac{u^{(2k+1)}(x)}{(2k+1)!}\right|^{-1/(2k+1)}\\&=\liminf_{k\to\infty}\left|\frac{\Delta^k u'(x)}{(2k+1)!}\right|^{-1/(2k+1)}\\&=\liminf_{k\to\infty}\left|\frac{\Delta^k u'(x)}{(2k)!}\right|^{-1/2k}=\frac{1}{m_1(x)}.\end{aligned}$$

By Theorem 1.2, Chapter II, the Laplacian order (and type) of u' on any domain are no greater than those of u. Thus, $\lambda\geqslant\lambda_1\geqslant\lambda_2$, where λ_1 and λ_2 are the local orders of u' and u''. Since $u''=\Delta u$, it must follow that $\lambda_2=\lambda$ and hence $\lambda=\lambda_1$. The same argument can now be made for type.

Since both R_0 and R_1 are continuous, both m and m_1 must be also, in which case $m=\mu$ and m_1 must be the local type of u' since $\mu(x)$ is finite at each x. By the result referred to, $\mu(x)=m_1(x)=\mu_2(x)$, where μ_2 is the local type of $u''=\Delta u$. It must follow that $R_1(x)=R_0(x)=1/m(x)=1/\mu(x)$, and each is the radius of convergence. ‖

This finishes the discussion of functions of one variable. The study of entire functions will be resumed in §2, Chapter VI.

The maximum and zero minimum principles show that low order and type values are taken on sets of positive measure only if certain high values occur nowhere. The next theorem shows that high values can occur on small sets in $\mathbb{R}^n$ only if they also occur on large sets.

Theorem 1.5. Let u be analytic on a domain $D\subset\mathbb{R}^n$.

(a) *If $l(x)\leqslant C$ for each x in an open set A such that $\sigma(D\setminus A)=0$, then $l(x)\leqslant\lambda(x)\leqslant C$ everywhere in D.*

(b) *If λ is constant on D, if μ is finite everywhere in D, and if $m(x) \leq C$ for each x in an open set A such that $\sigma(D \setminus A) = 0$, then $m(x) \leq \mu(x) \leq C$ everywhere in D.*

Proof

(a) If $C = \infty$ there is nothing to prove.

There exists a domain G in $\mathbb{C}^n$ on which a continuation of u is holomorphic. Let G' be a subdomain with compact closure in G and let $A' = A \cap G'$, $D' = D \cap G'$. Since

$$\limsup_{p\to\infty} \log \|\Delta^p u\|_{G'}^{1/\log(2p)!} \leq \sup_{G'} \log F = \sup_{G'} (1 - 1/\lambda) \leq 1$$

and since

$$\begin{aligned}\limsup_{p\to\infty} \log |\Delta^p u(x)|^{1/\log(2p)!} &\leq \sup_{G'} \log f(x) \\ &= \sup_{G'} (1 - 1/l(x)) \leq 1 - 1/C\end{aligned}$$

for each $x \in A'$, Theorem A5.3, Appendix, shows that

$$\limsup_{p\to\infty} \log \|\Delta^p u\|_{G'}^{1/\log(2p)!} \leq 1 - 1/C.$$

Thus, the local Laplacian type is at most C on G'.

(b) The same kind of argument can be applied, provided that G is constructed so small that μ is finite at each point of G. $\|$

In several examples the local type functions turned out to be the moduli of holomorphic functions. In fact several of the results so far suggest that local Laplacian order and type satisfy constraints closely related to those for moduli of holomorphic functions. In view of the fact that λ and μ are upper semi-continuous envelopes of limit suprema of such moduli the contrary would be surprising, but the question remains as to which constraints they must satisfy.

Some examples will be given in this section. Others will be constructed in §3, Chapter V.

Definition 1.2. If u is polyharmonic on an open set G in $\mathbb{R}^n$ or $\mathbb{C}^n$, let $P^\rho(u)$ (and $P^{\rho,\tau}(u)$) be the largest of the open sets on which u has Laplacian order ρ (and type τ).

Open Question 1.1. Clearly these sets are nested (compare Proposition 1.1, Chapter II) and $P^\rho(u)$ is the interior of the set on which $\lambda(x) \leq \rho$. If $P^\rho(u)$ is connected, then the maximum principle shows that either $\lambda \equiv \rho$ or $\lambda < \rho$ there. However, whether G being connected implies that $P^\rho(u)$ is connected remains an open question (see Open Question 1.2).

Similarly, if λ is constant on G then $P^{\lambda,\sigma}(u)$ is the interior of the set on which $\mu(x) \leq \sigma$, provided that σ is finite. The difference comes from the fact that type ∞ implies finite local type. If $P^{\lambda,\sigma}(u)$ is connected, then the maximum principle applies to show that either $\mu \equiv \sigma$ or $\mu < \sigma$ there, but in this case Ex. 1.1 shows that even G being connected does not imply that $P^{\lambda,\sigma}(u)$ is connected.

Example 1.8. Let $u(z_1, z_2, z_3) = \exp\{z_3 \sin(z_1 + iz_2)\}$ as in Ex. 1.1 where it is shown that $\lambda \equiv 1$ and $\mu(z_1, z_2, z_3) = |\sin(z_1 + iz_2)|$. Whether in $\mathbb{R}^3$ or in $\mathbb{C}^3$, the set of zeros of this function is not connected. In fact, if $0 < \varepsilon < 1$, the set $P^{1,\varepsilon}(u)$ on which $\mu < \varepsilon$ has infinitely many components.

Example 1.9. Let $u_1(z_1, z_2, z_3) = \exp\{z_3 \sin(z_1 + iz_2)\}$, let $u_2(z_1, z_2, z_3) = \exp\{z_3 \times (z_1 + iz_2)/2\pi\}$ and let $u = u_1 + u_2$. Both u_1 and u_2 have local order identically 1, with local types $\mu_1(z_1, z_2, z_3) = |\sin(z_1 + iz_2)|$ and $\mu_2(z_1, z_2, z_3) = |z_1 + iz_2|/2\pi$. By Proposition 1.3, Chapter II, and the fact that μ_1 and μ_2 agree on no open set it follows that u has local order identically equal to 1 and local type $\mu = \max\{\mu_1, \mu_2\}$. The local type has several local minima, at $z_1 + iz_2 = 0$ and near $z_1 + iz_2 = 0.9\pi$. The values there are zero and a number near $\frac{1}{2}$ respectively, but the minima are not assumed on open sets.

Open Question 1.2. Can it happen that μ has finite local minima of different values assumed on open sets? If $\mu = \infty$ is allowed the answer is known to be 'yes' (see Ex. 3.7, Chapter V). Of course the same question can be asked about local order, if local order allows disjoint sets of local minima. However, if the set of local minima for order is connected, or if the minimum value for either order or type is necessarily unique, then what is the significance of that uniqueness and the minimum value assumed?

Proposition 1.3. Let D be a domain in $\mathbb{R}^n$ or $\mathbb{C}^n$ and let u be polyharmonic on D.

(a) *If $\rho < \infty$ then $P^{\rho}(u)$ is the interior of $\cap \{P^{\sigma}(u) : \sigma > \rho\}$. If $\tau < \infty$ then $P^{\rho,\tau}(u)$ is the interior of $\cap \{P^{\rho,\sigma}(u) : \sigma > \tau\}$.*

(b) *The components of $P^{\rho}(u)$ are those of $\cup \{P^{\sigma}(u) : \sigma < \rho\}$ and those of the interior of the set on which $\lambda(x) = \rho$. If $\tau < \infty$ then the components of $P^{\rho,\tau}(u)$ are those of $\cup \{P^{\rho,\sigma}(u) : \sigma < \tau\}$ and those of the interior of the set on which $\mu(x) = \tau$.*

The proof follows at once from the definition and the maximum principle, as used above.

This section concludes with a kind of convexity result, showing how the existence of holomorphic capacity functions places constraints on all the local Laplacian order and type functions. Let G and G_0 be fixed domains in $\mathbb{C}^n$ with $G \subset G_0$. If $A' \subset G$ is a compact set with non-void interior and if G' is a domain containing A' and having compact closure in G_0, then the holomorphic capacity function $\alpha(z) = \alpha(A', G'; z)$ has positive infimum on compact subsets of G' (see Corollary A3.1 and Theorem A3.2,

Appendix). Define β on G_0 to be the supremum of all such α, and let γ be the lower semi-continuous envelope of β. Necessarily γ will have positive infimum on compact subsets of G_0.

Proposition 1.4. Let G and G_0 be domains in $\mathbb{C}^n$ with $G \subset G_0$, let γ be defined as above, and let u be holomorphic on G_0.

(a) *If u has constant local Laplacian order λ, with type M on G_0 and N on G, $N \leq M$, then its local type satisfies*

$$\mu(x) \leq N^{\gamma(x)} M^{1-\gamma(x)} \quad \text{for} \quad x \in G_0.$$

(b) *If u has Laplacian order S on G_0 and T on G, $T \leq S$, then its local Laplacian order satisfies*

$$F(x) = \exp\{1 - 1/\lambda(x)\} \leq \{\exp(1 - 1/T)\}^{\gamma(x)} \{\exp(1 - 1/S)\}^{1-\gamma(x)} \quad \text{for} \quad x \in G_0.$$

Proof. Let z be fixed in G_0, let β be defined as in the previous paragraph, and let $\eta > 0$. There exist A' and G' as in the previous paragraph such that

$$\alpha(z) = \alpha(A', G'; z) > \beta(z) - \eta.$$

(a) For each $\varepsilon > 0$ and large enough p,

$$\begin{aligned} |\Delta^p u(z)| &\leq \|\Delta^p u\|_{A'}^{\alpha(z)} \, |\Delta^p u\|_{G'}^{1-\alpha(z)} \\ &\leq \{(2p)!^{1-1/\lambda}(N+\varepsilon)^{2p}\}^{\alpha(z)} \{(2p)!^{1-1/\lambda}(M+\varepsilon)^{2p}\}^{1-\alpha(z)} \\ &\leq (2p)!^{1-1/\lambda}\{(N+\varepsilon)^{\alpha(z)}(M+\varepsilon)^{1-\alpha(z)}\}^{2p} \\ &\leq (2p)!^{1-1/\lambda}\{(N+\varepsilon)^{\beta(z)-\eta}(M+\varepsilon)^{1-\beta(z)+\eta}\}^{2p} \end{aligned}$$

since $(N+\varepsilon)/(M+\varepsilon) \leq 1$. The inequality holds for all ε, and so if p is large enough,

$$|\Delta^p u(z)| \leq (2p)!^{1-1/\lambda}\{N^{\beta(z)-\eta} M^{1-\beta(x)+\eta} + \varepsilon\}^{2p}.$$

Thus

$$m(z) \leq M^{1-\beta(z)+\eta} N^{\beta(z)-\eta}$$

for all $\eta > 0$, and so the inequality must hold after taking the limit and the upper semi-continuous envelope

$$m(z) \leq M\left(\frac{N}{M}\right)^{\beta(z)} \leq M\left(\frac{N}{M}\right)^{\gamma(z)}, \quad \text{for all } z \in G_0.$$

Since the last upper bound is upper semi-continuous, it follows that

$$\mu(x) \leq M\left(\frac{N}{M}\right)^{\gamma(x)}, \quad \text{for all } x \in G_0.$$

(b) The argument for order is similar. ∥

It might be said, then, that $\log \mu$ and $\log F$ are convex with respect to γ. In the next section more precise results in this direction will be obtained using domains with circular symmetry properties.

2. Convexity

This section concerns the logarithmic convexity of the functions μ, the local type, and $F=\exp(1-\lambda^{-1})$, derived from the local order, with respect to the homogeneous functions $|r|$, q, and s on $\mathbb{C}^n$, as defined in Exx. 3.3 and 3.4, Chapter II.

Definition 2.1. If g and h are positive real functions on a set D, then g is *logarithmically convex with respect to h on D* if, whenever $0<t_1\leq t\leq t_2$, the condition

$$(2.1)\qquad \begin{aligned} &\sup\{\log g(x): x\in D,\, h(x)=t\}\\ &\quad\leq \theta \sup\{\log g(x): x\in D,\, h(x)=t_1\}\\ &\qquad+(1-\theta)\sup\{\log g(x): x\in D,\, h(x)=t_2\}\end{aligned}$$

is satisfied where

$$(2.2)\qquad \theta=\frac{\log t_2-\log t}{\log t_2-\log t_1}.$$

That is, this supremum of log g is a convex function of $\log h$.

Phrased in this way, the Hadamard three-circle theorem (Hille [49]) states that on annuli in $\mathbb{C}^1$ the modulus of a holomorphic function is logarithmically convex with respect to distance from the origin in the modulus norm. The result has a generalization to $\mathbb{C}^n$, where any homogeneous function can be used in place of distance.

Lemma 1.1 (Three-shell lemma for $\mathbb{C}^n$). If h is a non-negative homogeneous function on $\mathbb{C}^n$, so that $h(\zeta x)=|\zeta|\, h(x)$ for $x\in\mathbb{C}^n$ and $\zeta\in\mathbb{C}$, and if u is holomorphic on a neighbourhood of

$$A=\{z\in\mathbb{C}^n: R_1\leq h(z)\leq R_2\}\quad \textit{where}\quad R_1>0$$

then $|u|$ is logarithmically convex with respect to h on A.

Proof. If z is a given point of A, then ζz generates a plane in $\mathbb{C}^n$. Of course $u(\zeta z)$ is a function of the single variable $\zeta\in\mathbb{C}$, holomorphic on $\mathbb{C}$ in a neighbourhood of the annulus defined by

$$R_1\leq|\zeta|\, h(z)=h(\zeta z)\leq R_2.$$

If

$$\theta=\frac{\log R_2-\log h(z)}{\log R_2-\log R_1},$$

then

$$\sup\{\log|u(\zeta z)|:|\zeta|=1\} \tag{2.3}$$
$$\leq \theta \sup\{\log|u(\zeta z)|:|\zeta|=R_1/h(z)\}$$
$$+(1-\theta)\sup\{\log|u(\zeta z)|:|\zeta|=R_2/h(z)\}$$
$$\leq \theta \sup\{\log|u(x)|:x\in\mathbb{C}^n,\ h(x)=R_1\} \tag{2.4}$$
$$+(1-\theta)\sup\{\log|u(x)|:x\in\mathbb{C}^n,\ h(x)=R_2\}.$$

Extending the supremum in (2.3) to the largest applicable set, $\{x\in\mathbb{C}^n : h(x)=h(z)\}$, shows that

$$\sup\{\log|u(x)|:x\in\mathbb{C}^n,\ h(x)=h(z)\}$$

is also bounded by the right-hand side of (2.4). Since the argument can be applied to all sub-annuli of A, the result follows. ||

The functions $|r|$, q, and s mentioned in the opening paragraph are all homogeneous. The first is

$$|r(\xi+i\eta)|=\left|\sum_{j=1}^{n}(\xi_j+i\eta_j)^2\right|^{\frac{1}{2}}, \tag{2.5}$$

the modulus of the analytic continuation of the distance function on $\mathbb{R}^n$, which vanishes on $V(0)$, the isotropic cone through the origin. The second and third are

$$q(\xi+i\eta)=[|\xi|^2+|\eta|^2+2\sqrt{\{|\xi|^2\,|\eta|^2-(\xi,\eta)^2\}}]^{\frac{1}{2}} \tag{2.6}$$

and

$$s(\xi+i\eta)=[|\xi|^2+|\eta|^2-2\sqrt{\{|\xi|^2\,|\eta|^2-(\xi,\eta)^2\}}]^{\frac{1}{2}} \tag{2.7}$$

which for $n=1$ reduce to $|r|$ but for $n\geq 2$ take the value R exactly on the boundary of $\tilde{B}_R(0)$ and $\tilde{C}_R(0)$, the harmonicity hulls of the ball and comball respectively (see Exx. 3.3 and 3.4, Chapter II).

The function q is a norm, the set $\tilde{B}_R(0)$ is convex, and $\partial\tilde{B}_R(0)$ is bounded. The function s is not a norm, $\tilde{C}_R(0)$ is not convex, and $\partial\tilde{C}_R(0)$ is not bounded. Nevertheless, both q and s participate in convexity theorems for Laplacian order and type. The first, which follows, is the prototype for a more powerful result on polyharmonic functions (Theorem 2.3).

Theorem 2.1 (Convexity in $\mathbb{C}^n$). Let h be a non-negative, continuous, and homogeneous function on $\mathbb{C}^n$, and let u be holomorphic on a neighbourhood of

$$A=\{z\in\mathbb{C}^n : R_1\leq h(z)\leq R_2\}\quad \textit{where}\quad R_1>0.$$

If $F=\exp(1-\lambda^{-1})$ where λ is the local Laplacian order of u and if μ is the local type (if defined), then F and μ are logarithmically convex with respect to h on A.

Thus, for $n\geq 2$ they are logarithmically convex with respect to q on $\tilde{B}_{R_2}(0)\setminus\tilde{B}_{R_1}(0)$; with respect to s on $\tilde{C}_{R_1}(0)\setminus\tilde{C}_{R_2}(0)$. (For $n=1$, h is necessarily a constant multiple of $|r|$ and A is the usual annulus.)

Proof. For any z in the interior of A define

$$\theta=\frac{\log R_2-\log h(z)}{\log R_2-\log R_1}.$$

By the three-shell lemma, each Laplacian $\Delta^k u$ satisfies

$$\log|\Delta^k u(z)|\leq\theta\sup\{\log|\Delta^k u(x)|:h(x)=R_1\}$$
$$+(1-\theta)\sup\{\log|\Delta^k u(x)|:h(x)=R_2\}.$$

First consider the case of local type, for which it is assumed that the local Laplacian order λ is constant. The action of the logarithm on powers shows that the following inequality is also satisfied:

$$\text{(2.8)}\quad \log\left(\frac{|\Delta^k u(z)|}{(2k)!^{1-1/\lambda}}\right)^{\frac{1}{2k}}\leq\theta\sup\left\{\log\left[\frac{|\Delta^k u(x)|}{(2k)!^{1-1/\lambda}}\right]^{\frac{1}{2k}}:h(x)=R_1\right\}$$
$$+(1-\theta)\sup\left\{\log\left[\frac{|\Delta^k u(x)|}{(2k)!^{1-1/\lambda}}\right]^{\frac{1}{2k}}:h(x)=R_2\right\}.$$

Taking the limit supremum as k increases now shows that the inequality is satisfied as well by the pointwise type m (Definition 1.1, Chapter IV):

$$\log m(z)\leq\theta\sup\{\log m(x):h(x)=R_1\}$$
$$+(1-\theta)\sup\{\log m(x):h(x)=R_2\}.$$

Because $m\leq\mu$,

$$\log m(z)\leq\theta\sup\{\log\mu(x):h(x)=R_1\}$$
$$+(1-\theta)\sup\{\log\mu(x):h(x)=R_2\}.$$

Since θ varies continuously with z, the right-hand side of the inequality defines a continuous upper bound for $\log m(z)$, and therefore for its upper semi-continuous envelope as well. This is $\log\mu$ wherever μ is finite (Theorem 1.2). The inequalities can be applied to all concentric sub-annuli of A where μ is finite. (This is always true if u is not polyharmonic.) If instead $\sup\{\log\mu(x):h(x)=R\}$ is infinite for some R, then the same is true for all $R'\geq R$ or all $R'\leq R$ by the maximum principle (Theorem 1.3). Therefore μ is logarithmically convex with respect to h on A. The argument proving the statement for local Laplacian order λ is virtually the same. ||

Theorem 2.2 (Constancy on $\mathbb{C}^n$). If u is entire, that is holomorphic on all of $\mathbb{C}^n$, then the local Laplacian order λ is constant. In addition, the local type μ is either constant or unbounded.

Proof. Because a real function defined on the whole real line and convex there is either constant or else unbounded, it follows that a logarithmically convex function bounded at both 0 and ∞ is constant. The function $F(x)=\exp\{1-\lambda^{-1}(x)\}$ is bounded by e on $\mathbb{C}^n\setminus\{0\}$ where q takes all positive values, and so

$$\sup\{F(x):q(x)=R\}$$

is a function of R which is bounded and logarithmically convex on $R>0$. Therefore the supremum of F on shells is constant and the same must be true of λ. That is, for each R

$$\begin{aligned}\sup\{\lambda(x):q(x)=R\}&=\sup\{\lambda(x):x\in\partial\tilde{B}_R(0)\}\\&=\sup\{\lambda(x):x\in\mathbb{C}^n\}.\end{aligned}$$

Since $\partial\tilde{B}_R(0)$ is compact and λ is upper semi-continuous, the supremum is actually assumed. By the maximum principle (Theorem 1.3) it follows that λ itself is constant.

If μ is bounded, the same argument shows that it, too, is constant. ||

Remark 2.1. The theorem shows that local Laplacian order attaches a single number to an entire function, just as the usual (exponential) order does (see §4, Chapter I). In Theorem 2.1, Chapter VI, it will be shown that entire functions are necessarily polyharmonic, but there do exist entire functions for which λ is not finite. Such a function is

$$u(x)=\sum_{k=0}^{\infty}\left[\frac{r(x)}{\log k}\right]^{2k}(2k)!^{-1/k}$$

given in Ex. 3.1, Chapter V. Examples of entire functions having finite Laplacian order are also given in §3, Chapter V. These may have constant local type, whether finite (Ex. 3.4) or infinite (Ex. 3.2), or unbounded local type, whether finite everywhere (Ex. 3.6) or finite on a disc only, where the local type may be bounded (Ex. 3.7) or unbounded (Ex. 3.9).

For the relationship between the Laplacian order and type of entire functions and the usual 'exponential' order and type see Theorem 1.1, Chapter II (for functions of one variable) and Theorem 2.1, Chapter VI (for the general case).

For functions which are analytic but not polyharmonic, $\lambda\equiv\infty$ and there is a relationship between $\mu(x)$ and the location of singularities of u in $\mathbb{C}^n$. For $n=1$ see Proposition 1.2. For the general case see §1, Chapter VI, in particular Corollary 1.2.

Theorem 2.3 (Convexity for polyharmonic functions). If u is a function polyharmonic on an open ball in $\mathbb{R}^n$, $n \geq 1$, or, for $n \geq 2$, on a comball or other annular domain

$$A = \{x \in \mathbb{R}^n : R_0 < r(x) < R_3\} \quad \textit{where} \quad r(x) = \left(\sum_j x_j^2\right)^{\frac{1}{2}},$$

then $F = \exp(1-\lambda^{-1})$ and μ (if defined) are logarithmically convex with respect to r there.

Proof. First consider a function u which is polyharmonic on a ball $B_{R_3}(0) \subset \mathbb{R}^n$. Here

$$u(y) = I(\partial B_R(0), u; y), \quad \text{if} \quad r(y) < R \quad \text{and} \quad 0 < R < R_3,$$

by the serio-integral representation (Theorem 2.4, Chapter II). It also follows that, for the continuation of u to $\hat{B}_R(0) = \tilde{B}_R(0)$,

$$\|\lambda\|_{\partial B_R(0)} \leq \|\lambda\|_{\partial \tilde{B}_R(0)} \leq \|\lambda\|_{B_t(0)} \quad \text{if} \quad R < t < R_3,$$

and taking the infimum on t shows that for such u

$$\|\lambda\|_{\partial B_R(0)} = \|\lambda\|_{\partial \tilde{B}_R(0)}. \tag{2.9}$$

This continuation of u must have the convexity property shown in Theorem 2.1. If $z \in \mathbb{R}^n$ and

$$0 < R_1 < r(z) < R_2 < R_3,$$

then $r(z) = q(z)$ and

$$z \in \{x \in \mathbb{C}^n : R_1 \leq q(x) \leq R_2\}$$

so that

$$\log F(z) \leq \theta \sup\{\log F(x) : x \in \mathbb{C}^n, q(x) = R_1\}$$
$$+ (1-\theta)\{\log F(x) : x \in \mathbb{C}^n, q(x) = R_2\}$$

where

$$\theta = \frac{\log R_2 - \log q(z)}{\log R_2 - \log R_1}.$$

By (2.9)

$$\sup\{F(x) : x \in \mathbb{C}^n, q(x) = R\} = \sup\{F(x) : x \in \mathbb{R}^n, r(x) = R\},$$

and so the required convexity is observed.

The same argument can be applied to μ.

Second, consider instead the local type of a function polyharmonic on $C_{R_0}(0)$ and having constant local Laplacian order λ there. By the serio-integral representation, u continues to

$$\tilde{C}_{R_0}(0) = \{x \in \mathbb{C}^n : s(x) > R_0\}$$

but the continuations may branch. Let $z \in \tilde{C}_{R_0}(0)$, let $0 < R_0 < R_1 < s(z) < R_2$, and let $\{\Delta^k \tilde{u}\}$ be the sequence of Laplacians of $\tilde{u}$, a continuation of u into a neighbourhood of z in $\mathbb{C}^n$.

In $\mathbb{C}^n$ the point z spans a plane of points $\{\tau z : \tau \in \mathbb{C}^1\}$ containing a plane annulus

$$\begin{aligned} D &= \{x \in \mathbb{C}^n : x = \tau z,\ \tau \in \mathbb{C}^1,\ R_1 \leq s(x) \leq R_2\} \\ &= \left\{z \in \mathbb{C}^n : x = \tau z,\ \tau \in \mathbb{C}^1,\ \frac{R_1}{s(z)} \leq |\tau| \leq \frac{R_2}{s(z)}\right\} \end{aligned}$$

because s is homogeneous by (3.16), Chapter II. Since D lies entirely in $\tilde{C}_{R_0}$, $\tilde{u}$ and all its Laplacians have unlimited analytic continuation upon it. To separate the continuations on D, for each ζ in a neighbourhood of the strip

$$E = \left\{\zeta \in \mathbb{C}^1 : \log \frac{R_1}{s(z)} \leq \mathcal{R}e\, \zeta \leq \log \frac{R_2}{s(z)}\right\}$$

we define

$$v_k(\zeta) = \Delta^k \tilde{u}(z e^{\zeta}).$$

This allows us to do the analysis on the interior of E, a simply connected plane domain. Letting

$$E_j = \{\zeta \in E : |\mathcal{I}m\, \zeta| \leq j\pi\}$$

gives compactness. To decompose ∂E_j let

$$H_j = \left\{\zeta \in \mathbb{C}^1 : \mathcal{R}e\, \zeta = \log \frac{R_2}{s(z)},\ |\mathcal{I}m\, \zeta| \leq j\pi\right\},$$

let J_j be the closure of $\partial E_j \backslash H_j$, and let ω_j be the harmonic measure of H_j on E_j. (See Hille [49, §18.3] for the notion of harmonic measure and for inequality (2.10) which follows.)

We shall consider the case in which u has constant local Laplacian order $\lambda > 0$ and show the logarithmic convexity of local type μ by making estimates of v_k on E_j.

$$\begin{aligned} \log\left\{\frac{|v_k(\zeta)|}{(2k)!^{1-1/\lambda}}\right\}^{\frac{1}{2k}} &\leq \omega_j(\zeta) \log\left\{\frac{\|v_k\|_{H_j}}{(2k)!^{1-1/\lambda}}\right\}^{\frac{1}{2k}} \\ &\quad + \{1 - \omega_j(\zeta)\} \log\left\{\frac{\|v_k\|_{J_j}}{(2k)!^{1-1/\lambda}}\right\}^{\frac{1}{2k}} \end{aligned} \tag{2.10}$$

for all ζ in the interior of E_j, by the defining property of ω_j. The limit supremum will be taken as k increases.

The points ζ in H_j correspond to points $x = z e^{\zeta}$ lying on a plane circular arc of length $j(2\pi R_2)$ in $\mathbb{C}^n$, lying also on the boundary of $\tilde{C}_{R_2}(0)$. By the same argument as at (2.9), the local type of $\tilde{u}$ satisfies

$$\|\mu\|_{\partial C_R(0)} = \|\mu\|_{\partial \tilde{C}_R(0)}, \qquad R_0 < R.$$

Thus,

$$\mu(x) \leq \sup\{\mu(\xi) : \xi \in \mathbb{R}^n, r(\xi) = R_2\}.$$

Since the arc on which x lies is compact, for each $\varepsilon > 0$ the arc has a neighbourhood U on which the continuation $\tilde{u}$ has type at most

$$\varepsilon + \sup\{\mu(\xi) : \xi \in \mathbb{R}^n, r(\xi) = R_2\}.$$

Since arc H_j is a compact subset of this neighbourhood U, it follows from the definition of type that

$$\limsup_{k\to\infty} \left(\frac{\|v_k\|_{H_j}}{(2k)!^{1-1/\lambda}}\right)^{\frac{1}{2k}} \leq \varepsilon + \sup\{\mu(\xi) : \xi \in \mathbb{R}^n, r(\xi) = R_2\}.$$

The set J_j does not correspond to a circular arc, but it does correspond to a compact set lying entirely in the closure of $\tilde{C}_{R_1}(0)$. Hence, the same argument given above can be made for it, using R_1 in place of R_2. Therefore, taking the limit supremum on both sides of (2.10) shows that for each $\varepsilon > 0$

(2.11) $$\log m(ze^{\zeta}) \leq \varepsilon + \omega_j(\zeta) \sup\{\log \mu(\xi) : \xi \in \mathbb{R}^n, r(\xi) = R_2\} + (1-\omega_j(\zeta))\{\log \mu(\xi) : \xi \in \mathbb{R}^n, r(\xi) = R_1\},$$

the inequality also holding in the limit $\varepsilon = 0$.

It is also the case that as j increases the harmonic measure for H_j in E_j converges to that for $\cup H_j$ in $E = \cup E_j$, that is

$$\begin{aligned}\lim_{j\to\infty} (1-\omega_j(\zeta)) &= 1 - \frac{\mathcal{Re}\,\zeta - \log\{R_2/s(z)\}}{\log\{R_2/s(z)\} - \log\{R_1/s(z)\}} \\ &= \frac{\log R_2 - \log s(ze^{\zeta})}{\log R_2 - \log R_1} \\ &= \theta,\end{aligned}$$

a continuous function of z. In the special case $\zeta = 0$, (2.11) implies that

$$\log m(z) \leq \theta \sup\{\log \mu(\xi) : \xi \in \mathbb{R}^n, r(\xi) = R_2\} + (1-\theta) \sup\{\log \mu(\xi) : \xi \in \mathbb{R}^n, r(\xi) = R_1\}.$$

The continuity of the right-hand side as a function of z shows that m can be replaced by its upper semi-continuous envelope, which is μ if μ is finite. This proves the convexity of finite μ. If instead $\sup\{\log \mu(x) : h(x) = R\}$ is infinite for some R, then the same is true for all $R' \geq R$, or all $R' \leq R$, by the maximum principle. This proves the convexity of μ. The convexity of F is proved similarly.

Finally, consider the local order of a function u which is polyharmonic on an annular domain

$$A = B_{R_3}(0) \cap C_{R_0}(0).$$

Here the function can be written as in Theorem 1.2, Chapter III

$$u = u_b + u_c,$$

with $u_b = A_u$ polyharmonic on $B_{R_3}(0)$ and $u_c = R_u$ polyharmonic on $\mathbb{R}^n \setminus \bar{B}_{R_0}(0)$. Let λ_b and λ_c be the local Laplacian orders of u_b and u_c respectively, let

$$M(R) = \max\{F(\xi) : \xi \in \mathbb{R}^n, \, r(\xi) = R\}$$

where $F(\xi) = \exp\{1 - \lambda^{-1}(\xi)\}$, and let M_b and M_c be defined similarly from λ_b and λ_c. Both $M_b(R)$ and $M_c(R)$ are logarithmically convex functions of R by the two previous parts of the proof, but to finish the proof it must be shown that M is logarithmically convex also.

Actually, $M_b(R)$ is an increasing function of R, while M_c is a decreasing function, by the serio-integral representation. Moreover, Corollary 1.1, Chapter III, implies that

$$\lambda_b(x) \leq \max\{\lambda(\xi) : \xi \in \mathbb{R}^n, \, r(\xi) = R\} = \max\{\lambda(\xi) : \xi \in \partial B_R(0)\}$$

for all x in $B_R(0)$. With the upper semi-continuity of λ this shows that

$$M_b(R_1) \leq M(R) \quad \text{if} \quad R_1 \leq R.$$

Similarly,

$$M_c(R_2) \leq M(R) \quad \text{if} \quad R_2 \geq R.$$

Passing to the special case $R_1 = R = R_2$ shows that both M_b and M_c are bounded by M. The opposite inequality is provided by Proposition 1.3, Chapter II, which implies that

$$M(R) \leq \max\{M_b(R), M_c(R)\}.$$

Thus

$$M(R) = \max\{M_b(R), M_c(R)\},$$

and since convexity is preserved under taking maxima it follows from the logarithmic convexity of M_b and M_c that M is logarithmically convex also.

If λ_b and λ_c are constant and equal, the same argument can be applied to type. ||

The seemingly exceptional case in this theorem corresponds to the fact that in $\mathbb{R}^1$ comballs are disconnected. Thus a polyharmonic function can be defined on the components of C_0 by completely unrelated formulas. From each component, however, the function continues to all of $\mathbb{C}^1$, including all of $\mathbb{R}^1$, notably the other component of C_0. To this continuation Theorem 2.2 applies, as does Theorem 2.1.

The analytic functions which are not polyharmonic have constant local Laplacian order, $\lambda \equiv \infty$. However, their local type does not satisfy a convexity property. For example, $u(x) = \mathcal{R}e(x-i)^{-1}$ on $\mathbb{R}^1$ has local type $\mu(x) = |x-i|^{-1}$, which has a maximum at $x = 0$. For many other examples of analytic functions which are not polyharmonic see §1, Chapter VI.

Theorem 2.4 (Constancy on $\mathbb{R}^n$). If u is analytic on all of $\mathbb{R}^n$, $n \geq 2$, except at a finite collection of points, then its local Laplacian order is constant. If u is polyharmonic as well then its local type is either constant or unbounded. Thus, for a polyharmonic function the principal part of u at a singular point isolated in $\mathbb{R}^n$ has a well-defined Laplacian order, the constant value of its local Laplacian order; its type, if not constant, is unbounded near the singularity.

Proof. If u is not polyharmonic then $\lambda \equiv \infty$.

Now consider a function u which is polyharmonic on $C_0 = \mathbb{R}^n \setminus \{0\}$. As was observed in the proof of Theorem 2.2, since

$$\max\{F(\xi) = \exp\{1 - \lambda^{-1}(\xi)\} : \xi \in \mathbb{R}^n, r(\xi) = R\}$$

is a logarithmically convex function of R, bounded on $R > 0$ by the number e, it must take a constant value, independent of R. Since u is polyharmonic on C_0, the maximum principle (Theorem 1.3) can be applied to show that $\lambda(x)$ is constant, just as was done in the proof of the constancy theorem for entire functions, since the spheres in $\mathbb{R}^n$ on which r is constant are compact. If μ is bounded, the same argument can be applied to it. However, the fact that serio-integral representation of a principal part gives values outside the region of integration means that

$$\max\{\mu(x) : r(x) = R\}$$

is non-increasing for $R > 0$. Hence the local type of a principal part is unbounded near the isolated singular point or else is constant.

The result now follows for u polyharmonic on $\mathbb{R}^n$ less a finite collection of singular points $x_1, \ldots, x_k$. For these, u has principal parts $u_1, \ldots, u_k$ such that

$$u = u_1 + \ldots + u_k + u_b \quad \text{on} \quad \mathbb{R}^n \setminus \{x_1, \ldots, x_k\},$$

u_b being entire. Each of the summands has constant local Laplacian order, and so the local Laplacian order of u is bounded by the largest of these constants. Conversely, each of the u_j is obtained from u by the serio-integral representation, and so the local Laplacian order of u must be at least that large at some point on some one of the compact surfaces of integration required, that is λ actually attains its supremum, and so it is constant, by the maximum principle. The same argument can also be applied to bounded type. ||

Polyharmonicity on the complement of a finite set in $\mathbb{R}^n$, $n \geqslant 2$, is enough to ensure that λ will be constant. Furthermore, the argument just given can be used to show that if $K_1, K_2, \ldots$ are disjoint compact sets in $\mathbb{R}^n$, each having the property that polyharmonicity on the complement entails having constant local order, then any finite union of these sets must have the same property.

Open Question. How large can a set K with connected complement in $\mathbb{R}^n$ be and still have the property that every function polyharmonic on $\mathbb{R}^n \setminus K$ will have constant local Laplacian order? Is belonging to the same class of sets either necessary or sufficient to ensure the constancy of bounded type on the complement?

V

THE ALMANSI EXPANSION

1. The Almansi expansion for holomorphic functions

In Proposition 1.3, Chapter I, it was shown that, if $u(x)$ is polyharmonic of finite degree p on a neighbourhood U of the origin in $\mathbb{R}^n$ and if $\bar{B}_R = \overline{B_R(0)} \subset U$, then $u(x)$ has a unique Almansi expansion

$$u(x) = \sum_{k=0}^{p-1} r^{2k} h_k(x)$$

on $\bar{B}_R$ where the functions $h_k(x)$ are harmonic on a neighbourhood of $\bar{B}_R$. From Theorem 2.5, Chapter II, the functions $u(x)$ and $h_k(x)$ can be continued analytically to $\hat{\bar{B}}_R$ and the above expansion remains valid there. This result will now be generalized to include all functions $u(x)$ holomorphic on a neighbourhood of the origin in $\mathbb{C}^n$, the proof yielding a formula of Cauchy type for holomorphic functions and an integral representation for harmonic functions. Next we shall see that a harmonic function on a neighbourhood of the origin in $\mathbb{C}^n$ is uniquely determined by its restriction to $V(0)$, the isotropic cone at the origin, and conversely a function analytic on a neighbourhood of the origin on $V(0)$ has a unique harmonic extension in a sufficiently small neighbourhood of the origin. This will enable us to define an algebra of germs of functions harmonic on a neighbourhood of the origin in $\mathbb{C}^n$. Furthermore, this algebra will determine an algebra of harmonic functions in B_R.

Theorem 1.1 (Almansi expansion). Let $u(x)$ be defined and holomorphic on a neighbourhood U of the origin in $\mathbb{C}^n$. There exists a unique Almansi expansion

$$u(x) = \sum_{k=0}^{\infty} r^{2k} h_k(x), \qquad r^2 = \sum_{j=1}^{n} x_j^2, \qquad h_k(x) \text{ harmonic,} \tag{1.1}$$

which is absolutely and uniformly convergent on every harmonicity hull $\hat{B}_{R_0}$ whose closure is contained in U. Conversely, any series of the form (1.1) *which is absolutely and uniformly convergent on a neighbourhood of the origin in $\mathbb{C}^n$ defines a holomorphic function in this neighbourhood.*

Remark 1.1. In the next section we shall derive formulae for the *Almansi coefficients* $h_k(x)$ in $\mathbb{R}^n$ and characterize the Laplacian order and type of $u(x)$ by means of these coefficients.

Proof. Let $\hat{\bar{B}}_{R_0} \subset U$. First suppose that $v(x)$ is polyharmonic of finite degree p on a neighbourhood of $\bar{\hat{B}}_{R_0}$, so that $v(x)$ has Almansi expansion

$$v(x) = \sum_{k=0}^{p-1} r^{2k} H_k(x)$$

on a neighbourhood of $\bar{\hat{B}}_{R_0}$. For $R < R_0$ and $x \in \hat{B}_R$, define

$$(1.2)\qquad \varphi_v(R, x) = \frac{1}{\omega_n R} \int_{S_R^{n-1}} \frac{R^2 - r(x)^2}{r(\theta - x)^n} v(\theta)\, d\sigma_R(\theta)$$

$$= \frac{R^{n-2}}{\omega_n} \int_{S_1^{n-1}} \frac{R^2 - r(x)^2}{r(R\theta - x)^n} v(R\theta)\, d\sigma(\theta).$$

(Here S_R^m is the m-sphere in $\mathbb{R}^{m+1}$ with the centre at the origin and radius R, $d\sigma_R$ is the measure on S_R^m, ω_n is the area of S_1^{n-1}, and $d\sigma \equiv d\sigma_1$.) Replacing $v(x)$ by its Almansi expansion in (1.2) gives

$$(1.3)\qquad \varphi_v(R, x) = \sum_{k=0}^{p-1} R^{2k} H_k(x).$$

Formula (1.3) shows that the real positive number R can be replaced by the complex variable ζ and the resulting function $\varphi_v(\zeta, x)$ is holomorphic on a neighbourhood of the set $\{(\zeta, x) : \zeta \in \mathbb{C}^1, x \in \bar{\hat{B}}_{R_0}\} \subset \mathbb{C}^1 \times \mathbb{C}^n$. It is now seen from (1.3) that $H_k(x)$ is the coefficient of ζ^{2k} in the Taylor expansion about $\zeta = 0$ of $\varphi_v(\zeta, x)$ as a function of ζ. Thus

$$(1.4)\qquad H_k(x) = \frac{1}{2\pi i} \int_{\Gamma_{R_0}} \frac{1}{\zeta^{2k+1}} \varphi_v(\zeta, x)\, d\zeta, \qquad x \in \bar{\hat{B}}_{R_0}$$

where Γ_{R_0} is the curve $|\zeta| = R_0$, $\zeta \in \mathbb{C}^1$, and

$$(1.5)\qquad v(x) = \frac{1}{2\pi i} \int_{\Gamma_{R_0}} \frac{\zeta}{\zeta^2 - r(x)^2} \varphi_v(\zeta, x)\, d\zeta, \qquad x \in \bar{\hat{B}}_{R_0}.$$

Returning to the given holomorphic function $u(x)$, define

$$(1.6)\qquad \varphi(\zeta, x) = \frac{\zeta^{n-2}}{\omega_n} \int_{S_1^{n-1}} \frac{\zeta^2 - r(x)^2}{r(\zeta\theta - x)^n} u(\zeta\theta)\, d\sigma(\theta)$$

$$0 < |\zeta| \le R_0, \qquad x \in \hat{B}_{|\xi|}.$$

Looking at the integrand in (1.6) and using Proposition 3.8, Chapter II, we see that $\varphi(\zeta, x)$ can be continued as a holomorphic function on $\{(\zeta, x) : |\zeta| \le R_0, x \in \hat{B}_{|\zeta|}\} \subset \mathbb{C}^1 \times \mathbb{C}^n$.

From Proposition 3.9, Chapter II, if $R_0' > R_0$ and $\bar{\hat{B}}_{R_0'} \subset U$, then $u(x)$ can be uniformly approximated on $\hat{B}_{R_0'}$ by a sequence $\{v_m(x)\}$ of polynomials. Since each polynomial $v_m(x)$ is polyharmonic of finite degree p_m,

(1.2), (1.4), and (1.5) imply that

$$(1.7)\qquad v_m(x)=\sum_{k=0}^{p_m-1} r(x)^{2k}h_{m,k}(x), \qquad x\in\bar{\hat{B}}_{R_0'}$$

$$(1.8)\quad \varphi_m(\zeta,x)=\frac{\zeta^{n-2}}{\omega_n}\int_{S_1^{n-1}}\frac{\zeta^2-r(x)^2}{r(\zeta\theta-x)^n}v_m(\zeta\theta)\,\mathrm{d}\sigma(\theta), \qquad x\in\hat{B}_{|\zeta|},\ |\zeta|\le R_0',$$

and also

$$(1.9)\qquad h_{m,k}(x)=\frac{1}{2\pi i}\int_{\Gamma_{R_0'}}\frac{1}{\zeta^{2k+1}}\varphi_m(\zeta,x)\,\mathrm{d}\zeta, \qquad x\in\bar{\hat{B}}_{R_0'}$$

$$(1.10)\quad v_m(x)=\frac{1}{2\pi i}\int_{\Gamma_{R_0'}}\frac{\zeta}{\zeta^2-r(x)^2}\varphi_m(\zeta,x)\,\mathrm{d}\zeta, \qquad x\in\hat{B}_{R_0'}$$

because the function $\varphi_m(\zeta,x)$ in (1.8) has analytic continuation to a neighbourhood of $\{(\zeta,x):\zeta\in\mathbb{C}^1,\ x\in\bar{\hat{B}}_{R_0'}\}$ with

$$(1.11)\qquad \varphi_m(\zeta,x)=\sum_{k=0}^{p_m-1}\zeta^{2k}h_{m,k}(x).$$

Formula (1.8) shows that for $\rho_0<R_0'$ the sequence $\{\varphi_m(\zeta,x)\}$ is uniformly convergent on $\{(\zeta,x):|\zeta|=R_0',\ x\in\bar{\hat{B}}_{\rho_0}\}$. Since each function $\varphi_m(\zeta,x)$ is holomorphic on a neighbourhood of $\{(\zeta,x):\zeta\in\mathbb{C}^1,\ x\in\bar{\hat{B}}_{R_0'}\}$, an application of the Cauchy integral formula for functions of one complex variable ζ yields uniform convergence on $\{(\zeta,x):|\zeta|\le R_0',\ x\in\bar{\hat{B}}_{\rho_0}\}$ to a function holomorphic on that set. From (1.6) and the sentence following it, this limit function provides analytic continuation of $\varphi(\zeta,x)$ to a neighbourhood of $\{(\zeta,x):|\zeta|\le R_0',\ x\in\bar{\hat{B}}_{R_0}\}$. By (1.9) and (1.10), as m approaches infinity the harmonic functions $h_{m,k}(x)$ converge uniformly to harmonic functions $h_k(x)$ given by

$$(1.12)\qquad h_k(x)=\frac{1}{2\pi i}\int_{\Gamma_{R_0}}\frac{1}{\zeta^{2k+1}}\varphi(\zeta,x)\,\mathrm{d}\zeta, \qquad x\in\bar{\hat{B}}_{R_0},$$

and

$$(1.13)\qquad u(x)=\frac{1}{2\pi i}\int_{\Gamma_{R_0}}\frac{\zeta}{\zeta^2-r(x)^2}\varphi(\zeta,x)\,\mathrm{d}\zeta, \qquad x\in\hat{B}_{R_0}.$$

Expanding $\varphi(\zeta,x)$ in powers of ζ and using the uniform convergence of $\{\varphi_m(\zeta,x)\}$ and (1.11), we obtain

$$(1.14)\qquad \varphi(\zeta,x)=\sum_{k=0}^{\infty}\zeta^{2k}h_k(x),$$

a uniformly and absolutely convergent series on $\{(\zeta,x):|\zeta|\le R_0,\ x\in\bar{\hat{B}}_{R_0}\}$.

Hence $\varphi(\zeta, x)$ is an even function in ζ and we can write

$$(1.15)\quad \varphi(z, x)=\frac{1}{2\pi i}\int_{\Gamma_{R_0}} \frac{\zeta}{\zeta^2-z^2}\varphi(\zeta, x)\,d\zeta \quad \text{for} \quad |z|<R_0, \qquad x\in \hat{B}_{R_0}.$$

Setting $z^2=r(x)^2$ in (1.14) and (1.15) and comparing with (1.13) yields the existence of the Almansi expansion.

To prove uniqueness of the expansion, consider again the function

$$\varphi(\zeta, x)=\frac{\zeta^{n-2}}{\omega_n}\int_{S_1^{n-1}} \frac{\zeta^2-r(x)^2}{r(\zeta\theta-x)^n}u(\zeta\theta)\,d\sigma(\theta)$$

defined in (1.6) and its analytic continuation described in the sentence following (1.6). Replacing $u(x)$ by its Almansi expansion, we see that

$$\varphi(\zeta, x)=\sum_{k=0}^{\infty}\zeta^{2k}h_k(x).$$

For each fixed x, this is a power series in ζ and is convergent for $|\zeta|\leq r(x)$. Since the analytic function defined by this power series is the continuation of $\varphi(\zeta, x)$ as given by the integral formula, it is uniquely determined by $u(x)$. However,

$$h_k(x)=\frac{1}{2\pi i}\int_{\Gamma_{R_0}} \frac{1}{\zeta^{2k+1}}\varphi(\zeta, x)\,d\zeta,$$

so that $h_k(x)$ is uniquely determined by $u(x)$ and the proof is complete. ‖

Remark 1.2. Notice that for $u(x)$ holomorphic on a neighbourhood of 0 in $\mathbb{C}^1$,

$$u(x)=\sum_{k=0}^{\infty}a_kx^k=\sum_{k=0}^{\infty}x^{2k}(a_{2k}+a_{2k+1}x)$$

is the Almansi expansion.

Remark 1.3. The Almansi expansion for analytic functions can be obtained locally by expanding $u(x)$ in spherical harmonics and using well-known characterizations of analytic functions. This was done by Baouendi, Goulaouic, and Lipkin [17]. The strength of Theorem 1.1 lies in the fact that the Almansi expansion is valid on all of $\tilde{\hat{B}}_{R_0}$. Moreover, it will be shown in the next section that if $u(x)$ is polyharmonic then the Almansi expansion is valid in any star domain about the origin on which $u(x)$ is holomorphic.

Using formulae (1.6) and (1.13) we obtain the following formula of Cauchy type.

Corollary 1.1. Let $u(x)$ be holomorphic on a neighbourhood U of the origin *in* $\mathbb{C}^n$. *If* $\bar{\hat{B}}_{R_0} \subset U$ *and* $x \in \hat{B}_{R_0}$, *then*

$$u(x) = \frac{1}{2\pi i\omega_n} \int_{\Gamma_{R_0}} \int_{S_1^{n-1}} \frac{\zeta^{n-1}}{r(\zeta\theta - x)^n} u(\zeta\theta)\, d\sigma(\theta)\, d\zeta. \tag{1.16}$$

Remark 1.4. The set $M^n_{R_0} = \{y = \zeta\theta : \zeta \in \mathbb{C}^1,\ |\zeta| = R_0,\ \theta \in S_1^{n-1}\}$ can be made into a real analytic n-dimensional manifold in a natural way, this manifold being a double covering of its geometric image in $\mathbb{C}^n$. It was pointed out in §3, Chapter II, that $\hat{B}_1$ is the classical domain $\mathcal{R}_{IV}$ of E. Cartan and that the boundary portion $\Lambda = \{e^{i\alpha}\theta : 0 \leq \alpha \leq \pi,\ \theta \in S_1^{n-1}\}$ is the Šilov (or characteristic) boundary. Harmonic analysis on the domain $\mathcal{R}_{IV}$ is discussed by Hua [51, Chapter IV]. The Cauchy kernel for $\mathcal{R}_{IV}$ is shown to be

$$H(x, \bar{\xi}) \equiv H(x, \alpha, \theta) = \frac{1}{Vr(\theta - e^{-i\alpha}x)^n}$$

where

$$V = \frac{2\pi^{n/2+1}}{\Gamma(n/2)}$$

is the volume of Λ, and the corresponding Cauchy formula is

$$u(x) = \int_\Lambda H(x, \bar{\xi}) u(\xi)\, d\alpha\, d\sigma(\theta)$$

where u is holomorphic on a neighbourhood of $\mathcal{R}_{IV}$ and $d\alpha\, d\sigma(\theta)$ is the volume element on Λ. It can be verified easily that this Cauchy formula is the same as that in (1.16).

By combining equations (1.6) and (1.12) we obtain formulae in $\mathbb{C}^n$ for the Almansi coefficients.

Corollary 1.2. Let $u(x)$ *be holomorphic on a neighbourhood* U *of the origin in* $\mathbb{C}^n$. *If* $\bar{\hat{B}}_{R_0} \subset U$ *then the Almansi coefficients of* $u(x)$ *are given on* $\hat{B}_{R_0}$ *by*

$$h_k(x) = \frac{1}{2\pi i\omega_n} \int_{\Gamma_{R_0}} \int_{S_1^{n-1}} \frac{\zeta^{n-2}}{\zeta^{2k+1}} \frac{\zeta^2 - r(x)^2}{r(\zeta\theta - x)^n} u(\zeta\theta)\, d\sigma(\theta)\, d\zeta. \tag{1.17}$$

Remark 1.5. Applying (1.17) to the case in which $u(x) \equiv h(x)$ is harmonic, so that $h_k(x) \equiv 0$ for $k = 1, 2, \ldots$, we obtain an integral representation on $\hat{B}_{R_0}$ for harmonic functions:

$$h(x) = \frac{1}{2\pi i\omega_n} \int_{\Gamma_{R_0}} \int_{S_1^{n-1}} \frac{\zeta^{n-3}(\zeta^2 - r(x)^2)}{r(\zeta\theta - x)^n} h(\zeta\theta)\, d\sigma(\theta)\, d\zeta. \tag{1.18}$$

Expressed in terms of $\mathcal{R}_{IV}$ as in Remark 1.4, with $R_0 = 1$, the kernel is written

$$A(x, \bar{\xi}) \equiv A(x, \alpha, \theta) = \frac{1 - r(e^{-i\alpha}x)^2}{V(\Lambda) r(\theta - e^{-i\alpha}x)^n}, \tag{1.19}$$

and the holomorphic harmonic functions on $\mathcal{R}_{IV}$ can be represented as

$$(1.18') \qquad h(x)=\int_{\Lambda} A(x,\bar{\xi})h(\xi)\,\mathrm{d}\alpha\,\mathrm{d}\sigma(\theta).$$

Corollary 1.3. If $h(x)$ is harmonic on a neighbourhood of the origin (in $\mathbb{R}^n$ or in $\mathbb{C}^n$) and if $h(x)/r^2$ is analytic, then $h(x)\equiv 0$.

Proof. Let $u(x)=h(x)/r^2$. Since $u(x)$ is analytic on a neighbourhood of the origin, it has a unique Almansi expansion

$$u(x)=\sum_{k=0}^{\infty} r^{2k}h_k(x)$$

on some $\hat{B}_R$. Therefore

$$h(x)=\sum_{k=0}^{\infty} r^{2k+2}h_k(x).$$

Since $h(x)$ is harmonic, the uniqueness of the Almansi expansion implies that $h_k(x)\equiv 0$ for all k. Thus $u(x)\equiv 0$ and $h(x)\equiv 0$. ‖

Recall that the isotropic cone at the origin in $\mathbb{C}^n$ is the set

$$V(0)\equiv\left\{x\in\mathbb{C}^n : r^2=\sum_{j=1}^{n} x_j^2=0\right\}.$$

For a point $x\in\mathbb{C}^n$, write $x=(x';x_n)$ where $x'=(x_1,\ldots,x_{n-1})$, and write $r'^2=\sum_{j=1}^{n-1}x_j^2$. With this notation $x_n^2=r^2-r'^2$, and the point x is on $V(0)$ if and only if $x=(x';\pm ir')$. This establishes a correspondence between subsets of $V(0)$ and subsets of $\mathbb{C}^{n-1}$. The next two lemmas demonstrate specific relations which will be of subsequent use.

Lemma 1.1. If $x'\in\hat{B}^{n-1}_{R/2}$ then $(x';\pm ir')\in\hat{B}^n_R\cap V(0)$, and $\hat{B}^{n-1}_{R/2}$ is the largest harmonicity hull of a ball which has this property.

Proof. Let $x'=\xi'+i\eta'$ and $x=(x';x_n)=\xi+i\eta$, so that $\xi=(\xi';\xi_n)$ and $\eta=(\eta';\eta_n)$. We know that $(x';\pm ir')\in V(0)$, and therefore $|\xi|^2=|\eta|^2$ and $(\xi,\eta)=0$. By Ex. 3.3, Chapter II, $x'\in\hat{B}^{n-1}_{R/2}$ if and only if

$$|\xi'|^2+|\eta'|^2+2\{|\xi'|^2\,|\eta'|^2-(\xi',\eta')^2\}^{\frac{1}{2}}<R^2/4.$$

To see that $(x';\pm ir')\in\hat{B}^n_R$, we note that $|\xi|=|\xi'|$ and

$$\begin{aligned}|\xi|^2+|\eta|^2+2(|\xi|^2\,|\eta|^2-(\xi,\eta)^2)^{\frac{1}{2}}&=4\,|\xi|^2\\&=2(|\xi|^2+|\eta|^2)\\&=2(|\xi'|^2+|\eta'|^2+|r'|^2)\\&\leqslant 2\{|\xi'|^2+|\eta'|^2+(|\xi'|^2+|\eta'|^2)\}\\&=4(|\xi'|^2+|\eta'|^2)\\&<R^2.\end{aligned}$$

The fact that $\hat{B}^{n-1}_{R/2}$ is the largest such harmonicity hull follows from observing that the point $x' = (R/2, 0, \ldots, 0) + i(0, \ldots, 0)$ is on $\partial\hat{B}^{n-1}_{R/2}$ and $(x'; \pm iR/2) \in \partial\hat{B}^n_R$. ‖

Lemma 1.2. $\hat{B}^n_R \subset \hat{B}^{n-1}_R \times \mathbb{C}^1$, *and* $\hat{B}^n_R$ *is the largest harmonicity hull of a ball which has this property.*

Proof. Because of the structure of the sets under consideration, it will suffice to prove that $\overline{\hat{B}^n_R} \subset \overline{\hat{B}^{n-1}_R \times \mathbb{C}^1}$. From Proposition 3.7, Chapter II, we know that both of these sets are closed and convex, and the extreme points of $\overline{\hat{B}^n_R}$ are the points $t\theta$ with $\theta \in \partial B^n_R$ and $|t| = 1$, $t \in \mathbb{C}^1$. To demonstrate the inclusion relation, it therefore suffices to show that all such points $t\theta$ lie in $\overline{\hat{B}^{n-1}_R \times \mathbb{C}^1}$, or that $t\theta'$ lies in $\overline{\hat{B}^{n-1}_R}$. If $t = a + ib$, then

$$|a\theta'|^2 + |b\theta'|^2 + 2\{|a\theta'|^2\,|b\theta'|^2 - (a\theta', b\theta')^2\}^{\frac{1}{2}} = |\theta'|^2 \leq R^2,$$

and so the inclusion is proved.

If $R_0 > R$, the point $(R_0, 0, \ldots, 0)$ is in $\overline{\hat{B}}{}^n_{R_0}$ but not in $\overline{\hat{B}^{n-1}_R \times \mathbb{C}^1}$. ‖

Definition 1.1. Let U be a neighbourhood of the origin in $\mathbb{C}^n$. A function defined on $U \cap V(0)$ will be called *analytic* if it is the restriction of a function holomorphic on some neighbourhood of the origin in $\mathbb{C}^n$.

Let $g(x)$ be analytic on $V(0)$ and let $u(x)$ be any function holomorphic on a neighbourhood of the origin in $\mathbb{C}^n$ whose restriction to $V(0)$ equals $g(x)$. Developing $u(x)$ locally in the variable x_n, we obtain

$$u(x) = \sum_{k=0}^{\infty} u_k(x')x_n^k. \tag{1.20}$$

Therefore on $V(0)$

$$\begin{aligned} g(x) = u(x) &= \sum_{k=0}^{\infty} u_{2k}(x')(-r'^2)^k + x_n \sum_{k=0}^{\infty} u_{2k+1}(x')(-r'^2)^k \\ &= g_1(x') + x_n g_2(x'), \qquad x_n = \pm ir'. \end{aligned} \tag{1.21}$$

The functions $g_1(x')$ and $g_2(x')$ are analytic on a neighbourhood of the origin in $\mathbb{C}^{n-1}$, and formula (1.21) shows that

$$g_1(x') = \frac{u(x'; x_n) + u(x'; -x_n)}{2} = \frac{g(x'; x_n) + g(x'; -x_n)}{2}, \qquad x_n = \pm ir' \tag{1.22}$$

$$g_2(x') = \frac{u(x'; x_n) - u(x'; -x_n)}{2x_n} = \frac{g(x'; x_n) - g(x'; x_n)}{2x_n}, \qquad x_n = \pm ir'.$$

Therefore the analytic functions $g_1(x')$ and $g_2(x')$ are uniquely determined by $g(x)$, and $g(x)$ is identically zero if and only if $g_1(x')$ and $g_2(x')$ are both identically zero.

Theorem 1.2. If $h(x)$ is harmonic on a neighbourhood of the origin in $\mathbb{C}^n$, then $h(x)$ is completely determined by its restriction to $V(0)$.

Proof. Equation (1.20) can be written (with $h(x)=u(x)$)

$$h(x)=\sum_{k=0}^{\infty} h_{2k}(x')(r^2-r'^2)^k + x_n \sum_{k=0}^{\infty} h_{2k+1}(x')(r^2-r'^2)^k. \tag{1.23}$$

If the restriction of $h(x)$ to $V(0)$ is identically zero, so that $h(x)=h(x)-h(x';ir')$, it follows from (1.23) that

$$\begin{aligned} h(x) &= \sum_{k=0}^{\infty} h_{2k}(x')\{(r^2-r'^2)^k-(-r'^2)^k\} \\ &\quad + x_n \sum_{k=0}^{\infty} h_{2k+1}(x')\{(r^2-r'^2)^k-(-r'^2)^k\}. \end{aligned} \tag{1.24}$$

Developing the terms in brackets in (1.24) we see that $h(x)$ is divisible by r^2. By Corollary 1.3, $h(x)\equiv 0$. ‖

Remark 1.6. Suppose that $u_1(x)$ and $u_2(x)$ are holomorphic on $\hat{B}^n_R$ and have the same restrictions to $V(0)$. On $\hat{B}^n_R$,

$$u_1(x)=\sum_{k=0}^{\infty} r^{2k}h_{1,k}(x)$$

and

$$u_2(x)=\sum_{k=0}^{\infty} r^{2k}h_{2,k}(x)$$

uniquely. The restrictions of $u_1(x)$ and $u_2(x)$ to $V(0)$ are the restrictions of $h_{1,0}(x)$ and $h_{2,0}(x)$, and by Theorem 1.2, $h_{1,0}(x)\equiv h_{2,0}(x)$ on $\hat{B}^n_R$. Therefore $u_1(x)-u_2(x)=r^2v(x)$ where $v(x)$ is holomorphic on $\hat{B}^n_R$. It now follows that $u_1(x)$ and $u_2(x)$ have the same restrictions to $V(0)$ if and only if $u_1(x)-u_2(x)=r^2v(x)$ where $v(x)$ is an arbitrary holomorphic function on $\hat{B}^n_R$. Furthermore, if $u_1(x)$ and $u_2(x)$ have equal restrictions to $V(0)$ and if $u_1(x)-u_2(x)$ is harmonic on $\hat{B}^n_R$, then $u_1(x)\equiv u_2(x)$.

Definition 1.2.

(a) Two functions g, g' analytic on neighbourhoods of the origin on $V(0)$ are *equivalent* if $g(x)\equiv g'(x)$ on some neighbourhood of the origin on $V(0)$. The equivalence classes are called *germs of analytic functions on $V(0)$ at* 0.

(b) Two functions h, h' harmonic on neighbourhoods of the origin in $\mathbb{C}^n$ are *equivalent* if $h(x)\equiv h'(x)$ on some neighbourhood of the origin in $\mathbb{C}^n$. The equivalence classes are called *germs of harmonic functions at* 0.

Since we shall only be interested in germs at the origin, the phrase 'at 0' will be suppressed.

Let $\mathscr{A}$ denote the set of germs of analytic functions on $V(0)$ and let $\mathscr{H}$ denote the set of germs of harmonic functions. Both $\mathscr{A}$ and $\mathscr{H}$ can be made into complex linear spaces by defining addition and scalar multiplication of germs by means of pointwise addition and scalar multiplication of representatives. We define a linear mapping T: $\mathscr{H} \to \mathscr{A}$ as follows. Let $H \in \mathscr{H}$ and let h be a representative of H. Define $T(H)$ to be the germ of the restriction of h to $V(0)$.

Theorem 1.3. The linear mapping T is an isomorphism between $\mathscr{H}$ and $\mathscr{A}$.

Proof. By Theorem 1.2, T is one to one. Let $A \in \mathscr{A}$ and let g be a representative of A with domain $\hat{B}^n_R \cap V(0)$. By Lemma 1.1 and the paragraph preceding Theorem 1.2, g uniquely determines a pair of holomorphic functions $g_1(x')$, $g_2(x')$ defined on $\hat{B}^{n-1}_{R/2}$, and $g(x) = g_1(x') + x_n g_2(x')$, $x_n = \pm ir'$. The function $u(x) = g_1(x') + x_n g_2(x')$ is holomorphic on $\hat{B}^{n-1}_{R/2} \times \mathbb{C}^1$, and by Lemma 1.2 $u(x)$ is holomorphic on $\hat{B}^n_{R/2}$. Therefore $u(x)$ has a unique Almansi expansion $\sum_{k=0}^{\infty} r^{2k} h_k(x)$ on $\hat{B}^n_{R/2}$ and the restriction of $u(x)$ to $V(0)$ is $g_1(x') \pm ir' g_2(x') = g(x)$. This is exactly the restriction of $h_0(x)$ to $V(0)$. If H_0 is the term corresponding to $h_0(x)$, then $T(H_0) = A$ and the theorem is proved. ||

Remark 1.7. For evaluating T^{-1} a formula for $h_0(x)$ is given by (1.17):

$$h_0(x) = \frac{1}{2\pi i \omega_n} \int_{\Gamma_{R'}} \int_{S_1^{n-1}} \frac{\zeta^{n-3}(\zeta^2 - r(x)^2)}{r(\zeta\theta - x)^n} u(\zeta\theta)\, d\sigma(\theta)\, d\zeta$$

where $u(x) = g_1(x') + x_n g_2(x')$, $x \in \hat{B}^n_{R'}$, $0 < R' < R/2$.

The linear space $\mathscr{A}$ also has a multiplication defined by means of pointwise multiplication of representatives. $\mathscr{A}$ is then the algebra of germs of analytic functions on $V(0)$. The isomorphism T allows us to define multiplication in $\mathscr{H}$ by $H_1H_2 = T^{-1}(T(H_1)T(H_2))$, so that $\mathscr{H}$ is now an algebra of germs of harmonic functions. In terms of representatives, multiplication in $\mathscr{H}$ is defined as follows. Let h, h' be representatives of H, H' with common domain $\hat{B}^n_R$. Restrict these functions to $V(0)$, multiply pointwise, and construct the harmonic function $h_0(x)$ in Theorem 1.3. Alternatively, if $u(x) = h(x)h'(x)$, then $u(x)$ is analytic on $\hat{B}^n_R$ and hence has unique Almansi expansion

$$u(x) = \sum_{k=0}^{\infty} r^{2k} h_k(x)$$

on $\hat{B}^n_R$. The product HH' is the germ H_0 of $h_0(x)$.

The second view of multiplication in $\mathscr{H}$ allows us to construct an algebra $\mathscr{H}(B^n_R)$ of functions harmonic on $B^n_R \subset \mathbb{R}^n$, $R > 0$ fixed. All functions in $\mathscr{H}(B^n_R)$ can be continued as single-valued harmonic functions on $\hat{B}^n_R$. If $h, h' \in \mathscr{H}(B_R)$, extend them to harmonic functions $\hat{h}$, $\hat{h}'$ on $\hat{B}^n_R$ and multiply them as above. The result $\hat{h}_0$ is harmonic on $\hat{B}^n_R$ and its restriction to B^n_R is defined to be the product hh' in the algebra $\mathcal{H}(B^n_R)$. For $R_0 < R$, harmonic functions on B^n_R can be considered (by restriction) as elements of $\mathcal{H}(B^n_{R_0})$. If h, $h' \in \mathcal{H}(B^n_R)$ are multiplied as elements of $\mathcal{H}(B^n_{R_0})$, their product is the restriction to $B^n_{R_0}$ of hh' multiplied as elements of $\mathcal{H}(B^n_R)$.

It is interesting to contrast multiplication in the algebra $\mathcal{H}(B^n_R)$ with another definition of multiplication of harmonic functions on B^n_R. On the linear space $\mathcal{L} = \mathcal{H}(B^n_R) \cap \mathcal{C}^0(\bar{B}^n_R)$ we can define a multiplication as follows. If h, h' are in $\mathcal{L}$, hh' is the element of $\mathcal{L}$ with boundary values $h(x)h'(x)$ on $\partial\bar{B}^n_R$. This multiplication does not preserve restrictions to $B^n_{R_0}$, $R_0 < R$, as does multiplication in $\mathcal{H}(B^n_R)$.

Notes

The Almansi expansion was proved for polyharmonic functions by Aronszajn [7]. Ciorănescu [26] proved it for real analytic functions of two variables. A proof of the theorem for n complex variables is given by Aronszajn [12] by quite a different method from that of this section. The Cauchy formula (1.16) is derived by means of a calculus of residues, and from it the Almansi expansion is deduced. A discussion of kernels, related to Remark 1.4, is also contained in that paper. Theorem 1.2 was proved for the case of two variables by Ciorănescu [27].

2. Characterization of order and type by the Almansi coefficients

In the preceding section we proved the existence and uniqueness of the Almansi expansion for holomorphic functions on the harmonicity hull of a ball. In this section we shall characterize the functions in the classes $\mathcal{P}^{\rho,\tau}(D)$ (see Definition 1.3, Chapter II) by means of the Almansi coefficients. As a consequence, we shall be able to construct functions exhibiting various types of behaviour with respect to order and type (see §3, Chapter V). Theorem 2.1 gives sufficient conditions on the coefficients that a function be in the class $\mathcal{P}^{\rho}(D)$, $\mathcal{P}^{\rho,\infty}(D)$, or $\mathcal{P}^{\rho,0}(D)$. In Theorem 2.2 we derive a serio-integral formula for the Almansi coefficients and show that the Almansi expansion for a polyharmonic function $u(x)$ converges in any star domain centred about the origin in which $u(x)$ is holomorphic. In Theorem 1.3 we complete the characterization of the polyharmonicity classes by showing that the sufficient conditions in Theorem 2.1 are also necessary and by deriving necessary and sufficient conditions for the class $\mathcal{P}^{\rho,\tau}(D)$.

Theorem 2.1. Let D be an open set in $\mathbb{R}^n$ or $\mathbb{C}^n$ and let $\{h_k(x)\}$ be a sequence of functions harmonic on D.

(a) *Let* $0 \leq \rho < \infty$. *Suppose that for every compact set* $K \subset D$ *and every* $\varepsilon > 0$ *there exists a constant* $C_{K,\varepsilon}$ *such that*

$$\|h_k\|_K \leq C_{K,\varepsilon}(2k)!^{-1/(\rho+\varepsilon)}.$$

(*Equivalently, for every compact set* $K \subset D$ *and for every* $\varepsilon > 0$,

$$\limsup_{k\to\infty} \|h_k\|_K^{1/2k} k^{1/(\rho+\varepsilon)} = 0.)$$

Then $\sum_{k=0}^{\infty} r^{2k} h_k(x)$ *defines a function in* $\mathcal{P}^{\rho}(D)$.

(b) *Let* $0 < \rho < \infty$. *Suppose that for every compact set* $K \subset D$ *there exist constants* C_K *and* τ_K *such that* $\|h_k\|_K \leq C_K \tau_K^{2k}(2k)!^{-1/\rho}$. (*Equivalently, for every compact set* $K \subset D$,

$$\limsup_{k\to\infty} \|h_k\|_K^{1/2k} k^{1/\rho} < \infty.)$$

Then $\sum_{k=0}^{\infty} r^{2k} h_k(x)$ *defines a function in* $\mathcal{P}^{\rho,\infty}(D)$.

(c) *Let* $0 < \rho < \infty$. *Suppose that for every compact set* $K \subset D$ *and every* $\varepsilon > 0$ *there exists a constant* $C_{K,\varepsilon}$ *such that*

$$|h_k\|_K \leq C_{K,\varepsilon}\varepsilon^{2k}(2k)!^{-1/\rho}.$$

(*Equivalently, for every compact set* $K \subset D$,

$$\limsup_{k\to\infty} \|h_k\|_K^{1/2k} k^{1/\rho} = 0.)$$

Then $\sum_{k=0}^{\infty} r^{2k} h_k(x)$ *defines a function in* $\mathcal{P}^{\rho,0}(D)$.

Proof. We first establish bounds for harmonic functions on subsets of $\mathbb{C}^n$. If the domain D is in $\mathbb{C}^n$ the bounds, given at (2.2) below, are automatic. Therefore we shall assume that D is a domain in $\mathbb{R}^n$.

Consider the collection $\mathcal{B}$ of all open balls $B_R(x)$ such that $x \in D$ and $\overline{B_R(x)} \subset D$. Clearly $D = \bigcup B_R(x)$ where the union is taken over $\mathcal{B}$. Define $G = \bigcup \hat{B}_R(x)$ where the union is taken over the family $\hat{\mathcal{B}}$ of harmonicity hulls of sets in $\mathcal{B}$. The set G is open in $\mathbb{C}^n$, $G \subset \hat{D}$, $G \cap \mathbb{R}^n = D$, and every function harmonic on D has unique analytic continuation as a harmonic function on G. Note also that if h is harmonic on a neighbourhood of a closed ball $\overline{B_{R_0}(x_0)}$, then the Poisson formula

$$h(x) = \frac{1}{\omega_n R_0} \int_{S_{R_0}(x_0)} \frac{R_0^2 - r(x-x_0)^2}{r(\theta-(x-x_0))^n} h(\theta)\, d\sigma_{R_0}(\theta) \tag{2.1}$$

is valid for $x \in B_{R_0}(x_0)$. By Proposition 3.8, Chapter II, the Poisson kernel can be continued analytically into all of $\hat{B}_{R_0}(x_0)$. When it is so continued, formula (2.1) represents the continuation of h to $\hat{B}_{R_0}(x_0)$.

Let K_0 be a compact subset of G. There exists a finite collection $\hat{B}_{R_1}(x_1),\ldots, \hat{B}_{R_l}(x_l)$ of sets in $\hat{\mathscr{B}}$ such that the sets $\hat{B}_{R_1/2}(x_1),\ldots, \hat{B}_{R_l/2}(x_l)$ cover K_0. If $x \in K_0$ then $x \in \hat{B}_{R_j/2}(x_j)$ for some $j \in \{1,\ldots, l\}$ and by ((2.1) we have

$$\begin{aligned}|h(x)| &\leq \frac{1}{\omega_n R_j}\left|\int_{S_{R_j}(x_j)} \frac{R_j^2 - r(x-x_j)^2}{r(\theta-(x-x_j))^n} h(\theta)\, \mathrm{d}\sigma_{R_j}(\theta)\right| \\ &\leq C_j \, \|h\|_{S_{R_j}(x_j)}.\end{aligned}$$

If

$$K_0' = \bigcup_{j=1}^{l} S_{R_j}(x_j),$$

the set K_0' is a compact subset of D depending only on K_0. Therefore we have the result: if K_0 is a compact subset of G there exists a constant C_{K_0} and a compact subset $K_0' \subset D$ such that every harmonic function $h(x)$ on D can be continued analytically to G and

$$\|h\|_{K_0} \leq C_{K_0} \|h\|_{K_0'}. \tag{2.2}$$

With the same notations as above, let $h(x)$ be harmonic on G. If K_1 is a compact subset of G, there exists a compact set K_0 such that K_0 is the closure of an open set and $K_1 \subset K_0^{\circ} \subset K_0 \subset G$. Applying the Cauchy bounds and (2.2), we see that there exists a constant L_{K_0} such that for every $p = 0, 1, 2,\ldots,$

$$\begin{aligned}\|\Delta^p(r^{2k}h)\|_{K_1} &\leq (2p)!\, L_{K_0}^{2p} \|r^{2k}h\|_{K_0} \\ &\leq (2p)!\, L_{K_0}^{2p} R_0^{2k} C_{K_0} \|h\|_{K_0'}\end{aligned} \tag{2.3}$$

where R_0 depends only on K_0.

We now prove (b). Let all of the harmonic functions $h_k(x)$ be continued analytically to $\hat{D}$, and hence to G. If K_1 is any compact subset of G we obtain from (2.3) for each p

$$\begin{aligned}\sum_{k=0}^{\infty} \|\Delta^p(r^{2k}h_k)\|_{K_1} &\leq \sum_{k=p}^{\infty} (2p)!\, L_{K_0}^{2p} R_0^{2k} C_{K_0} \|h_k\|_{K_0'} \\ &= (2p)!\, L_{K_0}^{2p} R_0^{2p} C_{K_0} \sum_{k=0}^{\infty} R_0^{2k} \|h_{k+p}\|_{K_0'}.\end{aligned} \tag{2.4}$$

By hypothesis in (b), the right-hand side of (2.4) is dominated by

$$\begin{aligned}&(2p)!\, L_{K_0}^{2p} R_0^{2p} C_{K_0} \sum_{k=0}^{\infty} R_0^{2k} C_{K_0'} \tau_{K_0'}^{2k+2p} (2k+2p)!^{-1/\rho} \\ &= (2p)!^{1-1/\rho} (L_{K_0} R_0 \tau_{K_0'})^{2p} C_{K_0} C_{K_0'} \sum_{k=0}^{\infty} (R_0 \tau_{K_0'})^{2k} \left\{\frac{(2p)!}{(2k+2p)!}\right\}^{1/\rho} \\ &\leq (2p)!^{1-1/\rho} \tau_{K_0}'^{2p} C'_{K_0} \sum_{k=0}^{\infty} R_0'^{2k} \left\{\frac{1}{(2k)!}\right\}^{1/\rho}.\end{aligned} \tag{2.5}$$

Since the numbers $\{1/(2k)!\}^{1/\rho}$ are coefficients of an entire function, combining (2.4) and (2.5) we see that $\sum_{k=0}^{\infty} r^{2k}h_k(x)$ is holomorphic on G and defines a function of class $\mathcal{P}^{\rho,\infty}(D)$.

To prove (c), let $\varepsilon' > 0$ be given, let K_0 be a compact set, and let $\varepsilon = \varepsilon'/M'_{K_0}$ where $M'_{K_0} = L_{K_0}R_0$ in (2.4). From (2.4) and the hypothesis of (c) we obtain

$$(2.6) \qquad \sum_{k=0}^{\infty} \|\Delta^p(r^{2k}h_k)\|_{K_1} \leq (2p)!\, M'^{2p}_{K_0} C_{K_0} \sum_{k=0}^{\infty} R_0^{2k} \|h_{k+p}\|_{K'_0}$$

$$\leq (2p)!\, M'^{2p}_{K_0} C_{K_0} \sum_{k=0}^{\infty} R_0^{2k} C_{K_0,\varepsilon}\varepsilon^{2p+2k}(2p+2k)!^{-1/\rho}$$

$$\leq (2p)!\, \varepsilon'^{2p} C_{K_0} C_{K_0,\varepsilon} \sum_{k=0}^{\infty} R_0^{2k}\left(\frac{\varepsilon'}{M'_{K_0}}\right)^{2k} (2p+2k)!^{-1/\rho}.$$

The same technique as in (2.5) yields

$$\sum_{k=0}^{\infty} \|\Delta^p(r^{2k}h_k)\|_{K_1} \leq (2p)!^{1-1/\rho}\varepsilon'^{2p} C'_{K_0,\varepsilon'}.$$

Therefore the function $\sum_{k=0}^{\infty} r^{2k}h_k(x)$ is holomorphic on G and defines a function of class $\mathcal{P}^{\rho,0}(D)$.

Statement (a) now follows easily from (c) by using the fact that $\mathcal{P}^{\rho}(D) = \bigcap_{\rho'>\rho} \mathcal{P}^{\rho',0}$. ‖

We now wish to derive formulas for the Almansi coefficients, to prove that the sufficient conditions in Theorem 2.1 are also necessary, and to derive necessary and sufficient conditions that a function u have Laplacian order ρ and positive finite type τ. In order to do this we shall make use of a version of Stirling's formula and a technique of dominating series.

Stirling's formula.

$$(2.7) \qquad \text{For } x > 0,\ \Gamma(x) = x^{x-1/2}e^{-x}(2\pi)^{1/2}e^{\theta(x)/12x}$$

where $0 < \theta(x) < 1$.

Technique. Let

$$S_k = \sum_{i=0}^{\infty} A_i a_{ik}, \qquad A_i > 0, \qquad a_{ik} > 0.$$

We shall want to obtain an inequality of the form $a_{ik} \leq C\gamma^k\delta^i$ where δ is chosen so that $\sum_{i=0}^{\infty} A_i\delta^i$ converges. Then γ will be chosen so that the inequality $a_{ik} \leq C\gamma^k\delta^i$ holds. If the A_i are coefficients of an entire function, any choice of δ will yield convergence of $\sum_{i=0}^{\infty} A_i\delta^i$; if the A_i are not coefficients of an entire function, the choice of δ will depend upon the radius of convergence. The construction of δ and γ will be achieved as follows. Consider any number $\alpha > 0$ and decompose the range of i into two parts: $i \leq \alpha k$ and $i > \alpha k$. Choose $\gamma = \gamma(\alpha)$ such that if $i \leq \alpha k$ then $a_{ik} \leq C_1 \gamma^k$, and choose $\delta = \delta(\alpha)$ such that if $i > \alpha k$ then $a_{ik} \leq C_2 \delta^i$. If both γ and δ are greater than or equal to unity then we shall obtain $a_{ik} \leq C\gamma^k \delta^i$.

Theorem 2.2. Let $u(x)$ be holomorphic on a neighbourhood of the origin in $\mathbb{C}^n$ and let

$$u(x) = \sum_{k=0}^{\infty} r^{2k} h_k(x)$$

be its Almansi expansion, valid on some neighbourhood U of the origin. If D is a star domain centred at the origin with $D \subset U$ then

$$h_0(x) = u(x) + \sum_{i=1}^{\infty} \frac{(-1)^i r(x)^{2i}}{2^{2i}(i-1)!\, i!} \times \int_0^1 \tau^{(n-2)/2} \tau^{i-1} (1-\tau)^{i-1} (\Delta^i u)(\tau x)\, \mathrm{d}\tau, \tag{2.8}$$

and for $k \geq 1$

$$h_k(x) = \sum_{i=0}^{\infty} \frac{1}{2^{2k} k!} \frac{(-1)^i r(x)^{2i}}{2^{2i} i!\, (k+i)!} \times \int_0^1 \tau^{(n-2)/2} \tau^{i-1} (1-\tau)^{k+i-1} (i+k\tau)(\Delta^{k+i} u)(\tau x)\, \mathrm{d}\tau \tag{2.9}$$

for each $x \in D$. ($(\Delta^p u)(\tau x)$ is the value of $\Delta^p u$ at $\tau x \in D$.) If $u(x)$ is polyharmonic on a neighbourhood of the origin in $\mathbb{C}^n$ then the Almansi expansion and formulae (2.8) and (2.9) are valid in any star domain centred at the origin in which $u(x)$ is holomorphic.

Proof. First we show that the series (2.8) and (2.9) are absolutely and uniformly convergent on compacts. Since $u(x)$ is holomorphic, for each compact set $K \subset D$ there exist constants M and C such that

$$|\Delta^i u(y)| \leq C(2i)!\, M^{2i}, \qquad i = 0, 1, \ldots, \quad y \in K. \tag{2.10}$$

Using (2.10) we majorize the series of absolute values of the terms in (2.9) by the series

$$\sum_{i=0}^{\infty} \frac{1}{2^{2k} k!} \frac{r^{2i}}{2^{2i} i!\, (k+i)!} \times \int_0^1 \tau^{(n-2)/2} \tau^{i-1} (1-\tau)^{k+i-1} (i+k\tau) C(2k+2i)!\, M^{2k+2i}\, \mathrm{d}\tau. \tag{2.11}$$

Split (2.11) into two series arising from the factor $i+k\tau$ in the integrand and use the relations

$$\int_0^1 \tau^{(n-2)/2+i-1}(1-\tau)^{k+i-1}\,d\tau$$

$$=B\left(\frac{n-2}{2}+i,\, k+i\right)=\frac{\Gamma((n-2)/2+i)\Gamma(k+i)}{\Gamma((n-2)/2+k+2i)}$$

and

$$\int_0^1 \tau^{(n-2)/2+i}(1-\tau)^{k+i-1}\,d\tau$$

$$=B\left(\frac{n-2}{2}+i+1,\, k+i\right)=\frac{\Gamma(n/2+i)\Gamma(k+i)}{\Gamma(n/2+k+2i)}.$$

The series (2.11) is equal to

$$\frac{CM^{2k}}{2^{2k}k!}\sum_{i=0}^{\infty}\left(\frac{rM}{2}\right)^{2i}\frac{(2k+2i)!}{i!\,(k+i)!} \times\left\{\frac{i\Gamma((n-2)/2+i)\Gamma(k+i)}{\Gamma((n-2)/2+k+2i)}+\frac{k\Gamma(n/2+i)\Gamma(k+i)}{\Gamma(n/2+k+2i)}\right\}. \tag{2.12}$$

Since (2.12) is the sum of two power series in $(rM/2)^2$ and each of the summands has positive coefficients, the radius of convergence of (2.12) will be the minimum of the two radii of convergence. The summands have coefficients

$$\frac{(2k+2i)!\,i\Gamma((n-2)/2+i)\Gamma(k+1)}{i!\,(k+i)!\,\Gamma((n-2)/2+k+2i)}$$

and

$$\frac{(2k+2i)!\,k\Gamma(n/2+i)\Gamma(k+i)}{i!\,(k+i)!\,\Gamma(n/2+k+2i)}.$$

Using Stirling's formula (2.7), we can show each radius of convergence to be unity. Therefore, the series (2.12) is convergent for $rM/2<1$, and consequently the series (2.9) is absolutely and uniformly convergent for $r\leq R'<2/M$ and $x\in D$. If $u(x)$ is polyharmonic, then in (2.10) M can be taken arbitrarily so that the series defining $h_k(x)$ converges in any star domain centred at the origin where $u(x)$ is holomorphic. (Recall that if $u(x)$ is holomorphic on D and polyharmonic on an open subset of D, then $u(x)$ is polyharmonic on D. See Corollary 1.4, Chapter IV.) The same kind of estimates can be carried out for $h_0(x)$ as defined in (2.9).

Next we show that each function $h_k(x)$ is harmonic. Note that each term in the series expansion for $h_k(x)$ is holomorphic and that the Laplacian can be taken under the integral sign. If N is any positive integer,

$$(2.13)\quad \Delta\left\{\sum_{i=0}^{N}\frac{1}{2^{2k}k!}\frac{(-1)^i r(x)^{2i}}{2^{2i}i!\,(k+i)!}\right.$$
$$\left.\times\int_0^1 \tau^{(n-2)/2}\tau^{i-1}(1-\tau)^{k+i-1}(i+k\tau)(\Delta^{k+i}u)(\tau x)\,\mathrm{d}\tau\right\}$$
$$=\sum_{i=0}^{N}\frac{1}{2^{2k}k!}\frac{(-1)^i}{2^{2i}i!\,(k+i)!}$$
$$\times\int_0^1 \tau^{(n-2)/2}\tau^{i-1}(1-\tau)^{k+i-1}(i+k\tau)\Delta\{r(x)^{2i}(\Delta^{k+i}u)(\tau x)\}\,\mathrm{d}\tau.$$

From the formula (equation (1.1), Chapter I)

$$(2.14)\qquad \Delta(r^{2i}v)=2i(n+2i-2)r^{2i-2}v+r^{2i}\Delta v+4ir^{2i-1}\frac{\partial v}{\partial r},$$

it follows that

$$(2.15)\quad \Delta\{r(x)^{2i}(\Delta^{k+i}u)(\tau x)\}=2i(n+2i-2)r(x)^{2i-2}(\Delta^{k+i}u)(\tau x)$$
$$+r(x)^{2i}\tau^2(\Delta^{k+i+1}u)(\tau x)+4ir(x)^{2i-2}\tau\frac{\partial}{\partial\tau}(\Delta^{k+i}u)(\tau x).$$

Using (2.15) we can write (2.13) as a sum of three expressions:

$$(2.16)\quad \Delta\left[\sum_{i=0}^{N}\dots\right]=\sum_{i=0}^{N}\frac{(-1)^i 2i(n+2i-2)r^{2i-2}}{2^{2k}k!\,2^{2i}i!\,(k+i)!}$$
$$\times\int_0^1 \tau^{(n-2)/2}\tau^{i-1}(1-\tau)^{k+i-1}(i+k\tau)(\Delta^{k+i}u)(\tau x)\,\mathrm{d}\tau$$
$$+\sum_{i=1}^{N+1}\frac{(-1)^{i-1}r^{2i-2}}{2^{2k}k!\,2^{2(i-1)}(i-1)!\,(k+i-1)!}$$
$$\times\int_0^1 \tau^{(n-2)/2}\tau^{i}(1-\tau)^{k+i-2}(i-1+k\tau)(\Delta^{k+i}u)(\tau x)\,\mathrm{d}\tau$$
$$+\sum_{i=0}^{N}\frac{(-1)^i 4ir^{2i-2}}{2^{2k}k!\,2^{2i}i!\,(k+i)!}$$
$$\times\int_0^1 \tau^{(n-2)/2}\tau^{i}(1-\tau)^{k+i-1}(i+k\tau)\frac{\partial}{\partial\tau}(\Delta^{k+i}u)(\tau x)\,\mathrm{d}\tau.$$

For $k>0$ integrate by parts in the last sum in (2.16) and obtain

$$(2.17)\quad -\sum_{i=1}^{N} \frac{(-1)^i 4ir^{2i-2}}{2^{2k}k!\,2^{2i}i!\,(k+i)!} \int_0^1 \left\{\left(\frac{n-2}{2}+i\right)\tau^{(n-2)/2}\tau^{i-1}(1-\tau)^{k+i-1}(i+k\tau)\right.$$

$$-(k+i-1)\tau^{(n-2)/2}\tau^{i}(1-\tau)^{k+i-2}(i+k\tau)$$

$$\left.+k\tau^{(n-2)/2}\tau^{i}(1-\tau)^{k+i-1}\right\}(\Delta^{k+i}u)(\tau x)\,\mathrm{d}\tau.$$

In the integrand above, abbreviate $\{\ldots\}$ by $A-B+C$ and note that the portion of (2.17) corresponding to $\int_0^1 A(\Delta^{k+i}u)(\tau x)\,\mathrm{d}\tau$ cancels the first sum in (2.16). Next,

$$-B+C=-\tau^{(n-2)/2}\tau^{i}(1-\tau)^{k+i-2}(k+i)(i-1+k\tau),$$

so the portion of (2.17) corresponding to $\int_0^1(-B+C)(\Delta^{k+i}u)(\tau x)\,\mathrm{d}\tau$ cancels the second sum in (2.16) except for the term with $i=N+1$. What remains is

$$\frac{(-1)^N r^{2N}}{2^{2k}k!\,2^{2N}N!\,(k+N)!}$$
$$\times\int_0^1 \tau^{(n-2)/2}\tau^{N+1}(1-\tau)^{k+N-1}(N+k\tau)(\Delta^{k+N+1}u)(\tau x)\,\mathrm{d}\tau.$$

Use (2.10) to majorize the absolute value of this expression by

$$(2.18)\quad \frac{M^{2k+2}C}{2^{2k}k!}\frac{\{2(k+N+1)\}!}{N!\,(N+k-1)!}\left(\frac{rM}{2}\right)^{2N}\int_0^1 \tau^{(n-2)/2+N+1}(1-\tau)^{k+N-1}\,\mathrm{d}\tau,$$

and then apply the technique following expression (2.12) to see that, for each fixed k, (2.18) converges to zero uniformly for $rM/2\leq R'<1$. Therefore $\Delta(\sum_{i=0}^{N}\ldots)$ in (2.13) converges uniformly to zero for $r\leq R''<2/M$, and hence the functions $h_k(x)$ defined in (2.9) are harmonic. For the function $h_0(x)$ defined in (2.8) the procedure is the same; the only difference in detail is that upon integration by parts one of the boundary terms is $-\Delta u(x)$, and this cancels with $\Delta u(x)$ in the expression for the Laplacian of the Nth partial sum.

Finally we shall demonstrate the convergence of the series $\sum_{k=0}^{\infty} r^{2k}h_k$ to u, which will establish the theorem (by uniqueness of the Almansi expansion).

Let

$$S_q(x)=\sum_{k=0}^{q} r^{2k}h_k(x),$$

where the functions $h_k(x)$ are defined by (2.8) and (2.9). Group the terms in this double series according to the same power of Δ and set $p=k+i$:

(2.19)
$$S_q(x)=u(x)+\sum_{p=1}^{\infty}\int_0^1(\Delta^p u)(\tau x)\sum_{k=0}^{\min(p,q)}\frac{(-1)^{p-k}r^{2p}}{2^{2p}k!\,(p-k)!\,p!}\tau^{(n-2)/2+p-k-1}(1-\tau)^{p-1}\times\{p-k(1-\tau)\}\,d\tau.$$

Each integrand is a sum of differences, because of the factor $p-k(1-\tau)$. It can be rewritten as a difference of sums:

$$(\Delta^p u)(\tau x)\frac{(-1)^p r^{2p}\tau^{(n-2)/2+p-1}(1-\tau)^{p-1}}{2^{2p}p!}\times\left\{p\sum_{k=0}^{\min(p,q)}\frac{(-1)^k\tau^{-k}}{k!\,(p-k)!}-(1-\tau)\sum_{k=0}^{\min(p,q)}\frac{k(-1)^k\tau^{-k}}{k!\,(p-k)!}\right\}.$$

If $p\leqslant q$, the above sums run from $k=0$ to $k=p$, and the expression in brackets becomes

$$\{\ldots\}=\frac{1}{(p-1)!}\left(1-\frac{1}{\tau}\right)^p+\frac{1-\tau}{\tau}\frac{1}{(p-1)!}\left(1-\frac{1}{\tau}\right)^{p-1}=0.$$

For $p>q$

$$\{\ldots\}=\frac{1}{(p-1)!}\sum_{k=0}^{q}\binom{p}{k}(-\tau)^{-k}+\frac{1-\tau}{\tau}\frac{1}{(p-1)!}\sum_{k=0}^{q-1}\binom{p-1}{k}(-\tau)^{-k}.$$

The second summand above can be written

$$-\frac{1}{(p-1)!}\sum_{k=0}^{q-1}\binom{p-1}{k}(-\tau)^{-k}-\frac{1}{(p-1)!}\sum_{k=0}^{q-1}\binom{p-1}{k}(-\tau)^{-k-1}.$$

In the second term of this expression, replace k by $k-1$ and use the equation

$$\binom{p-1}{k}+\binom{p-1}{k-1}=\binom{p}{k}$$

to obtain finally

$$\begin{aligned}\{\ldots\}&=\frac{1}{(p-1)!}\left\{\binom{p}{q}(-\tau)^{-q}-\binom{p-1}{q-1}(-\tau)^{-q}\right\}\\&=\frac{1}{(p-1)!}\binom{p-1}{q}(-\tau)^{-q}\\&=\frac{(-\tau)^{-q}}{q!\,(p-q-1)!}.\end{aligned}$$

Therefore

$$S_q(x)-u(x)=\sum_{p=q+1}^{\infty}\frac{(-1)^{p+q}r^{2p}}{2^{2p}p!\,q!\,(p-q-1)!}$$
$$\times\int_0^1 \tau^{(n-2)/2}\tau^{p-q-1}(1-\tau)^{p-1}(\Delta^p u)(\tau x)\,d\tau,$$

and setting $p=q+i$ shows that

$$S_q(x)-u(x)=\sum_{i=1}^{\infty}\frac{(-1)^i r^{2i+2q}}{2^{2q+2i}q!\,(i-1)!\,(q+i)!}\tag{2.20}$$
$$\times\int_0^1 \tau^{(n-2)/2+i-1}(i-\tau)^{q+i-1}(\Delta^{q+i}u)(\tau x)\,d\tau.$$

Applying (2.10) to (2.20) and using the formula

$$\int_0^1 y^{s-1}(1-y)^{t-1}\,dy=\frac{\Gamma(s)\Gamma(t)}{\Gamma(s+t)}$$

we obtain

(2.21)
$$|S_q(x)-u(x)|\leqslant 2C\left(\frac{rM}{2}\right)^{2q}\sum_{i=1}^{\infty}\frac{\Gamma(2q+2i)\Gamma((n-2)/2+i)}{\Gamma(i)\Gamma(q+1)\Gamma((n-2)/2+q+2i)}\left(\frac{rM}{2}\right)^{2i}.$$

To study the series in (2.21) we use the technique described prior to the statement of Theorem 2.2. Let

$$A_i=\frac{\Gamma((n-2)/2+i)}{\Gamma(i)}\left(\frac{rM}{2}\right)^{2i}$$

and

$$a_{iq}=\frac{\Gamma(2q+2i)}{\Gamma(q+1)\Gamma(q+2i)}.$$

Then

$$|S_q(x)-u(x)|\leqslant 2C\left(\frac{rM}{2}\right)^{2q}\sum_{i=1}^{\infty}A_i a_{iq},\tag{2.22}$$

and the A_i are coefficients of a power series with radius of convergence $(2/rM)^2$. For fixed q, a_{iq} is increasing in i and for fixed i, a_{iq} is increasing in q. Let $0<\alpha<\alpha_0$ where α_0 is a fixed large number. If $i\leqslant\alpha q$, from Stirling's formula (2.7) we find that there is a constant C_0, independent of

i, q, α, α_0, such that

$$\begin{aligned} a_{iq} &= \frac{\Gamma(2q+2i)}{\Gamma(q+1)\Gamma(q+2i)} \\ &\leqslant \frac{\Gamma(2q+2\alpha q)}{\Gamma(q+1)\Gamma(q+2\alpha q)} \\ &\leqslant C_0 \left\{ \frac{(2+2\alpha)^{2+2\alpha}}{(1+2\alpha)^{1+2\alpha}} \right\}^q \end{aligned}$$

If $i > \alpha q$, that is $q < i/\alpha$, there is a constant C_0', depending only on α_0, such that

$$\begin{aligned} a_{iq} &= \frac{\Gamma(2q+2i)}{\Gamma(q+1)\Gamma(q+2i)} \\ &\leqslant \frac{\Gamma((2/\alpha)i+2i)}{\Gamma((1/\alpha)+1)\Gamma((1/\alpha)+2i)} \\ &\leqslant C_0' \left\{ \frac{(2+2/\alpha)^{2+2/\alpha}}{(1/\alpha)^{1/\alpha}(1/\alpha+2)^{1/\alpha+2}} \right\}^i. \end{aligned}$$

Set

$$\gamma(\alpha) = \frac{(2+2\alpha)^{2+2\alpha}}{(1+2\alpha)^{1+2\alpha}}$$

and

$$\delta(\alpha) = \frac{(2+2/\alpha)^{2+2/\alpha}}{(1/\alpha)^{1/\alpha}(1/\alpha+2)^{1/\alpha+2}}.$$

Then $\gamma(\alpha)$ is an increasing function of α and $\gamma(\alpha) \geqslant 4$; for $\alpha \geqslant 1$, $\delta(\alpha)$ is a decreasing function and $\delta(\alpha) \geqslant 1$. Thus there is a constant C_0'' depending only on α_0 such that if $1 \leqslant \alpha \leqslant \alpha_0$ then

$$a_{iq} \leqslant C_0'' \gamma(\alpha)^q \delta(\alpha)^i$$

for all i, q. From (2.22) we obtain

$$|S_q(x) - u(x)| \leqslant 2CC_0'' \left\{ \left(\frac{rM}{2}\right)^2 \gamma(\alpha) \right\}^q \sum_{i=1}^{\infty} A_i \delta(\alpha)^i.$$

Choose $\alpha = 1$; then $\gamma(1) = \delta(1)$. If

$$r < \frac{2}{\{\delta(1)\}^{1/2} M},$$

the series $\sum_{i=1}^{\infty} A_i\delta(1)^i$ converges and

$$\left(\frac{rM}{2}\right)^2\gamma(1)=\left(\frac{rM}{2}\right)^2\delta(1)<1.$$

As q approaches infinity, $|S_q(x)-u(x)|$ approaches zero, so the series $\sum_{k=0}^{\infty} r^{2k}h_k(x)$ converges to $u(x)$ for $x\in D$ with

$$r<\frac{2}{\{\delta(1)\}^{1/2}M}.$$

If $u(x)$ is polyharmonic, M can be chosen arbitrarily small so that the series converges in any star domain centred at the origin on which $u(x)$ is holomorphic. ‖

To treat functions of order ρ and finite type we shall need estimates for

$$\text{(2.23)}\qquad \Delta^p(r^{2k}h_k(x))\equiv r^{2k-2p}h_{k-p}^{(p)}(x),\qquad p=0,1,\ldots,k.$$

Lemma 2.1. Let $u(x)$ be polyharmonic of order ρ and finite type τ on a neighbourhood of the origin, and let $\sum_{k=0}^{\infty} r^{2k}h_k(x)$ be its Almansi expansion. For each $\varepsilon>0$, each $\gamma>1$, and each compact set $K\subset\mathbb{C}^n$ which is in the largest star domain centred at the origin on which $u(x)$ is holomorphic, there exists a constant $C_{K,\varepsilon,\gamma}$ such that $h_{k-p}^{(p)}$, defined by (2.23), satisfies

$$\text{(2.24)}\qquad |h_{k-p}^{(p)}(x)|\leq C_{K,\varepsilon,\gamma}\frac{\{(\tau+\varepsilon)\gamma\}^{2k}(2k)!^{1-1/\rho}\Gamma(n/2)}{2^{2(k-p)}(k-p)!\Gamma(n/2+k-p)},$$

all $x\in K$, $p=0,1,\ldots,k$.

Proof. Using (2.14) for a harmonic function $v=h$,

$$\Delta(r^{2k}h)=r^{2k-2}\left\{2k(2k+n-2)h+4kr\frac{\partial h}{\partial r}\right\}$$

and $r\,\partial h/\partial r$ is harmonic since h is harmonic. Consequently,

$$h_{k-1}^{(1)}=2k(2k+n-2)h_k+4kr\frac{\partial h_k}{\partial r}$$

is harmonic, and by induction one shows that $h_{k-p}^{(p)}$ is harmonic.

The holomorphic function $\Delta^p u$ has a unique Almansi expansion

$$\Delta^p u(x)=\sum_{i=0}^{\infty} r^{2i}g_i^{(p)}(x),$$

valid in a neighbourhood of the origin. Taking Laplacians termwise in the

Almansi expansion for $u(x)$ we obtain

$$\Delta^p u(x) = \sum_{k=p}^{\infty} r^{2k-2p} h_{k-p}^{(p)}(x),$$

and therefore $h_{k-p}^{(p)}(x) = g_{k-p}^{(p)}(x)$. For $p \leqslant k$ we use fomulae (2.8) and (2.9) to obtain

$$(2.25)\quad h_{k-p}^{(p)}(x) = \sum_{i=0}^{\infty} \frac{1}{2^{2(k-p)}(k-p)!} \frac{(-1)^i r^{2i}}{2^{2i} i!\,(k-p+1)!} \times \int_0^1 \tau^{(n-2)/2+i-1}(1-\tau)^{k-p+i-1}\{i+(k-p)\tau\}(\Delta^{k+i}u)(\tau x)\,d\tau.$$

If $u(x)$ is of order ρ and finite type τ, for every $\varepsilon > 0$ and every compact set K as in the hypothesis, there exists a constant $C'_{K,\varepsilon}$ such that

$$(2.26)\qquad |\Delta^i u(y)| \leqslant C'_{K,\varepsilon}(2i)!^{1-1/\rho}(\tau+\varepsilon)^{2i}, \qquad i = 0, 1, \ldots, \quad y \in K.$$

Together (2.25) and (2.26) yield

$$|h_{k-p}^{(p)}(x)| \leqslant C'_{K,\varepsilon}(\tau+\varepsilon)^{2k} \frac{1}{2^{2(k-p)}(k-p)!} \times \sum_{i=0}^{\infty} \left\{\frac{r(\tau+\varepsilon)}{2}\right\}^{2i} \frac{(2k+2i)!^{1-1/\rho}}{i!\,(k-p+i)!}[\ldots]$$

where

$$[\ldots] = \frac{i\Gamma((n-2)/2+i)\Gamma(k-p+i)}{\Gamma((n-2)/2+k-p+2i)} + \frac{(k-p)\Gamma(n/2+i)\Gamma(k-p+i)}{\Gamma(n/2+k-p+2i)}.$$

Simplifying and enlarging [...], we obtain

$$(2.27)\qquad |h_{k-p}^{(p)}(x)| \leqslant C'_{K,\varepsilon} \frac{2(\tau+\varepsilon)^{2k}(2k)!^{1-(1/\rho)}\Gamma(n/2)}{2^{2(k-p)}(k-p)!\,\Gamma(n/2+k-p)}\{\ldots\}$$

where

$$(2.28)\quad \{\ldots\} = 1 + \sum_{i=1}^{\infty} \left\{\frac{r(\tau+\varepsilon)}{2}\right\}^{2i} \left\{\frac{(2k)!}{(2k+2i)!}\right\}^{1/\rho} \frac{\Gamma(n/2+i)}{i!\,\Gamma(n/2)} \frac{(2k+2i)!\,\Gamma(n/2+k-p)}{\Gamma(n/2+k-p+2i)(2k)!}.$$

To study the series in (2.28) we again use the technique described prior to Theorem 2.2. Since

$$\left\{\frac{(2k)!}{(2k+2i)!}\right\}^{1/\rho} \leqslant (2i)!^{-1/\rho}$$

and

$$\frac{\Gamma((n/2)+k-p)}{\Gamma((n/2)+k-p+2i)} \leqslant \frac{\Gamma(n/2)}{\Gamma(2i+1)},$$

if we set

$$A_i = \left\{\frac{r(\tau+\varepsilon)}{2}\right\}^{2i} (2i)!^{-1/\rho} \frac{\Gamma(n/2+i)}{i!}$$

and

$$a_{ik} = \frac{\Gamma(2k+2i+1)}{\Gamma(2k+1)\Gamma(2i+1)},$$

the series in (2.28) is dominated by $\sum_{i=1}^{\infty} A_i a_{ik}$. The numbers A_i are coefficients of an entire function, and a_{ik} is increasing in i for fixed k and increasing in k for fixed i. As in the proof of Theorem 2.2, let $\alpha > 0$ and separate the a_{ik} into two classes, those for which $i \leqslant \alpha k$ and those for which $i > \alpha k$. Use Stirling's formula to obtain the existence of a constant C_0 such that for $i \leqslant \alpha k$

$$a_{ik} \leqslant C_0 \left\{\frac{(2+2\alpha)^{2+2\alpha}}{2^2(2\alpha)^{2\alpha}}\right\}^k$$

and for $i > \alpha k$

$$a_{ik} \leqslant C_0 \left\{\frac{(2/\alpha+2)^{2/\alpha+2}}{(2/\alpha)^{2/\alpha}2^2}\right\}^i.$$

Set

$$\{\gamma(\alpha)\}^2 = \frac{(2+2\alpha)^{2+2\alpha}}{2^2(2\alpha)^{2\alpha}}$$

and

$$\delta(\alpha) = \gamma(1/\alpha).$$

Observe that $\gamma(\alpha)$ is an increasing function of α and

$$\lim_{\alpha \to 0} \gamma(\alpha) = 1.$$

Thus $\delta(\alpha)$ is decreasing in α and $\gamma(\alpha) \geqslant 1$, $\delta(\alpha) \geqslant 1$. It now follows that

$$\sum_{i=1}^{\infty} A_i a_{ik} \leqslant C_0 \gamma(\alpha)^{2k} \sum_{i=1}^{\infty} A_i \delta(\alpha)^{2i} \tag{2.29}$$

and the series on the right converges for any value of $\delta(\alpha)$. From (2.27), (2.28), and (2.29) we obtain the desired conclusion of the lemma. ‖

We are now in a position to state necessary and sufficient conditions on the Almansi coefficients of a polyharmonic function which determine its Laplacian order and type. We use the notation in Definition 1.3, Chapter II.

Theorem 2.3. Let $G \subset \mathbb{C}^n$ (or $\mathbb{R}^n$) be a star domain centred at the origin and let $u(x)$ be polyharmonic on G with Almansi expansion

$$u(x) = \sum_{k=0}^{\infty} r^{2k} h_k(x),$$

valid in G (Theorem 2.2).

(a) *Let $0 \leq \rho < \infty$. $u \in \mathcal{P}^{\rho}(G)$ if and only if for every compact set $K \subset G$ and every $\varepsilon > 0$ there exists a constant $C_{K,\varepsilon}$ such that $\|h_k\|_K \leq C_{K,\varepsilon}(2k)!^{-1/(\rho+\varepsilon)}$. (Equivalently, for every compact set $K \subset G$ and every $\varepsilon > 0$,*

$$\limsup_{k \to \infty} \|h_k\|_K^{1/2k} \; k^{1/(\rho+\varepsilon)} = 0.)$$

(b) *Let $0 < \rho < \infty$. $u \in \mathcal{P}^{\rho,\infty}(G)$ if and only if for every compact set $K \subset G$ there exist constants C_K and M_K such that*

$$\|h_k\|_K \leq C_K M_K^{2k} (2k)!^{-1/\rho}.$$

(Equivalently, for every compact set $K \subset G$,

$$\limsup_{k \to \infty} \|h_k\|_K^{1/2k} \; k^{1/\rho} < \infty.)$$

(c) *Let $0 < \rho \leq \infty$. $u \in \mathcal{P}^{\rho,0}(G)$ if and only if for every compact set $K \subset G$ and every $\varepsilon > 0$ there exists a constant $C_{K,\varepsilon}$ such that*

$$\|h_k\|_K \leq C_{K,\varepsilon} \varepsilon^{2k} (2k)!^{-1/\rho}.$$

(Equivalently, for every compact $K \subset G$,

$$\limsup_{k \to \infty} \|h_k\|_K^{1/2k} \; k^{1/\rho} = 0.)$$

(d) *Let $0 < \rho < \infty$, $0 < \tau < \infty$. $u \in \mathcal{P}^{\rho,\tau}(G)$ if and only if for every compact set $K \subset G$ and every $\varepsilon > 0$ there exists a constant $C_{K,\varepsilon}$ such that*

$$(2.30) \qquad \|r^{2p-2k} \Delta^p (r^{2k} h_k(x))\|_K \leq C_{K,\varepsilon} (\tau + \varepsilon)^{2k} \frac{(2k)!^{1-1/\rho}}{(2k-2p)!}, \quad 0 \leq p \leq k.$$

The condition (2.30) is equivalent to (2.24) for any given compact K.

Proof. We first show that the condition (2.30) is equivalent to (2.24). Let (2.24) hold. Given $\varepsilon > 0$ choose $\gamma > 1$ so that $\tau(1-\gamma) + \varepsilon > 0$, set $\varepsilon' = \{\tau(1-\gamma) + \varepsilon\}/\gamma$, and apply (2.24) with the choices of ε' and γ. The right-hand side of (2.24) can then be written

$$C_{K,\varepsilon,\gamma}\frac{(\tau+\varepsilon)^{2k}(2k)!^{1-1/\rho}}{(2k-2p)!}\frac{(2k-2p)!\,\Gamma(n/2)}{2^{2(k-p)}(k-p)!\,\Gamma(n/2+k-p)},$$

and the second factor is bounded. Thus the desired condition follows from (2.24).

Conversely, suppose that the condition (2.30) holds and let $\varepsilon>0$ and $1<\gamma<2$ be given. The right-hand side in (2.30) can be written

$$C_{K,\varepsilon}\frac{(\tau+\varepsilon)^{2k}(2k)!^{1-1/\rho}\Gamma(n/2)}{2^{2(k-p)}(k-p)!\,\Gamma(n/2+k-p)}\frac{2^{2(k-p)}(k-p)!\,\Gamma(n/2+k-p)}{(2k-2p)!\,\Gamma(n/2)}.$$

If the first factor is multiplied by $\gamma^{2(k-p)}$ and the second by $\gamma^{-2(k-p)}$, then the second factor above becomes

$$\frac{(2/\gamma)^{2(k-p)}(k-p)!\,\Gamma((n/2)+k-p)}{\Gamma(n/2)(2k-2p)!}$$

with $1<2/\gamma<2$. Using Stirling's formula (2.7), we find that this expression is bounded by a constant depending upon γ. Thus (2.24) follows.

By Lemma 2.1, we can now conclude that the condition in (d) is a necessary condition. Moreover, using the relations between the classes $\mathscr{P}^{\rho,\tau}$ and $\mathscr{P}^{\rho}$ (Proposition 1.1, Chapter II) and setting $p=0$ in (d) we find that the conditions in (a), (b), and (c) are necessary.

Finally we show that the condition in (d) is sufficient. Apply (d) to the Almansi expansion and re-index by setting $k=p+i$:

$$(2.31)\quad |\Delta^p u(x)|\leq C_{K,\varepsilon}\tau^{2p}(2p)!^{1-1/\rho}\sum_{i=0}^{\infty}(\tau r)^{2i}\frac{(2p+2i)!\,(2p)!^{1/\rho}}{(2p+2i)!^{1/\rho}(2i)!\,(2p)!}.$$

We again use the technique described prior to Theorem 2.2. The factors $\{(2p)!/(2p+2i)!\}^{1/\rho}$ are majorized by $\{1/(2i)!\}^{1/\rho}$. With $A_i=(\tau r)^{2i}(2i)!^{-1/\rho}$ and

$$a_{ip}=\frac{(2p+2i)!}{(2i)!\,(2p)!},$$

the series in (2.31) is dominated by $\sum_{i=0}^{\infty}A_i a_{ip}$ and the A_i are coefficients of an entire function. In exactly the same way as in the proof of Lemma 2.1, we obtain

$$|\Delta^p u(x)|\leq C'_{K,\varepsilon}(2p)!^{1-1/\rho}(\tau\gamma)^{2p}$$

where γ is arbitrarily close to unity. Thus $u\in\mathscr{P}^{\rho,\tau}(G)$, and the theorem is proved. ‖

Corollary 2.1. Let G be a star domain centred at the origin in $\mathbb{C}^n$ *and let u be holomorphic in G with Almansi expansion*

$$u(x)=\sum_{k=0}^{\infty} r^{2k}h_k(x)$$

about the origin. For each compact set $K\subset G$ *define the formal power series in one complex variable*

$$f_K(z)=\sum_{k=0}^{\infty} \|h_k\|_K z^{2k}.$$

(a) $u\in\mathcal{P}^{\infty,0}(G)$, *that is u is polyharmonic, if and only if each* f_K *is entire.*

(b) $u\in\mathcal{P}^{\rho}(G)$ *and* $u\notin\mathcal{P}^{\rho'}(G)$ *for every* $\rho'<\rho$ *if and only if each* f_K *has exponential order* ρ.

(c) *If each* f_K *has positive finite exponential order* ρ *and finite type (or type 0) then* $u\in\mathcal{P}^{\rho,\infty}(G)$ *(or* $\mathcal{P}^{\rho,0}(G)$*).*

(d) *Let u have least order* λ *on G. If* $u\in\mathcal{P}^{\lambda}(G)\backslash\mathcal{P}^{\lambda,\infty}(G)$, *then each* f_K *has exponential order* λ *and type* ∞. *If* $u\in\mathcal{P}^{\lambda,\tau}(G)$, $0<\tau\leq\infty$, *but* $u\notin\mathcal{P}^{\lambda,\tau'}(G)$ *for every* $\tau'<\tau$, *then each* f_K *has exponential order* λ *and finite non-zero type. If* $u\in\mathcal{P}^{\lambda,0}(G)$, *then each* f_K *has exponential order* λ *and type* 0.

The proof consists of applying Theorem 2.3 above, Theorem 4.1, Chapter I, Proposition 1.1, Chapter II, and Definition 1.4, Chapter II. The essential reason that parts (c) and (d) must be separated is that exponential order is exclusive while Laplacian order is inclusive. To see that the implication in part (c), for example, could not be reversed as it stands, let

$$h_k(x)\equiv\frac{(\log k)^{2k}}{(2k)!}.$$

The function

$$u(x)=\sum_{k=0}^{\infty} r^{2k}h_k(x)$$

is in $\mathcal{P}^{\rho,0}(\mathbb{C}^n)$ for every $\rho>1$, for instance for $\rho=\frac{3}{2}$. The corresponding entire function has exponential order 1 and infinite type.

Theorem 2.3 will enable us to construct examples showing that the classes $\mathcal{P}^{\rho,\tau}$ are distinct, that is the inclusions given in Proposition 1.1(a), Chapter II, are all proper. These examples comprise the first portion of the next section.

3. Examples

Three groups of polyharmonic functions will be constructed using the Almansi expansion and Theorem 2.3 to illustrate the various combinations of local Laplacian order and type discussed. They will show, successively, constant local Laplacian order (and type), growth as a power of distance from the origin in $\mathbb{R}^n$, and arbitrary logarithmically convex growth. Piecewise constant type will also be shown.

Group 1. Constant local Laplacian order (and type). An entire function with Almansi expansion

$$u(x)=\sum_{k=0}^{\infty} r^{2k}(x)h_k(x)$$

with $\lambda(x)\equiv\rho$ (and $\mu(x)\equiv\tau$ where applicable) will be defined on $\mathbb{C}^n$.

Example 3.1. $\rho=\infty, \tau=0$:

$$h_k(x)\equiv\left(\frac{1}{\log k}\right)^{2k}(2k)!^{-1/k}$$

Example 3.2. $0<\rho<\infty$, *with no finite type at any point:*

$$h_k(x)\equiv(\log k)^{2k}(2k)!^{-1/\rho}$$

Example 3.3. $\rho=0$, *no finite degree (type not applicable):*

$$h_k(x)\equiv(2k)!^{-k}$$

Example 3.4. $0<\rho<\infty$, $0\leqslant\tau<\infty$:

$$h_k(x)\equiv\tau^{2k}(2k)!^{-1/\rho}$$

The required inequality in one direction stems from the identity

$$r^{2p-2k}\Delta^p(r^{2k})=\frac{4^pk!}{(k-p)!}\left(k-1+\frac{n}{2}\right)\left(k-2+\frac{n}{2}\right)\dots\left(k-p+\frac{n}{2}\right).$$

For the opposite inequality consider the special case $p=0$ when taking the maximum over $0\leqslant p\leqslant k$ in Theorem 2.3.

Lemma 3.1.

$$\limsup_{k\to\infty}\|h_k\|_K^{1/\log(2k)!}=e^{-1/\rho}$$

if and only if for each $\varepsilon>0$ *and no* $\varepsilon<0$, *for some constant* $C_{K,\varepsilon}$,

$$\|h_k\|_K\leqslant C_{K,\varepsilon}(2k)!^{-1/(\rho+\varepsilon)}\quad\text{for all } k.$$

Lemma 3.2.

$$\limsup_{k\to\infty}\max_{0\leqslant p\leqslant k}\left\{\frac{(2k-2p)!}{(2k)!^{1-1/\rho}}\left\|\frac{\Delta^p(r^{2k}h_k)}{r^{2k-2p}}\right\|_K\right\}^{\frac{1}{2k}}=\tau$$

if and only if for each $\varepsilon>0$ and no $\varepsilon<0$, for some constant $C_{K,\varepsilon}$,

$$\|r^{2p-2k}\Delta^p(r^{2k}h_k)\|_K \leq C_{K,\varepsilon}(\tau+\varepsilon)^{2k}\frac{(2k)!^{1-1/\rho}}{(2k-2p)!} \quad \text{for all } k, \qquad 0\leq p\leq k.$$

The examples in Groups 2 and 3 are stated for $n=2$ but can be generalized by making functions constant in the remaining variables.

Group 2. *Monomial growth of* $F=\exp(1-\lambda^{-1})$ *or* μ (see §2, Chapter IV). A function with Almansi expansion

$$u(x)=\sum_{k=0}^{\infty} r^{2k}(x)h_k(x), \qquad x=(x_1,x_2)$$

will be defined on balls in $\mathbb{R}^2$, hence on Lie balls in $\mathbb{C}^2$.

Example 3.5. $F(x)/\mathrm{e}=\{r(x)/b\}^t$, *where* $b>0$, $t>0$, *and* u *is defined for* $r(x)=(x_1^2+x_2^2)^{\frac{1}{2}}<b$. Let

$$v(x)=x_1+ix_2 \quad \text{and} \quad h_k(x)=\left\{\frac{v(x)}{b}\right\}^{[t\log(2k)!]}$$

where $[y]$ is the largest integer in y. If K is a compact ball of radius $R<b$ about the origin in $\mathbb{R}^2$, then

$$\lim_{k\to\infty}\|h_k(x)\|_K^{1/\log(2k)!}=\lim_{k\to\infty}\left(\frac{R}{b}\right)^{t+1/\log(2k)!}=\left(\frac{R}{b}\right)^t.$$

Now Theorem 2.3(d) and Lemma 3.1 combine to show that $\mathrm{e}^{-1/\rho}=(R/b)^t$ on the disc of radius R where ρ is the least order there. Hence $F(x)/\mathrm{e}=\{r(x)/b\}^t$ for $r(x)<b$. (Clearly absolute and uniform convergence of the series for $r(x)>b$ would be inconsistent with the theorems of the previous section.)

Example 3.6. $\mu(x)=\alpha\{r(x)\}^t$ *where* $\alpha>0$, $t>0$, *and* u *is defined on all of* $\mathbb{R}^2$ *with constant local Laplacian order* ρ, $0<\rho<\infty$. Let

$$v(x)=x_1+ix_2, \qquad \beta=\alpha(1+2t)^{-\frac{1}{2}}, \qquad t_k=[2kt]/2k,$$

and

$$h_k(x)=\beta^{2k}v^{2kt_k}(2k!)^{-1/\rho}, \qquad k=1,2,\dots.$$

Direct calculation shows that

$$r^{2p-2k}\Delta^p(r^{2k}h_k)=h_k\frac{4^p(k!)^2}{\{(k-p)!\}^2}(1+2t_k)^p,$$

so that

$$\left\{\frac{(2k-2p)!}{(2k)!^{1-1/\rho}}\left\|\frac{\Delta^p(r^{2k}h_k)}{r^{2k-2p}}\right\|_K\right\}^{\frac{1}{2k}}=\beta\,\|v\|_K^{t_k}\left(\frac{4^p(k!)^2}{\{(k-p)!\}^2}\frac{(2k-2p)!}{(2k)!}(1+2t_k)^p\right)^{\frac{1}{2k}}.$$

For very large k the radicand on the right-hand side is greater for $0<p\leq k$ than it

is for $p=0$. Consequently Stirling's formula allows the simplification

$$\max_{0\leq p\leq k}\left\{\frac{(2k-2p)!}{(2k)!^{1-1/\rho}}\left\|\frac{\Delta^p(r^{2k}h_k)}{r^{2k-2p}}\right\|_K\right\}^{\frac{1}{2k}}\sim\beta\,\|v\|_K^{t_k}\max_{0\leq p\leq k}\left\{\left(\frac{k-p}{k}\right)^{\frac{1}{2}}(1+2t_k)^p\right\}^{\frac{1}{2k}}.$$

The maximum is taken for an integer value of p near $k-\{1/\log(1+2t_k)\}$. Thus the limit as k increases exists and equals $\beta\,\|v\|_K^t(1+2t)^{\frac{1}{2}}=\alpha\,\|v\|_K^t$. Therefore the least type on the ball of radius R is αR^t, and the result follows by rotational symmetry.

The following example is the limit of the previous one as t increases.

Example 3.7.

$$\mu(x)=\begin{cases}0, & r(x)<R\\ \infty, & r(x)\geq R\end{cases}$$

where $R>0$, and u is defined on all of $\mathbb{R}^2$ with constant local Laplacian order ρ, $0<\rho<\infty$. Let

$$v(x)=v(x_1,x_2)=(x_1+ix_2)/R$$
$$h_k(x)=(1+2k)^{2k}v^{2k^2}(2k)!^{-1/\rho},\qquad k=0,1,2,\ldots.$$

The same calculations as in the previous example show that if $K=\bar{B}_{R_0}$ then

$$\lim_{k\to\infty}\max_{0\leq p\leq k}\left\{\frac{(2k-2p)!}{(2k)!^{1-1/\rho}}\left\|\frac{\Delta^p(r^{2k}h_k)}{r^{2k-2p}}\right\|_K\right\}^{\frac{1}{2k}}=\lim_{k\to\infty}(1+2k)^{-\frac{1}{2}}\left\|\frac{x_1+ix_2}{R}\right\|_K^k(1+2k)^{\frac{1}{2}}$$
$$=\begin{cases}0, & R_0<R\\ 1, & R_0=R\\ \infty, & R_0>R\end{cases}$$

By symmetry it follows that $\mu(x)$ is as required.

Ex. 3.7 exhibits a local type which is discontinuous. Since $\max\{\mu(x):r(x)=R\}$ is logarithmically convex, its only discontinuities must be jumps from a finite value up to infinity, the way the example shows. Such jumps are not possible for local Laplacian order because of the maximum principle (Theorem 1.3, Chapter IV).

It will be important for the examples of Group 3 that for each example constructed so far every infinite subseries has the same growth rate.

Group 3. Arbitrary logarithmically convex growth of $F(x)$ or $\mu(x)$. A function with Almansi expansion

$$u(x)=\sum_{k=0}^{\infty}r^{2k}(x)h_k(x),\qquad x=(x_1,x_2)$$

will be defined on balls in $\mathbb{R}^2$ and Lie balls in $\mathbb{C}^2$. Let $C(R)$ be a non-decreasing logarithmically convex function of R, possibly taking $C(R)=\infty$ for R greater than or equal to some R_1. (Necessarily $C(R)$ is not zero on any interval, lest it be zero everywhere and the problems reduce to Group 1.) The examples will have $F(x)=\exp\{1-\lambda^{-1}(x)\}=C(r(x))$, u being defined on $\{x\in\mathbb{R}^2: C(r(x))<\mathrm{e}\}$, or will be defined on all of $\mathbb{R}^2$, having constant local Laplacian order ρ, $0<\rho<\infty$, with $\mu(x)=C(r(x))$.

In each example the construction is based on the fact that C has a sequence $\{c_s\}_{s=0}^{\infty}$ of monomial support functions

$$c_s(R)=\alpha_s R^{t_s}=(R/b_s)^{t_s}, \qquad C(R)=\sup_s\{c_s(R)\},$$

that polyharmonic functions with growth rate c_s can be chosen as in the previous examples, and that the growth of the sum of an Almansi expansion is the supremum of the growth rates of its subseries.

Example 3.8. $F(x)=C(r(x))$, u defined on $B=\{x\in\mathbb{R}^2: C(r(x))<\mathrm{e}\}$. For each support function $c_s(R)=(R/b_s)^{t_s}$ let v_s be constructed as in Group 1 or Group 2, defined on a neighbourhood of B and having local Laplacian order λ_s such that $\exp\{1-1/\lambda_s(x)\}=c_s(r(x))$. Each v_s has an Almansi expansion

$$v_s(x)=\sum_{k=0}^{\infty} r^{2k}(x)h_{s,k}(x)$$

valid on $\tilde{B}$.

From the double sequence $\{h_{s,k}\}$ and the zero function there can be chosen a single sequence h_k as follows.

(a) If h_k is not identically zero then $h_k=h_{s,k}$ for some s.
(b) For each s, infinitely many of the $h_{s,k}$ appear in the sequence $\{h_k\}$.
(c) The non-zero terms of $\{h_k\}$ are so sparse that the series

$$u(x)=\sum_{k=0}^{\infty} r^{2k}(x)h_k(x)$$

converges absolutely and uniformly on compact subsets of $\tilde{B}$.

The resulting series is the Almansi expansion for a function u which has the desired properties because

$$\limsup_{k\to\infty}\|h_k(x)\|_{\tilde{B}_R}^{1/\log(2k)!}=\sup_s\left\{\frac{c_s(R)}{\mathrm{e}}\right\}=\frac{C(R)}{\mathrm{e}}$$

and because rotational symmetry shows that

$$F(x)=C(R) \quad \text{if} \quad r(x)=R.$$

Example 3.9. $\mu(x)=C(r(x))$, u defined on all of $\mathbb{R}^2$ with constant local Laplacian order ρ, $0<\rho<\infty$. For each support function $c_s(R)=\alpha_s R^{t_s}$ let v_s be constructed as in Group 1 or Group 2, defined on $\mathbb{R}^2$ with constant local Laplacian order ρ and local type $\mu_s(x)=c_s(r(x))$. The construction of u is carried out as in the previous example, except that the non-zero terms must be so sparse that the series is absolutely and uniformly convergent on all compact subsets of $\mathbb{C}^2$. The sum of the series

$$u(x)=\sum_{k=0}^{\infty} r^{2k}(x)h_k(x)$$

has the required local Laplacian order and type because

$$\limsup_{k\to\infty} \max_{0\le p\le k} \left\{\frac{(2k-2p)!}{(2k)!^{1-1/\rho}} \left\|\frac{\Delta^p(r^{2k}h_k)}{r^{2k-2p}}\right\|_{\bar{B}_R}\right\}^{\frac{1}{2k}}$$

$$= \sup_s \{c_s(R)\}$$

$$= C(R).$$

Notice that it is possible for u to have finite positive local type on an open disc but $\mu(x) \equiv \infty$ on its complement. In fact, $C(R)$ may be continuous approaching the value infinity, or discontinuous, but if it is discontinuous at R_0 then $C(R) = \infty$ for $R > R_0$.

Thus the examples show that Theorem 2.3, Chapter IV, is sharp on balls in $\mathbb{R}^2$, for an arbitrary non-decreasing logarithmically convex function describes the growth of F or μ for some polyharmonic function. The examples generalize readily to $\mathbb{R}^n$, $n \ge 2$. (F and μ are constant for $n = 1$.)

Remark 3.1. The functions in Group 1 are entire and have constant local Laplacian order (and type). For the functions in Group 2 and Group 3 we now calculate the local Laplacian order (and type) of their continuations from $\mathbb{R}^2$ into $\mathbb{C}^2$.

Each example is based on the function

$$v(x_1, x_2) = x_1 + ix_2$$

which is entire. An inequality

$$|v(x_1, x_2)| < R$$

defines an open subset of $\mathbb{C}^2$, a star domain in fact. It follows that in Ex. 3.5, the function u has analytic continuation to the set

$$\{(x_1, x_2) \in \mathbb{C}^2 : |v(x_1, x_2)| < b\}$$

on which

$$F(x_1, x_2) = \mathrm{e}(|x_1 + ix_2|/b)^t.$$

Similarly, in Ex. 3.8, u has analytic continuation to

$$\{(x_1, x_2) \in \mathbb{C}^2 : |v(x_1 + ix_2)| < \mathrm{e}^{-1}\}$$

where

$$F(x_1, x_2) = C(|x_1 + ix_2|).$$

Exx. 6, 7, and 9 give entire functions u for which the local Laplacian order is constant and the local type is given by

$$\mu(x_1, x_2) = \alpha\, |x_1 + ix_2|^t,$$

$$\mu(x_1, x_2) = \lim_{k\to\infty} \left(\frac{|x_1 + ix_2|}{R}\right)^k,$$

or

$$\mu(x_1, x_2) = C(|x_1 + ix_2|)$$

respectively.

To give functions on which the local Laplacian order (and type) take specified values not on the level sets of $|v|$ but on Lie balls is now a simple matter, for

$$\partial\tilde{B}_R(0) = \{(x_1, x_2) : \max\{|x_1 + ix_2|, |x_1 - ix_2|\} = R\}$$

by Ex. A2.5, Appendix. It is only necessary to construct a function w by the same procedure as that for u but using $x_1 - ix_2$ in place of $v(x_1, x_2) = x_1 + ix_2$, and then adding it to u. The sum $u + w$ will have the prescribed local Laplacian order and type, by Proposition 1.3, Chapter II. For example, if u and w are constructed according to Ex. 3.9, then $u + w$ will have constant local Laplacian order ρ and

$$\mu(x) = C(q(x))$$

where q is the function taking value R exactly on $\partial\tilde{B}_R(0)$ (see Ex. A2.5, Appendix).

All the previous examples can be restricted to a smaller domain by adding a harmonic function with natural boundary on that domain.

VI

EXTENSIONS AND APPLICATIONS

1. Harmonic extension to $\mathbb{R}^{n+1}$

The starting point of this section is the following well-known result.

Theorem 1.1. Let D be a domain in $\mathbb{R}^n$. *If u and v are arbitrary functions analytic on D, then there exist a domain* $D' \subset \mathbb{R}^{n+1}$ *such that* $D \subset D' \cap \mathbb{R}^n$ *and a unique function U harmonic on D' such that*

$$U(x, 0) = u(x), \qquad \frac{\partial U}{\partial x_{n+1}}(x, 0) = v(x), \qquad x \in D.$$

In fact, the series for U in powers of x_{n+1} *is*

$$(1.1)\quad U(x, x_{n+1}) = \sum_{k=0}^{\infty} (-1)^k \Delta^k u(x) \frac{x_{n+1}^{2k}}{(2k)!} + \sum_{k=0}^{\infty} (-1)^k \Delta^k v(x) \frac{x_{n+1}^{2k+1}}{(2k+1)!}.$$

(The operator Δ *in* (1.1) *is the n-dimensional Laplacian.) Of the two sums in* (1.1), *the first defines a function* U_e *which is even in* x_{n+1} *and determined by u, the second defines a function* U_o *which is odd in* x_{n+1} *and determined by v. Both* U_e *and* U_o *are harmonic on D'.*

The first portion of this section consists of a discussion of the relation between the Laplacian order and type of u (or v) and the growth rate of U_e (or U_o) as a function of the single variable x_{n+1}. This discussion will rely on the results in §1, Chapter IV, on local and pointwise order and type, and on holomorphic capacity (presented in the Appendix and already used in §1, Chapter IV). In the second portion, examples will be constructed of functions which are analytic, but not polyharmonic, and which exhibit partially specified local type. Finally, the theorem on separation of singularities (§2, Chapter III) will be generalized to all analytic functions.

Proof of Theorem 1.1. The analytic function u continues to an analytic function on a domain $G \subset \mathbb{C}^n$. From the Cauchy bounds (§4, Chapter I), for each compact polydisc $K \subset G$ there exists $\delta_K > 0$ such that

$$(1.2)\qquad \|\Delta^k u\|_K \leq (2k)!\, \delta_K^{-2k}, \qquad k = 0, 1, \ldots.$$

Thus the series

$$(1.3)\qquad U_e(x, x_{n+1}) = \sum_{k=0}^{\infty} (-1)^k \Delta^k u(x) \frac{x_{n+1}^{2k}}{(2k)!}$$

is uniformly absolutely convergent on $K \times \mathscr{B}_{\delta_K}(0)$, and hence on compact subsets of an open set $G' \subset \mathbb{C}^{n+1}$. The series (1.3) defines a holomorphic function on G' and differentiations can be computed termwise. The same argument applies to $U_o(x, x_{n+1})$, and thus $U(x, x_{n+1})$ is defined on a domain $G'' \subset \mathbb{C}^{n+1}$. Termwise differentiation shows that U_e, U_o, and U are all harmonic on G''. For the uniqueness, if $V(x, x_{n+1})$ is any extension satisfying the desired conditions, then $V = V_e + V_o$ where

$$V_e(x, x_{n+1}) = \sum_{k=0}^{\infty} f_k(x) \frac{x_{n+1}^{2k}}{(2k)!}$$

and

$$V_o(x, x_{n+1}) = \sum_{k=0}^{\infty} g_k(x) \frac{x_{n+1}^{2k+1}}{(2k+1)!}.$$

By termwise differentiation, it is seen that the harmonicity of V implies that of V_e and V_o. Since

$$0 = \Delta_{n+1} V_e(x, x_{n+1}) = \sum_{k=0}^{\infty} \{\Delta f_k(x) + f_{k+1}(x)\} \frac{x_{n+1}^{2k}}{(2k)!},$$

it follows that

$$f_k(x) = (-1)^k \Delta^k f_0(x) = (-1)^k \Delta^k u(x).$$

The same argument applies to V_o, and we conclude that V_e, V_o, and V are uniquely determined. ‖

The Laplacian order (and type) of u, and the corresponding local and pointwise functions, are related to properties of U_e through the rate of convergence of the series (1.3). The first of these relations is pointwise and is proved by comparing the definitions (Definition 1.1, Chapter IV) of pointwise Laplacian order $l(x)$ and pointwise type $m(x)$ with the definitions (Definitions 4.3 and 4.4, Chapter I) of exponential order and type for entire functions. Recall (Corollary 1.2, Chapter IV) that if $l(x_0) = \infty$ for even a single point x_0 then $\lambda(x) \equiv \infty$ so that pointwise type is defined. Moreover, $l(x) = \infty$ almost everywhere.

Corollary 1.1. Let u be holomorphic on a domain $G \subset \mathbb{C}^n$, let U_e be defined by (1.3), and let $x \in G$.

(a) *If $l(x) = \infty$ then the series (1.3) has radius of convergence exactly $1/m(x)$. If $l(x) = \infty$ and $m(x) = 0$, then $U_e(x, x_{n+1})$ is entire in x_{n+1}.*

(b) *If $l(x) = \infty$ and if*

$$\limsup_{p \to \infty} \left| \frac{\Delta^p u(y)}{(2p)!} \right|^{1/2p} = 0$$

for each y in a subset A of $G \cap \mathbb{R}^n$ with positive measure there, then $U_e(x, x_{n+1})$ is entire in x_{n+1}.

(c) $l(x)<\infty$ *if and only if* $U_e(x, x_{n+1})$ *is entire in* x_{n+1} *of exponential order* $l(x)$. *If* $0<l(x)<\infty$, *then the exponential type is*

$$\tau=\frac{n(x)^{l(x)}}{l(x)}$$

where

$$n(x)=\limsup_{p\to\infty}\left\{\frac{|\Delta^p u(x)|}{(2p)!^{1-1/l(x)}}\right\}^{1/2p}.$$

The proofs of (a) and (c) follow as explained above and the proof of (b) follows from Theorem 1.4, Chapter IV.

The next three results are reminiscent of theorems of Hartogs in which pointwise convergence of a series is made to imply uniform convergence. Indeed, versions of Hartogs' lemmas (Propositions A5.2 and A5.4, Appendix) will be used in the proof of Theorem 1.2 below. A proof of Theorem 1.2 could also be constructed using the semi-continuity results of §1, Chapter IV, but since an argument of this type will be used to prove Theorem 1.3 and Corollary 1.2, we leave to the interested reader the details for Theorem 1.2. The semi-continuity results, of course, ultimately rely on the theory of holomorphic capacity.

Theorem 1.2. Let u be analytic on a domain $D\subset\mathbb{R}^n$ *and let* U_e *be defined by* (1.3).

(A) *The following properties are equivalent:*

- (a) *u is of Laplacian order* ∞ *and type* τ *on D;*
- (b) *the series* (1.3) *converges uniformly absolutely on compact subsets of* $G\times\mathcal{B}_{1/\tau}(0)\subset\mathbb{C}^n\times\mathbb{C}^1$ *where G is some neighbourhood of D in* $\mathbb{C}^n$*;*
- (c) *there exists a set A open in D such that* $\sigma(D\backslash A)=0$ *and such that if x is any point of A then the series* (1.3) *has radius of convergence at least* $1/\tau$.

(B) *The following properties are equivalent:*

- (a) *u is polyharmonic on D;*
- (b) *the series* (1.3) *is uniformly absolutely convergent on compact subsets of* $\hat{D}\times\mathbb{C}$ *if u is continued analytically to* $\hat{D}$*;*
- (c) *there exists a measurable set* $A'\subset D$ *with* $\sigma(A')>0$ *such that if* $x\in A'$ *then the series* (1.3) *has radius of convergence* $1/\tau=\infty$.

Proof

(A) Corollary 1.1 shows that (a) implies (c). To prove that (b) implies (a), first note that by the uniform convergence of the series, U_e is holomorphic on $G\times\mathcal{B}_{1/\tau}(0)$. The Cauchy bounds show that if K is any compact subset of $G\times\mathcal{B}_{1/\tau}(0)$ and if $\varepsilon>0$ then there exists $C_{K,\varepsilon}>0$ such that

$$\frac{|\Delta^k u(x)|}{(2k)!} = \left|(-1)^k \frac{\partial^{2k} U_e}{\partial x_{n+1}^{2k}}(x, 0)\right| \leq C_{K,\varepsilon}(\tau+\varepsilon)^{2k}, \qquad x \in K \cap D.$$

By the definition of order ∞ and type τ, this last inequality implies (a).

It remains to prove that (c) implies (b), but this is the content of the version of Hartogs' lemma in Proposition A5.4, Appendix, with $R = 1/\tau$ because each point $x \in D$ has a polydisc neighbourhood in $\mathbb{C}^{n+1}$ on which (1.3) is uniformly absolutely convergent.

(B) The equivalence of (a) and (c) of part (A) shows that (b) $\Rightarrow$ (a) $\Rightarrow$ (c). It remains to show that (c) implies (b). By Theorem 1.1, each $x \in D$ has a polydisc neighbourhood in $\mathbb{C}^{n+1}$ on which the series (1.3) is uniformly absolutely convergent. The version of Hartogs' lemma in Proposition A5.2, Appendix, can be applied showing that the series (1.3) is uniformly absolutely convergent on compact subsets of a set $G \times \mathbb{C} \subset \mathbb{C}^{n+1}$ where G is some neighbourhood of D in $\mathbb{C}^n$. Thus the sum of the series is holomorphic on $G \times \mathbb{C}$, and part (A) shows that u is polyharmonic on G, and therefore on D, and therefore on $\hat{D}$ by analytic continuation. The defining inequalities for polyharmonicity of u on $\hat{D}$ show the uniform absolute convergence of the series (1.3) on $\hat{D} \times \mathbb{C}$. ||

Local and pointwise Laplacian order and type are characterized by the two previous results and the next. The characterization of type is only partial.

Theorem 1.3. Let u be analytic on a domain $D \subset \mathbb{R}^n$ and for each $x \in D$ let $U_e(x, \cdot)$ be defined by (1.3).

(A) *The following properties are equivalent:*

(a) *u is of Laplacian order $\rho < \infty$ on D;*

(b) *for each $x \in D$, $U_e(x, \cdot)$ is an entire function of exponential order no greater than $\rho < \infty$;*

(c) *there exists an open set $A \subset D$ such that $\sigma(D \backslash A) = 0$ and for each $x \in A$, $U_e(x, \cdot)$ is an entire function of exponential order no greater than $\rho < \infty$;*

(B) *If the local Laplacian order of u is positive, finite, and constant with value ρ on D, then the following properties* (a), (b′), *and* (c′) *are equivalent and each implies both* (b) *and* (c):

(a) *u is of Laplacian order ρ and type $\tau < \infty$ on D;*

(b′) *u is of Laplacian order ρ and type ∞ on D, and for each $x \in D$, $U_e(x, \cdot)$ is an entire function of exponential order no greater than ρ and type no greater than τ^ρ/ρ;*

(c′) *u is of Laplacian order ρ and type ∞ on D, and there exists an open subset $A' \subset D$ such that $\sigma(D \backslash A') = 0$ and for each $x \in A'$,*

$U_e(x, \cdot)$ is an entire function of exponential order no greater than ρ and type no greater than τ^ρ/ρ;

(b) *for each $x \in D$, $U_e(x, \cdot)$ is an entire function of exponential order no greater than ρ and type no greater than τ^ρ/ρ;*

(c) *there exists an open set $A' \subset D$ such that $\sigma(D \backslash A') = 0$ and for each $x \in A'$, $U_e(x, \cdot)$ is an entire function of exponential order no greater than ρ and type no greater than τ^ρ/ρ.*

Remark 1.1. The phrase 'no greater than' is needed in the statements above for the same reason as mentioned at the end of §2, Chapter V, namely Laplacian order is inclusive but exponential order is not. Even replacing 'Laplacian order ρ' by 'least Laplacian order ρ' in Theorem 1.3 would not avoid this difficulty (see Corollary 1.1(c)). Instead of adopting the terminology 'exponential growth (ρ, τ)' to indicate that the order is no greater than ρ and the type no greater than τ, which is not completely standard (Boas [18, Chapter 2]), we prefer to make explicit mention of this phenomenon in the few places in which it occurs.

Proof

(A) By Corollary 1.1(c), (a) implies (b) and certainly (b) implies (c). Under the assumption that (c) holds, Corollary 1.1(c) shows that $l(x) \leq \rho$ on A, and from Theorem 1.5, Chapter IV, it follows that $l(x) \leq \lambda(x) \leq \rho$ on all of D. Thus (a) follows from (c).

(B) The equivalence of (a), (b′), and (c′) is proved exactly as in part (A). Certainly (b′) $\Rightarrow$ (b) $\Rightarrow$ (c). ||

It is an open question whether (c), or even (b), implies that u has order ρ and type ∞ on D.

Remark 1.2. In the preceding theorem, D was a domain in $\mathbb{R}^n$, where $\mathbb{R}^n$ is considered to be the hyperplane $\{x_{n+1} = 0\}$ in $\mathbb{R}^{n+1}$. The same result holds if D is a domain in any of the hyperplanes $\{x_{n+1} = c\}$, since the required properties of entire functions are invariant under translation. There are similar modifications available for Theorem 1.2 and Corollary 1.1. In part (A) of those results, if the function is to be expanded in powers of x_{n+1}, then τ and $m(x)$ must be taken to be zero. If the expansion is in powers of $x_{n+1} - c$, the results are identical.

Remark 1.3. Corollary 1.1, Theorems 1.2 and 1.3, and Remark 1.2 concerned the function U_e which is even in x_{n+1}. The same results hold for U_o since $(\partial/\partial x_{n+1})U_o(x, \cdot)$ is even in x_{n+1} and has the same exponential order and type (and radius of convergence if not entire) as $U_o(x, \cdot)$. Thus if $U = U_e + U_o$, the radius of convergence of the series for $U(x, \cdot)$ is the smaller of those for $U_e(x, \cdot)$ and $U_o(x, \cdot)$, while the exponential order (and type) is the larger of those of $U_e(x, \cdot)$ and $U_o(x, \cdot)$.

Remark 1.4. The harmonicity (and uniqueness) of U, U_e, and U_o has not yet been used. In Corollary 1.2, which follows, the harmonicity of U_e is used only to guarantee continuation to the harmonicity hull, so that any polyharmonic extension of $u(x)$ would suffice. While there are many analytic extensions of $u(x)$ and $v(x)$, the advantage of the harmonic extensions is the ease of extracting the desired information.

In Theorem 1.2(A) it was shown that the restriction to $\mathbb{R}^n$ of U_e is of Laplacian order ∞ and type τ on D if and only if the series for U_e converges uniformly absolutely on compact subsets of $G\times\mathscr{B}_{1/\tau}(0)\subset\mathbb{C}^n\times\mathbb{C}^1$. In the next result, type τ is replaced by local type $\mu(x)$, and the harmonicity of U_e will be assumed on a specified open ball in $\mathbb{R}^{n+1}$. There will not be equivalence as there was above, but the result is sharp. The usefulness and sharpness will be seen in Ex. 1.2.

Corollary 1.2. If U_e (or U_o) is harmonic on $B_R(x;0)\subset\mathbb{R}^{n+1}$, where $x\in\mathbb{R}^n$, and if u (or v) has constant local order and is not polyharmonic in $\mathbb{R}^n$, then $\mu(x)$, the local type at x of u (or of v), satisfies

$$\mu(x)\leq 1/R.$$

Proof. The function U_e has analytic continuation to the reduced harmonicity hull $\hat{B}_R(x;0)\subset\mathbb{C}^{n+1}$. By using translation and the fact that $\hat{B}_R(x;0)$ is a complete circular domain (Remark 3.3, Chapter II), we see that $(x;\zeta)\in\hat{B}_R(x;0)$ if and only if $(0;\zeta)\in\hat{B}_R(0;0)$ if and only if $|\zeta|<R$. From Corollary 1.1(a) it follows that $m(x)\leq 1/R$. Now if ξ is any point in $\mathbb{R}^n\cap B_R(x;0)$, then $B_{R-|\xi-x|}(\xi;0)\subset B_R(x;0)$, so that

$$m(\xi)\leq\frac{1}{R-|\xi-x|},$$

by the argument above. Since $\lambda(x)\equiv\infty$, μ is the upper semi-continuous envelope of m and so

$$\mu(\xi)\leq\frac{1}{R-|\xi-x|}$$

for each $\xi\in\mathbb{R}^n\cap B_R(x;0)$. In particular, $\xi=x$ gives $\mu(x)\leq 1/R$. The same argument applies to $\partial U_o/\partial x_{n+1}$, and hence the result is valid for U_o. ‖

Remark 1.5. From the definition of local type μ, it must be the case that if $\lambda\equiv\infty$ then $\mu(x)<\infty$ always, even though τ may be infinity. The inequality in Corollary 1.2 gives still another view of the local nature of μ.

Example 1.1. In this example we investigate the local type of some functions which are analytic but not polyharmonic on domains $D\subset\mathbb{R}^1$. The domain $D'\subset\mathbb{R}^2=\mathbb{R}^{n+1}$ of the harmonic extension will be constructed first, as will the harmonic extension itself, and u will be the restriction to D.

First consider the strip $D' = \{(x, y) \in \mathbb{R}^2 : |y| < 1/\tau\}$, $\tau > 0$. Let F be a conformal mapping of D' onto the unit disc $\mathcal{B} \subset \mathbb{C}^1$ and let f be a holomorphic function on $\mathcal{B}$ with natural boundary exactly $\partial\mathcal{B}$. Define $U(x, y) = \mathcal{R}e\, f(F(x+iy))$ so that U is harmonic on D' with natural boundary exactly $\partial D'$. If $u(x) = U(x, 0)$, then u cannot be polyharmonic on D, a fact which can be seen either by observing that u is not the restriction of an entire function (Theorem 1.1, Chapter II) or by noting that $U(x, \cdot)$ is not entire. By Corollary 1.1, $1/m(x) = 1/\tau$, and by Corollary 1.2, $m(x) \leq \mu(x) \leq \tau$. Thus $\tau = m(x) = \mu(x)$ for all $x \in \mathbb{R}^1$.

Next, let $0 < \tau_1 < \tau_0 < \infty$ and let

$$D' = \left\{(x, y) \in \mathbb{R}^2 : |y| < \frac{1}{\tau_0} \text{ if } x \leq 0,\ |y| < \frac{1}{\tau_1} \text{ if } x > 0\right\}.$$

Define $U(x, y)$ and $u(x) = U(x, 0)$ in the same way as above, and note that u has local type $\mu(x) = \tau_0$ if $x \leq 0$, $\mu(x) = \tau_1$ if $x \geq (\tau_1^{-2} - \tau_0^{-2})^{\frac{1}{2}}$, and $\tau_1 \leq \mu(x) \leq (\tau_0^{-2} + x^2)^{-\frac{1}{2}}$ if $0 < x < (\tau_1^{-2} - \tau_0^{-2})^{\frac{1}{2}}$, by Corollaries 1.1 and 1.2. Notice that the non-constant local type of this non-polyharmonic function assumes its maximum (compare Theorem 1.3, Chapter IV).

By the same technique, a function u can be constructed for which $\mu(x)$ has a unique strict maximum. To do this, let $D' = \{(x, y) \in \mathbb{R}^2 : |y| < 2$ if $x \neq 0$, $|y| < 1$ if $x = 0\}$ and proceed as above. The function $u(x) = \log(x^2 + 1)$ could also be used. In this case $\mu(x) = (x^2 + 1)^{-\frac{1}{2}}$.

If we take $D' = \{(x, y) \in \mathbb{R}^2 : |y| < \frac{1}{2}(1 + \cos^2 x)\}$, then the corresponding function u has local type with strict minimum at each point $x = k\pi$ and strict maximum at each point $x = (2k+1)\pi/2$, $k = 0, \pm 1, \ldots$.

If D is not all of $\mathbb{R}^1$ and if $\lambda(x) \equiv \infty$, then the local type $\mu(x)$ is the reciprocal of $R(x)$, the radius of convergence of the power series of u about x (see Proposition 1.2, Chapter IV). This says that $\mu(x)$ must be unbounded as x approaches a boundary point of D which is a singular point of u. For instance, let $D' = \{(x, y) \in \mathbb{R}^2 : |x| < 2, |y| < 1\}$ and let u be defined as above. As before, if $1 < |x| < 2$, then $1 \leq \mu(x) \leq 1/(2 - |x|)$ and $1/\mu(x) = R(x) \leq 2 - |x|$. Thus $\mu(x) = 1/(2 - |x|)$ if $1 < |x| < 2$ and $\mu(x) = 1$ if $|x| \leq 1$. The points $x = \pm 2$ are boundary points of D and singular points of u.

If $n \geq 2$, the phenomenon of unbounded type just exhibited need not occur. To see this, let D_1 and D_2 be domains in $\mathbb{R}^n$, $n \geq 2$, let $D_1 \cap D_2 \neq \phi$, let u_1 be harmonic on D_1 with natural boundary ∂D_1, and let u_2 be analytic on D_2. The function $u_1 + u_2$ is defined on $D_1 \cap D_2$ and its local Laplacian order and type are the same as those of u_2. However, $u_1 + u_2$ is singular at least on all of $\partial D_1 \cap \partial(D_1 \cap D_2)$, and its local type need not be unbounded near this set.

Example 1.2. Let $n \geq 2$,

$$r^2(x) = \sum_{j=1}^{n} x_j^2,$$

and let $\zeta_{n+1} = \xi_{n+1} + i\eta_{n+1}$ be a fixed complex number. Define

$$u(x) = \{r^2(x) + \zeta_{n+1}^2\}^{1-(n+1)/2}.$$

The function u is analytic on $\mathbb{R}^n$ except at those points x such that $r^2(x) = -\zeta_{n+1}^2$, if there are any such x. First it will be shown that u is not polyharmonic, and its pointwise and local types will be computed. Part of the interest here centres on the fact that u is a root of a polynomial, and if n is odd it is a rational function.

Next, the inequality in Corollary 1.2 will be shown to be sharp, and finally it will be shown that Corollary 1.2 cannot be improved to an equivalence.

One harmonic extension of u to $\mathbb{R}^{n+1}$ is given by

$$U(x, x_{n+1}) = \{r^2(x) + (x_{n+1} - \zeta_{n+1})^2\}^{1-(n+1)/2}. \tag{1.4}$$

Since $\{U(x, x_{n+1}) + U(x, -x_{n+1})\}/2$ is harmonic in $\mathbb{R}^{n+1}$ and an even function in x_{n+1}, it is the unique even harmonic extension of u guaranteed by Theorem 1.1. If $(x; x_{n+1})$ varies in a domain in $\mathbb{C}^{n+1}$, U and the even harmonic extension of u can now be considered as holomorphic functions on some domain in $\mathbb{C}^{n+1}$. If y_0 is a fixed point in $\mathbb{R}^n$, the function $U(y_0, \cdot)$ has singular points at the two roots of the polynomial $P(\zeta) = r^2(y_0) + (\zeta - \zeta_{n+1})^2$, $\zeta \in \mathbb{C}^1$. By Corollary 1.1, u is not polyharmonic on any domain in $\mathbb{R}^n$ and has local order $\lambda(x) \equiv \infty$ on $\mathbb{R}^n$. If ζ is a root of P with minimal modulus, Corollary 1.1 also implies that $1/m(y_0) = |\zeta|$. The two roots of P are $\zeta_{n+1} \pm i|y_0| = \xi_{n+1} + i(\eta_{n+1} \pm |y_0|)$, and the one of minimal modulus is

$$\zeta = \begin{cases} \xi_{n+1} + i\,|y_0|, & \eta_{n+1} = 0 \\ \xi_{n+1} + i(|\eta_{n+1}| - |y_0|)\eta_{n+1}/|\eta_{n+1}|, & \eta_{n+1} \neq 0. \end{cases} \tag{1.5}$$

Thus

$$\left\{\frac{1}{m(y_0)}\right\}^2 = |\zeta|^2 = |\xi_{n+1}|^2 + |\eta_{n+1}|^2 - 2\,|y_0|\,|\eta_{n+1}| + |y_0|^2, \tag{1.6}$$

and the continuity of this expression as a function of y_0 shows that $\mu(y_0) = m(y_0) = 1/|\zeta|$ is the local type of u at y_0. Replacing y_0 by x and $|y_0|$ by $r(y_0)$ in (1.6), we see that the local type at $x \in \mathbb{R}^n$ is

$$\mu(x) = \{r(x)^2 + |\zeta_{n+1}|^2 - 2r(x)\,|\mathcal{I}m\ \zeta_{n+1}|\}^{-\frac{1}{2}}. \tag{1.7}$$

If ζ_{n+1}^2 is real, then

$$\mu(x) = \begin{cases} \{r(x)^2 + \zeta_{n+1}^2\}^{-\frac{1}{2}}, & \zeta_{n+1}^2 \geq 0 \\ \{r(x) - |\zeta_{n+1}|\}^{-1}, & \zeta_{n+1}^2 < 0. \end{cases}$$

In the case $n = 1$, the function $u(x) = \log(x^2 + \zeta_2^2)$ can be used to obtain the same result as above.

To show that the inequality in Corollary 1.2 is sharp we must show that the distance R from $(y_0; 0) \in \mathbb{R}^{n+1}$ to the $(n-1)$-sphere $\mathbb{R}^{n+1} \cap V(0; \zeta_{n+1})$ is exactly $|\zeta|$, given by the square root of the expression in (1.6). By a translation in $\mathbb{R}^{n+1}$ by $(-y_0; 0)$, R is the distance from the origin in $\mathbb{R}^{n+1}$ to $\mathbb{R}^{n+1} \cap V(-y_0; \zeta_{n+1})$. This last set is the $(n-1)$-sphere in $\mathbb{R}^{n+1}$ with centre $(-y_0; \xi_{n+1})$ and radius $|\eta_{n+1}|$, lying in the hyperplane perpendicular to the vector $(0; \eta_{n+1})$. A simple calculation shows that $R^2 = |y_0|^2 + |\eta_{n+1}|^2 - 2\,|y_0|\,|\eta_{n+1}| + |\xi_{n+1}|^2$, and this is $|\zeta|^2$.

Finally we modify u to produce an example of a function w whose local type at y_0 satisfies $\mu(y_0) \leq 1/R$ but whose harmonic extension W_e is not harmonic on $B_R(y_0) \subset \mathbb{R}^{n+1}$. To this end, let $w(x) = u(x + i\eta_0)$, where $\eta_0 = (\eta_1, \ldots, \eta_n) \in \mathbb{R}^n$, and let x vary in a domain in $\mathbb{C}^n$ so that $w(x)$ is an analytic continuation of the given function. The local type of w at $y_0 \in \mathbb{R}^n$ is the same as that of (the continuation of) u at $y_0 + i\eta_0 \in \mathbb{C}^n$. The harmonic extension of w into $\mathbb{R}^{n+1}$, corresponding to U defined in (1.4), is

$$W(x, x_{n+1}) = \{r(x + i\eta_0)^2 + (x_{n+1} - \zeta_{n+1})^2\}^{1-(n+1)/2},$$

which has unlimited continuation on the complement of the isotropic cone $V(-i\eta_0; \zeta_{n+1})$. We shall compute the local type of w at a fixed $y_0 \in \mathbb{R}^n$ and also compute the radius R of the largest ball in $\mathbb{R}^{n+1}$ about y_0 which does not intersect $V(-i\eta_0; \zeta_{n+1})$.

As before, $1/m(y_0) = |\zeta|$ where ζ is a root of minimal modulus of the polynomial $P(\zeta) = r(y_0 + i\eta_0)^2 + (\zeta - \zeta_{n+1})^2$. The roots are

$$\zeta = \zeta_{n+1} \pm ir(y_0 + i\eta_0) \tag{1.8}$$

so that ζ is a continuous function of y_0. Thus the local type is $\mu(y_0) = 1/|\zeta|$, where $|\zeta|$ is the smaller of the values given by (1.8).

The distance R from $(y_0; 0)$ to $\mathbb{R}^{n+1} \cap V(-i\eta_0; \zeta_{n+1})$ is the same as that from the origin to $\mathbb{R}^{n+1} \cap V(y + i\eta)$ where $y = (-y_0; \xi_{n+1})$ and $\eta = (-\eta_0; \eta_{n+1})$. By (3.19) Chapter II,

$$R^2 = |y|^2 + |\eta|^2 - 2\{|y|^2\,|\eta|^2 - (y, \eta)^2\}^{\frac{1}{2}} \tag{1.9}$$

where (y, η) is the inner product in $\mathbb{R}^{n+1}$.

Now let $y_0 = (1, 0, \ldots, 0, 1)$, $\eta_0 = (1, 0, \ldots, 0, -1)$, and $\zeta_{n+1} = 1 + i$. Since $r(y_0 + i\eta_0) = 0$, it follows from (1.8) that $\zeta = \zeta_{n+1} = 1 + i$ and $|\zeta|^2 = 2$. From (1.8) we obtain $R^2 = 6 - 4\sqrt{2}$, so that $R^2 < |\zeta|^2$. The conclusion is that $\mu(y_0) = 1/|\zeta| < 1/R$, but $B_R(y_0)$ is the largest ball about y_0 on which W_e is harmonic.

The separation of singularities theorem (Theorem 2.1, Chapter III) was proved for polyharmonic functions. By the use of harmonic extension to $\mathbb{R}^{n+1}$, the theorem can be generalized to analytic functions.

Theorem 1.4 (Separation of singularities). Let G be an open dense set in $\mathbb{R}^n$ and let u be analytic on G. If F_1 and F_2 are closed subsets of $\mathbb{R}^n$ such that $F_1 \cup F_2 = \mathbb{R}^n \setminus G$, then there exist functions u_1 and u_2, analytic on $\mathbb{R}^n \setminus F_1$ and $\mathbb{R}^n \setminus F_2$, respectively, such that $u = u_1 + u_2$ on G.

Proof. Let U_e be the unique even harmonic extension of u into $\mathbb{R}^{n+1}$, given by Theorem 1.1, and let G_e be the largest open subset of $\mathbb{R}^{n+1}$ such that

(1) U_e is harmonic on G_e, and

(2) if $(x; x_{n+1}) \in G_e$ with $x \in \mathbb{R}^n$ then $(x, t) \in G_e$ for all $t \in \mathbb{R}^1$ such that $|t| \leq |x_{n+1}|$.

On the open set $\mathbb{R}^{n+1} \setminus \bar{G}_e$, define U_e to be identically zero. Define $\mathcal{F} = \partial G_e$,

$$\mathcal{F}_1 = \{(x; x_{n+1}) \in \mathcal{F} : x \in \mathbb{R}^n \text{ and } d(x, F_1) \leq d(x, F_2)\}$$
$$\mathcal{F}_2 = \{(x; x_{n+1}) \in \mathcal{F} : x \in \mathbb{R}^n \text{ and } d(x, F_2) \leq d(x, F_1)\}.$$

Each $\mathcal{F}_k$ is closed, $\mathcal{F}_1 \cup \mathcal{F}_2 = \mathcal{F}$, $\mathbb{R}^{n+1} \setminus \mathcal{F}$ is dense in $\mathbb{R}^{n+1}$, and $\mathcal{F}_k \cap \mathbb{R}^n = F_k$, $k = 1, 2$. By Theorem 2.1, Chapter III, the singularities of U_e can be separated, that is there exist functions U_1 and U_2, harmonic on $\mathbb{R}^{n+1} \setminus \mathcal{F}_1$ and $\mathbb{R}^{n+1} \setminus \mathcal{F}_2$ respectively, such that $U_e = U_1 + U_2$ on $\mathbb{R}^{n+1} \setminus \mathcal{F}$. The restrictions of U_1 and U_2 to $\mathbb{R}^n$ are analytic on $\mathbb{R}^n \setminus F_1$ and $\mathbb{R}^n \setminus F_2$ respectively, and give the desired decomposition. ‖

The functions u_1 and u_2 in the preceding theorem, which are the restrictions of the harmonic functions U_1 and U_2, can be analysed further. To do this, assume that u has local Laplacian order ∞ everywhere on G. This assumption merely says that u is not polyharmonic, by Corollary 1.2, Chapter IV, unless it is of type 0.

For each $x\in G$, let $m(x)$ and $\mu(x)$ be the pointwise and local types of u, and let

$$R_e(x)=\begin{cases}\sup\{|t|: t\in\mathbb{R}^1, (x;t)\in G_e\}, & x\in G\\ 0, & x\notin G.\end{cases}$$

Here G_e is the set constructed in the proof of the preceding theorem. The inequality

$$\mu(x)\geqslant m(x)\geqslant\frac{1}{R_e(x)}$$

gives an estimate of the size of G_e from which will be derived estimates of μ_1 and μ_2, the local types of u_1 and u_2. If $x\in\mathbb{R}^n\backslash F_1$ is identified with $(x;0)\in\mathbb{R}^{n+1}\backslash\mathscr{F}_1$ and if $R=d(x,\mathscr{F}_1)$, then U_1 is harmonic on $B_R(x)\subset\mathbb{R}^{n+1}$ and Corollary 1.2 implies that

(1.10)
$$\mu_1(x)\leqslant\frac{1}{d(x,\mathscr{F}_1)}=\sup\{[(x-\xi)^2+R_e(\xi)^2]^{-\frac{1}{2}}:\xi\in\mathbb{R}^n, d(\xi,F_1)\leqslant d(\xi,F_2)\}$$

and similarly for μ_2, by the definition of $\mathscr{F}_1$ and $\mathscr{F}_2$.

If u has type τ on G, so that

$$\frac{1}{\tau}\leqslant\frac{1}{\mu(x)}\leqslant R_e(x)$$

for all $x\in G$, then the right-hand side of (1.10) is bounded by

$$\sup\{[(x-\xi)^2+R_e(\xi)^2]^{-\frac{1}{2}}:\xi\in G\cup F_1\}\leqslant\max\{\tau, 1/d(x,F_1)\},$$

and the same is true for μ_2. Thus,

$$\text{(1.11)}\qquad \mu_k(x)\leqslant\max\{\tau, 1/d(x,F_k)\},\qquad x\in G,\qquad k=1,2.$$

It is always true that $\mu(x)\leqslant\max\{\mu_1(x),\mu_2(x)\}$ and that (Proposition 1.3, Chapter II) equality occurs if $\mu_1(x)\neq\mu_2(x)$. Since $\mu(x)\leqslant\tau$, if $\mu_1(x)>\tau$ inequality (1.11) implies that

$$\mu_1(x)=\mu_2(x)\leqslant\min\{1/d(x,F_1), 1/d(x,F_2)\}.$$

Thus, for each $x\in G$

$$\mu_k(x)\leqslant\max\{\tau,\min\{1/d(x,F_1), 1/d(x,F_2)\}\},$$

and by upper semi-continuity the inequality holds for μ_k on $\mathbb{R}^n\backslash F_k$. Therefore we have the following corollary.

Corollary 1.3. If the function u in Theorem 1.4 is of Laplacian order ∞ *and type* $\tau<\infty$ *on G, then* u_1 *and* u_2 *can be taken to have Laplacian order* ∞ *and local types*

$$\mu_k(x)\leq \max\{\tau, \min\{1/d(x,F_1), 1/d(x,F_2)\}\} \quad \text{for } x\in\mathbb{R}^n\backslash F_k,$$

$k=1,2$.

2. Relations between Laplacian order and exponential order of entire functions

The relations between Laplacian order and type and exponential order and type for functions of one variable were completely determined in Theorem 1.1, Chapter II. In this section we derive relations for $n\geq 2$ and also prove a multiplier theorem. The notions of exponential order and type and the Cauchy bounds to be used were discussed in §4, Chapter I. Recall that $P(x;R)$ is the polydisc with centre x and polyradius $(R,\ldots,R)$ and that $M(u;P(x;R))$ is the supremum of $|u|$ on the polydisc.

Theorem 2.1

(a) *If u is analytic on* $\mathbb{R}^n$ *and is the restriction of an entire function on* $\mathbb{C}^n$, *then u is polyharmonic on* $\mathbb{R}^n$ *(that is it has Laplacian order* ∞ *and type* 0).

(b) *If u is entire of exponential order* ρ *and type* τ *on* $\mathbb{C}^n$, *then u is polyharmonic of Laplacian order* ρ *and type* $(\tau\rho)^{1/\rho}\sqrt{n}$. *If u is entire of exponential order* ρ, *then u is polyharmonic of Laplacian order* ρ.

Proof

(a) Let K be a compact set. For all R sufficiently large, if $x\in K$ then K is contained in the polydisc $P(x;R)$. Let

$$U(R)=\bigcup_{x\in K} P(x;R).$$

Using the Cauchy bounds we obtain for all $x\in K$

$$\begin{aligned}
|\Delta^p u(x)| &\leq \sum_{|\alpha|=p}\frac{p!}{\alpha!}M(D^{2\alpha}u;P(x;R))\\
&\leq \sum_{|\alpha|=p}\frac{p!}{\alpha!}\frac{(2\alpha)!}{R^{2p}}M(u;P(x;R))\\
&\leq M(u;U(R))n^p\frac{1}{R^{2p}}(2p)!
\end{aligned} \tag{2.1}$$

where 2α denotes the multi-index $(2\alpha_1,\ldots,2\alpha_n)$. Since R can be chosen arbitrarily large, (2.1) shows that u is polyharmonic.

(b) Let u be entire of exponential order ρ and type τ, let K be a compact set and let $\varepsilon > 0$ be given. From the definition of exponential order and type, there is a constant $C_{K,\varepsilon}$ such that for all $x \in K$, $M(u; P(x; R)) \leq C_{K,\varepsilon} \exp\{(\tau+\varepsilon)R^\rho\}$. From the second inequality in (2.1) it follows that

$$(2.2) \qquad |\Delta^p u(x)| \leq n^p (2p)! \frac{1}{R^{2p}} C_{K,\varepsilon} \exp\{(\tau+\varepsilon)R^\rho\}, \qquad x \in K.$$

Fix p and consider the function

$$g(R) = R^{-2p} \exp\{(\tau+\varepsilon)R^\rho\}.$$

This function achieves its minimum at

$$R = \left\{\frac{2p}{\rho(\tau+\varepsilon)}\right\}^{1/\rho},$$

and the minimum value is $\{\rho(\tau+\varepsilon)\}^{2p/\rho}\{(2p/e)^{2p}\}^{-1/\rho}$. Thus for p sufficiently large and all $x \in K$,

$$(2.3) \qquad |\Delta^p u(x)| \leq C'_{K,\varepsilon}(2p)!^{1-1/\rho} n^p (2p)^{1/2\rho}\{\rho(\tau+\varepsilon)\}^{2p/\rho}.$$

Inequality (2.3) shows that u is of Laplacian order ρ and type $(\rho\tau)^{1/\rho}\sqrt{n}$.

For the case u of exponential order ρ, notice that if $\rho' > \rho$ and $\varepsilon > 0$, then $M(u; P(x; R)) \leq C_{K,\varepsilon} \exp(\varepsilon R^{\rho'})$ for all sufficiently large R. This allows us to carry through the preceding proof with $\tau = 0$ and obtain the result that u is of Laplacian order ρ' and type 0 for every $\rho' > \rho$. Therefore u is of Laplacian order ρ, and the theorem is proved. ‖

For the case of two variables, Theorem 2.1(a) was proved by Ciorănescu [25].

If $n > 1$, the converse of Theorem 2.1(b) does not hold, and even the most extreme case can occur. For example, let $(x_1, x_2) \in \mathbb{R}^2$, let $z = x_1 + ix_2$, and define $u(x_1, x_2) = \mathcal{R}e \exp(e^z) = \exp(e^{x_1} \cos x_2)\cos(e^{x_1} \sin x_2)$. The function u is harmonic on $\mathbb{R}^2$ (hence Laplacian order 0), and the function $u(x_1, x_2) = \exp(e^{x_1} \cos x_2)\cos(e^{x_1} \sin x_2)$, $(x_1, x_2) \in \mathbb{C}^2$, is its continuation as an entire function. Evaluating $u(R, 0) = \exp(e^R)$ for all $R > 0$, we see that $u(x_1, x_2)$ is of exponential order ∞.

Remark 2.1. Every function polyharmonic on all of $\mathbb{R}^n$ is the restriction of an entire function since the reduced harmonicity hull of $\mathbb{R}^n$ is $\mathbb{C}^n$. The example given above shows that even for harmonic functions the Laplacian and exponential orders may be unequal. For a discussion of the exponential order for harmonic functions in terms of gradients at the origin, see for example Armitage [5], Fryant [36], and Fugard [37].

Theorem 2.2.

(a) *If u is polyharmonic (that is, it has Laplacian order ∞ and type 0) on $D\subset\mathbb{C}^n$, and if f is entire on $\mathbb{C}^n$ then the function uf is polyharmonic on D.*

(b) *If u has Laplacian order ρ on $D\subset\mathbb{C}^n$ and f is entire with exponential order ρ_1, then uf has Laplacian order $\rho_0=\max\{2\rho_1,\rho\}$ on D. If u has positive finite Laplacian order ρ and type σ on D and if f is entire with positive finite exponential order ρ_1 and type τ, then uf has Laplacian order $\rho_0=\max\{2\rho_1,\rho\}$ and type*

$$3^{1/\rho_0}\left(nt^2+\frac{4^{1+1/\rho_0}}{R_0'}t+\sigma^2\right)^{\frac{1}{2}}.$$

where $t=(\rho_1\tau)^{1/\rho_1}$, on any domain $D'\subset D$ such that $P(x;R_0')\subset D$ for each $x\in D'$.

Proof

(a) Let K be a compact subset of D and let $\varepsilon>0$ be given. For each $x\in K$ there is an open polydisc $P(x;R)\subset D$. Let $R_0'=\sup\{R\colon \text{for all } x\in K, P(x;R)\subset D\}$ and choose R_0, $0<R_0<R_0'$. For each $x\in K$ the Cauchy bounds yield

$$(2.4)\qquad |D^\alpha\Delta^k u(x)|\leqslant\frac{\alpha!}{R_0^{|\alpha|}}M(\Delta^k u;P(x;R_0))\leqslant\frac{\alpha!}{R_0^{|\alpha|}}C_{K,\varepsilon}(2k)!\,\varepsilon^{2k}.$$

Since f is entire, for any $R>0$ we can apply the Cauchy bounds for f on the polydisc $P(x;R)$ and obtain, for all $x\in K$,

$$(2.5)\qquad \begin{aligned}|D^\alpha\Delta^l f(x)|&\leqslant n^l(2l+|\alpha|)!\left(\frac{1}{R}\right)^{2l+|\alpha|}M(f;P(x;R))\\&\leqslant C_{K,R}n^l(2l+|\alpha|)!\left(\frac{1}{R}\right)^{2l+|\alpha|}.\end{aligned}$$

From Lemma 1.1, Chapter I, we have the formula

$$(2.6)\qquad \Delta^p(u(x)f(x))=\sum_{\substack{k,l,\alpha\\k+l+|\alpha|=p}}2^{|\alpha|}\frac{p!}{k!\,l!\,\alpha!}\{D^\alpha\Delta^k u(x)\}\{D^\alpha\Delta^l f(x)\}.$$

Substituting (2.4) and (2.5) into (2.6), we obtain for $x\in K$

$$(2.7)\qquad \begin{aligned}&|\Delta^p(u(x)f(x))|\\&\leqslant\sum 2^{|\alpha|}\frac{p!}{k!\,l!\,\alpha!}\frac{\alpha!}{R_0^{|\alpha|}}C_{K,\varepsilon}(2k)!\,\varepsilon^{2k}C_{K,R}n^l(2l+|\alpha|)!\left(\frac{1}{R}\right)^{2l+|\alpha|}\end{aligned}$$

where the sum is over k, l, α with $k+l+|\alpha|=p$. Still summing over these indices, from (2.7) we obtain

(2.8) $|\Delta^p(u(x)f(x))|$

$$\leqslant C_{K,\varepsilon}C_{K,R}\sum 2^{|\alpha|}\frac{p!}{k!\,l!\,|\alpha|!}|\alpha|!\,(2k)!\,(2l+|\alpha|)!$$

$$\times n^l(\varepsilon^2)^k\left(\frac{1}{R^2}\right)^l\left(\frac{1}{RR_0}\right)^{|\alpha|}$$

$$\leqslant C_{K,\varepsilon}C_{K,R}2^p n^p(2p)!\sum\frac{p!}{k!\,l!\,|\alpha|!}(\varepsilon^2)^k\left(\frac{1}{R^2}\right)^l\left(\frac{1}{RR_0}\right)^{|\alpha|}.$$

For each fixed integer m, $0\leqslant m\leqslant p$, there are

$$\binom{n+m-1}{m}$$

multi-indices α with $|\alpha|=m$. Thus for each integer m, $0\leqslant m\leqslant p$, the sum (2.8) has

$$\binom{n+m-1}{m}$$

identical terms, and this binomial coefficient is at most 2^{n-1+m}. Therefore from (2.8) and these remarks it follows that

(2.9) $|\Delta^p(u(x)f(x))|$

$$\leqslant C_{K,\varepsilon}C_{K,R}2^{p+n-1}n^p(2p)!\sum_{\substack{k,l,m\\ k+l+m=p}}\frac{p!}{k!\,l!\,m!}(\varepsilon^2)^k\left(\frac{1}{R^2}\right)^l\left(\frac{2}{RR_0}\right)^m$$

$$\leqslant C_{K,\varepsilon}C_{K,R}2^{n-1+p}n^p(2p)!\left(\varepsilon^2+\frac{1}{R^2}+\frac{2}{RR_0}\right)^p.$$

If $\varepsilon'>0$ is given, choose $\varepsilon>0$ so small and $R>0$ so large that

$$|\Delta^p(u(x)f(x))|\leqslant C_{K,\varepsilon'}(2p)!\,(\varepsilon')^{2p}.$$

Thus uf is polyharmonic.

(b) We prove the second statement. A minor modification proves the first. For a given compact set $K\subset D'$ and a number $\varepsilon>0$, (2.4) becomes

$$|D^\alpha\Delta^k u(x)|\leqslant\frac{\alpha!}{R_0^{|\alpha|}}C_{K,\varepsilon}(2k)!^{1-1/\rho}(\sigma+\varepsilon)^{2k} \tag{2.10}$$

for all $x\in K$. For each $R>0$, (2.5) becomes

$$|D^\alpha\Delta^l f(x)|\leqslant n^l(2l+|\alpha|)!\left(\frac{1}{R}\right)^{2l+|\alpha|}C'_{K,\varepsilon}\exp\{(\tau+\varepsilon)R^{\rho_1}\}. \tag{2.11}$$

Minimizing $R^{-2l-|\alpha|}\exp\{(\tau+\varepsilon)R^{\rho_1}\}$ as in the proof of the preceding theorem, we obtain

$$(2.12)\quad |D^\alpha \Delta^l f(x)| \leqslant n^l(2l+|\alpha|)!\, C''_{K,\varepsilon}(2l+|\alpha|)^{1/2\rho_1} \times\{\rho_1(\tau+\varepsilon)\}^{(2l+|\alpha|)/\rho_1}(2l+|\alpha|)!^{-1/\rho_1}.$$

Substituting (2.10) and (2.12) into (2.6) gives the estimate

$$(2.13)\quad |\Delta^p(u(x)f(x))| \leqslant \sum_{\substack{k,l,\alpha\\ k+l+|\alpha|=p}} 2^{|\alpha|}\frac{p!}{k!\,l!\,\alpha!}\frac{\alpha!}{R_0^{|\alpha|}}C_{K,\varepsilon}(2k)!^{1-1/\rho}(\sigma+\varepsilon)^{2k} \times C''_{K,\varepsilon}n^l(2l+|\alpha|)^{1/2\rho_1}\{\rho_1(\tau+\varepsilon)\}^{(2l+|\alpha|)/\rho_1}(2l+|\alpha|)!^{1-1/\rho_1}.$$

In (2.13), cancel the $\alpha!$ which appears as both a numerator and denominator and insert the factor $|\alpha|!/|\alpha|!$. Let us now collect factors and examine the factor

$$(2.14)\quad b(k,l,|\alpha|)\equiv|\alpha|!\,(2l+|\alpha|)!^{1-1/\rho_1}(2k)!^{1-1/\rho}$$

which appears in each term of (2.13). Since $k+l+|\alpha|=p$,

$$b(k,l,|\alpha|)\leqslant(2p)!\,(2k)!^{-1/\rho}(2l)!^{-1/\rho_1}\,|\alpha|!^{-2/2\rho_1}.$$

If $\rho_0=\max\{2\rho_1,\rho\}$, then

$$(2.15)\quad \begin{aligned} b(k,l,|\alpha|)&\leqslant(2p)!\,(2k)!^{-1/\rho_0}(|\alpha|!^2)^{-1/\rho_0}(2l)!^{-1/\rho_0}\\ &\leqslant(2p)!\,(2k)!^{-1/\rho_0}(2l)!^{-1/\rho_0}(2\,|\alpha|)!^{-1/\rho_0}(2^{2\,|\alpha|})^{1/\rho_0}\\ &\leqslant(2p)!^{1-1/\rho_0}(3^{2p}2^{2\,|\alpha|})^{1/\rho_0}.\end{aligned}$$

Substituting (2.15) into (2.13) we obtain

$$(2.16)\quad |\Delta^p(u(x)f(x))| \leqslant C'''_{K,\varepsilon}(2p)!^{1-1/\rho_0}(3^{1/\rho_0})^{2p}(2p)^{1/2\rho_1}\sum_{\substack{k,l,\alpha\\ k+l+|\alpha|=p}}\frac{p!}{k!\,l!\,|\alpha|!}\{(\sigma+\varepsilon)^2\}^k \times\{n(\rho_1(\tau+\varepsilon))^{2/\rho_1}\}^l\left[\frac{2^{1+2/\rho_0}}{R_0}\{\rho_1(\tau+\varepsilon)\}^{1/\rho_1}\right]^{|\alpha|}.$$

Using our remark following (2.8) we obtain from (2.16)

$$(2.17)\quad |\Delta^p(u(x)f(x))|\leqslant C'''_{K,\varepsilon}(2p)^{1/2\rho_1}(2p)!^{1-1/\rho_0}(3^{1/\rho_0})^{2p} \times\left[(\sigma+\varepsilon)^2+n\{\rho_1(\tau+\varepsilon)\}^{2/\rho_1}+\frac{4^{1+1/\rho_0}}{R_0}\{\rho_1(\tau+\varepsilon)\}^{1/\rho_1}\right]^p.$$

If t is set equal to $(\rho_1\tau)^{1/\rho_1}$, (2.17) shows that uf has Laplacian order $\rho_0=\max\{2\rho_1,\rho\}$ and type $3^{1/\rho_0}\{nt^2+(4^{1+1/\rho_0}/R_0)t+\sigma^2\}^{\frac{1}{2}}$. Since this holds for all $R_0<R_0'$, the conclusion follows. ‖

Remark 2.2. The order $\rho_0 = \max\{2\rho_1, \rho\}$ is the best possible, although the expression for type is not known to be sharp. To see that order ρ_0 is the best, let D be a bounded domain in $\mathbb{C}^1$ and let $f(z) = f(x+iy)$ be holomorphic on D with ∂D as its natural boundary. The function $h(x, y) = f(x+iy)$ is harmonic in D, and hence is of Laplacian order $\rho = 0$. Multiply $h(x, y)$ by the function e^x, entire of exponential order $\rho_1 = 1$. Change the variables to z, $\bar{z}$ and write $v = f(z)\exp\{(z+\bar{z})/2\}$. With these variables, $\Delta = \partial^2/\partial z\,\partial\bar{z}$ and

$$\Delta^p v = \frac{d^p}{dz^p}\{e^{z/2}f(z)\}\frac{d^p}{d\bar{z}^p}e^{\bar{z}/2} = \frac{d^p}{dz^p}\{e^{z/2}f(z)\}(\tfrac{1}{2})^p e^{\bar{z}/2}.$$

The Cauchy bounds on a small disc of radius R contained in D give

$$\left|\frac{d^p}{dz^p}e^{z/2}f(z)\right| \leq \frac{1}{R^p}\,p!\,M(e^{z/2}f(z); P(z; R))$$

so that

$$|\Delta^p v| \leq \left(\frac{1}{2R}\right)^p p!\,M(e^{z/2}f(z); P(z; R))M(e^{\bar{z}/2}; D).$$

This shows that v is of Laplacian order 2 with type $1/\sqrt{(2R)}$ on every compact K whose minimum distance to ∂D is R. However, v cannot have type 0, for this would require

$$\left|\frac{d^p}{dz^p}e^{z/2}f(z)\right| \leq C_{K,\varepsilon}(2p)!^{\frac{1}{2}}\varepsilon^{2p}$$

which, using Stirling's formula on $(2p)!^{\frac{1}{2}}$, would yield convergence of the Taylor series for $e^{z/2}f(z)$ outside of D.

3. Bounded polyharmonic functions

It is known that if u is polyharmonic of order 1 and type 0 and bounded on all of $\mathbb{R}^n$, then u is constant (Friedman [34, pp. 86–87]). This is the best statement as far as type is concerned, for $\cos x$ is of order 1 and finite non-zero type and is bounded. In this section we improve the above statement and derive relations between the Almansi coefficients of entire functions bounded on $\mathbb{R}^n$ having arbitrary finite Laplacian order ρ with type τ.

We begin with some preliminary notations, definitions, and observations.

Denote the type of an entire or polyharmonic function by the symbol τ which will signify only three values: zero, finite non-zero, or ∞ (hence we shall not distinguish between different finite non-zero types). We shall say that an entire function has *inclusive* exponential order ρ and type τ if it has order $\rho' < \rho$, or has order ρ and type $\tau' \leq \tau$. (Such a function is sometimes said to have *growth* (ρ, τ).) With this convention, Corollary 2.1(a), (c), (d), Chapter V, implies that if

$$u(x) = \sum_{k=0}^{\infty} r^{2k} h_k(x)$$

is the Almansi expansion of an analytic function and if u is polyharmonic (of order ρ and type τ), then for every compact set K the series $\sum_{k=0}^{\infty} \|h_k\|_K z^{2k}$ represents an entire function (of inclusive order ρ and type τ).

If $h_k(x; x_0)$, $k = 0, 1, \ldots$, are the Almansi coefficients of a polyharmonic function (of order ρ and type τ) expanded about the point x_0, then the function

$$h(x; x_0; R) = \sum_{k=0}^{\infty} h_k(x; x_0) R^{2k}$$

is harmonic in x for each fixed R, and for fixed x it is entire in R (of inclusive order ρ and type τ). Notice also that if

$$u(x) = \sum_{k=0}^{\infty} r(x - x_0)^{2k} h_k(x; x_0)$$

is polyharmonic of order ρ and type τ on $\bar{B}_{R_0}$ and if μ is a finite Borel measure supported on $\bar{B}_{R_0}$, then

$$f(R) = \sum_{k=0}^{\infty} \int h_k(x; x_0) \, \mathrm{d}\mu(x) R^{2k}$$

is entire of inclusive order ρ and type τ.

For a compact set $X \subset \mathbb{R}^n$, denote by $\mathscr{C}(X)$ the Banach space of continuous functions on X (with the supremum norm) and by $\mathscr{C}(X)^*$ its dual space, that is $\mathscr{C}(X)^*$ is the space of finite Borel measures μ on X where $\|\mu\|$ is the absolute total mass on X. Let $\mathscr{H}(B_R)$ be the linear space of functions harmonic on B_R and let $\mathscr{H}(B_R) \cap \mathscr{C}(\bar{B}_R)$ be the closed subspace of $\mathscr{C}(\bar{B}_R)$ formed by functions in $\mathscr{C}(\bar{B}_R)$ which are harmonic on B_R. This subspace is isometrically isomorphic to the space $\mathscr{C}(S_R)$, $S_R = \partial B_R$. Let $A : \mathscr{C}(S_R) \to \mathscr{H}(B_R) \cap \mathscr{C}(\bar{B}_R)$ be this isomorphism, and let $A_1 : \mathscr{H}(B_R) \cap \mathscr{C}(\bar{B}_R) \to \mathscr{H}(B_{R_1}) \cap \mathscr{C}(\bar{B}_{R_1})$, $R_1 < R$, be the linear mapping which restricts $h \in \mathscr{H}(B_R) \cap \mathscr{C}(\bar{B}_R)$ to $\bar{B}_{R_1}$. The mapping A_1 is one-to-one and $\|A_1\| = 1$. Finally, let $T_1 = A_1 A$ so that T_1 is a one-to-one linear mapping of $\mathscr{C}(S_R)$ into $\mathscr{H}(B_{R_1}) \cap \mathscr{C}(\bar{B}_{R_1})$ with $\|T_1\| = 1$.

Definition 3.1. Let $\{m_k\}$ be a sequence of integers, either finite or infinite, with $0 < m_1 < m_2 < \ldots$. Define $\Omega_{\rho,\tau}$ to be the collection of all such sequences such that every entire function $f(z)$ (of one complex variable) which is of inclusive exponential order ρ and type τ, which has the form

$$f(z) = \sum_{k \geq 1} a_{m_k} z^{2m_k}$$

and which is bounded on the real line is identically zero.

Applications of the Phragmén–Lindelöf theorem show that $\{1, 2, 3, \ldots\}$ is in $\Omega_{1,0}$ and that if a and p are integers, $a \geq 0$, $p \geq 1$, then $\{a + pk\}_{k=1}^{\infty}$ is in $\Omega_{p,0}$.

Theorem 3.1. Let u be polyharmonic and bounded on $\mathbb{R}^n$, and of Laplacian order ρ and type τ for $|x - x_0| < R_0$. Let $\{m_k\} \in \Omega_{\rho,\tau}$ and let $\{m_k'\}$ be the complementary sequence of $\{m_k\}$ in the non-negative integers. In the Almansi expansion of u about x_0 the coefficients h_{m_k} are uniform limits on compacts of linear combinations of the coefficients $h_{m_k'}$.

Proof. We assume that x_0 is the origin. The Almansi expansion of u is valid on $\mathbb{R}^n$ and the Almansi coefficients are harmonic on all of $\mathbb{R}^n$ (see Theorem 2.2, Chapter V). If the theorem were not true, for some $R_1 > 0$ the closed linear span of the functions $\{h_{m_k'}\}$ in the space $\mathscr{C}(\bar{B}_{R_1})$ would omit at least one function $h_{m_{k_0}}$. By the Hahn–Banach Theorem, there exists a finite Borel measure μ on $\bar{B}_{R_1}$ such that

$$\int h_{m_k'}(x)\, d\mu(x) = 0$$

for all k and

$$\int h_{m_{k_0}}(x)\, d\mu(x) \neq 0.$$

As remarked earlier, the function

$$h(R) = \sum_{k=0}^{\infty} \int h_k(x)\, d\mu(x) R^{2k} = \sum_{k \geq 1} \int h_{m_k}(x)\, d\mu(x) R^{2m_k}$$

is entire of inclusive exponential order ρ and type τ, and since

$$\int h_{m_{k_0}}(x)\, d\mu(x) \neq 0,$$

this entire function is not identically zero. If we can prove that $h(R)$ is bounded on the real line, then it will follow from the definition of $\Omega_{\rho,\tau}$ that $h(R)$ is identically zero, a contradiction.

To see that $h(R)$ is bounded on the real line, let R be fixed with

$R \geqslant R_1$. If the Borel measure $\nu_R \in \mathscr{C}(S_R)^*$ is defined by

$$\nu_R(f) \equiv \int_{S_R} f(x)\,\mathrm{d}\nu_R(x) \equiv \mu(T_1 f) \equiv \int_{B_{R_1}} (T_1 f)(x)\,\mathrm{d}\mu(x),$$

then $\|\nu_R\| \leqslant \|\mu\|$. Letting $|u(x)| \leqslant K$ on $\mathbb{R}^n$ we see that

$$\begin{aligned}
|h(R)| &= \left| \sum_{k=0}^{\infty} \int_{B_{R_1}} h_k(x)\,\mathrm{d}\mu(x) R^{2k} \right| \\
&= \left| \sum_{k=0}^{\infty} \int_{S_R} h_k(x)\,\mathrm{d}\nu_R(x) R^{2k} \right| \\
&= \left| \sum_{k=0}^{\infty} \int_{S_R} h_k(x)\,|x|^{2k}\,\mathrm{d}\nu_R(x) \right| \\
&= \left| \int_{S_R} u(x)\,\mathrm{d}\nu_R(x) \right| \\
&\leqslant K\,\|\nu_R\| \\
&\leqslant K\,\|\mu\|.
\end{aligned}$$

Hence $h(R)$ is bounded on the real line, and the theorem is proved. ||

The improvement of Friedman's result is contained in the following theorem.

Theorem 3.2. If u is polyharmonic and bounded on all of $\mathbb{R}^n$, *of Laplacian order* 1 *and type* 0 *on some domain* $D \subset \mathbb{R}^n$, *then u is harmonic and hence constant.*

Proof. Let $x_0 \in D$, and choose $R_0 > 0$ so that $\overline{B_{R_0}(x_0)} \subset D$. Choose a constant b so that $u(x_0) + b \neq 0$. The function $u(x) + b$ satisfies the hypotheses (and the conclusion) of the theorem if and only if $u(x)$ does, so we can assume that $u(x_0) \neq 0$. Let μ_R be the mean-value measure on the sphere $S_R(x_0) \equiv \partial B_R(x_0)$. The function u has Almansi expansion

$$u(x) = \sum_{k=0}^{\infty} r(x - x_0)^{2k} h_k(x; x_0)$$

about the point x_0, and the function

$$\begin{aligned}
h(R; x_0) &= \sum_{k=0}^{\infty} \int_{S_R(x_0)} h_k(x; x_0)\,\mathrm{d}\mu_R(x) R^{2k} \\
&= \sum_{k=0}^{\infty} h_k(x_0; x_0) R^{2k} \\
&= \int_{S_R(x_0)} u(x)\,\mathrm{d}\mu_R(x)
\end{aligned}$$

is entire of inclusive order 1, type 0, and is bounded on the real line. By the Phragmén–Lindelöf theorem, $h(R; x_0)$ is constant, and hence $h(R; x_0) \equiv h_0(x_0; x_0)$, $h_k(x_0; x_0) = 0$ for $k = 1, 2, \ldots$.

If $\{m_k\} = \{k : k = 1, 2, \ldots\}$ then $\{m_k\} \in \Omega_{1,0}$, and by Theorem 3.1 it follows that $h_k(x; x_0) = \alpha_k h_0(x; x_0)$ for all $k = 1, 2, \ldots$, and constants α_k. Since $h_0(x_0; x_0) = u(x_0) \neq 0$ and $h_k(x_0; x_0) = 0$ for all $k \geq 1$, we now have $\alpha_k = 0$ for all $k \geq 1$. Therefore $h_k(x; x_0) \equiv 0$ for $k \geq 1$ and $u(x) = h_0(x; x_0)$. ‖

Notes

Theorem 3.2 is a theorem of the Liouville type (of a rather strong form) for the whole class of functions of Laplacian order 1 and type 0. Friedman's result was first proved in [33], but he obtained additional theorems for solutions of the meta-harmonic equation which are not, in general, of order 1 and type 0. See the Notes at the end of §5 for further references.

4. Comparison with Gevrey classes

Let G be an open set in $\mathbb{R}^n$, and let $s > 1$ be fixed. The corresponding *Gevrey class* is the class of functions $u \in \mathscr{C}^\infty(G)$ which satisfy the following: for every compact $K \subset G$ there exist constants C and L (depending upon u and K) such that

$$\max_{x \in K} |D^\alpha u(x)| \leq C L^{|\alpha|} (|\alpha|!)^s$$

for every derivative D^α, $\alpha = (\alpha_1, \ldots, \alpha_n)$. These and more general classes are discussed by Lions and Magenes [65, Chapter 7]. Each of these classes contains non-analytic functions.

The construction of these classes can be generalized in the following way. Let A be a linear elliptic differential operator of order $2q$ whose coefficients are sufficiently regular in G. Consider the class of functions $u \in \mathscr{C}^\infty(G)$ which satisfy the following: for every compact $K \subset G$ there exist constants C and L (depending upon u and K) such that

$$\max_{x \in K} |A^m u(x)| \leq C L^{2m} (2qm)!^s.$$

It is proved by Lions and Magenes [65] that the same classes of functions are obtained with this definition as with the preceding; that is, the Gevrey classes can be defined using any such elliptic operator A. If $A = \Delta$, the above inequalities become

$$\max_{x \in K} |\Delta^m u(x)| \leq C L^{2m} (2m)!^s.$$

The inequalities defining the class of functions of Laplacian order ρ and type ∞ can therefore be viewed as extensions of these Gevrey inequalities by setting $s=1-1/\rho$ and letting s range from $-\infty$ to 1.

The question now arises as to whether the class of functions of Laplacian order ρ and type ∞ is independent of the elliptic operator used in the definition. To see that the answer is 'no', let A be a second-order linear homogeneous elliptic operator with constant coefficients and suppose that A is not a constant multiple of Δ. There is a non-singular linear transformation $x=L\xi$ of $\mathbb{R}^n$ which carries A (with the variable x) into Δ (with the variable ξ). The function $|\xi|^{2-n}$ is a fundamental solution for Δ (with $n>2$), and so $A\,|L^{-1}x|^{2-n}=0$. The role of the isotropic cone issuing from the origin is now played by

$$\left\{x\in\mathbb{C}^n : \sum_{j=1}^{n} \xi_j^2=0,\ \xi=(\xi_1,\ldots,\xi_n)=L^{-1}x\right\},$$

and L is not a scalar multiple of the identity. Hence the isotropic cones corresponding to A differ from those corresponding to Δ, and therefore the 'harmonicity' hulls are different. However, the harmonicity hull is precise, in the sense that all functions in the class can be continued into it and there are functions in the class which cannot be continued anywhere beyond it. This shows that the classes determined by the operators A and Δ are different. For the case $n=2$ the argument is clearly the same and so is the result.

5. Solutions of metaharmonic and related partial differential equations

Let $P(t)$ be a polynomial in the single variable t and form the differential operator $P(\Delta)$. The partial differential equation

$$P(\Delta)u=0$$

is called the *metaharmonic equation* and its solutions are called *metaharmonic functions.*

As pointed out in Ex. 1.2, Chapter II, if u is a non-trivial solution of $(\Delta-\lambda)u=0$, $\lambda\neq 0$, on an open set $D\subset\mathbb{R}^n$, then u is of least Laplacian order $\rho=1$ and least type $|\lambda|^{\frac{1}{2}}$ on D. We shall need the following generalization of this result.

Lemma 5.1. If $(\Delta-\lambda)u=w$ and w is of Laplacian order $\rho\geq 1$ and type τ $(1\leq\rho\leq\infty,\ 0\leq\tau\leq\infty)$ on an open set D, then u is of Laplacian order ρ and type $N=\max\{|\lambda|^{\frac{1}{2}},\tau\}$.

Proof. $(\Delta-\lambda)u=w$ implies that $\Delta u=\lambda u+w$. The successive Laplacians are

$$\Delta^p u=\lambda^p u+\sum_{k=0}^{p-1}\lambda^{p-1-k}\Delta^k w.$$

Suppose that $\tau<\infty$. Let K be a compact subset of D and let $\varepsilon>0$ be given. The function $|u|$ is bounded on K by a constant C_K, and from the assumptions on w it follows that

$$|\Delta^p u|\leqslant|\lambda|^p\, C_K+\sum_{k=0}^{p-1}|\lambda|^{p-1-k}\, C_{K,\varepsilon}(2k)!^{1-1/\rho}\left(\tau+\frac{\varepsilon}{2}\right)^{2k}$$

on K. Let $N=\max\{|\lambda|^{\frac{1}{2}},\tau\}$, set $s=1-1/\rho$, and note that $s\geqslant0$. If $s>0$ then

$$\begin{aligned}|\Delta^p u|&\leqslant N^{2p}C_K+N^{2(p-1)}C_{K,\varepsilon}\sum_{k=0}^{p-1}\left(\frac{N+\varepsilon/2}{N}\right)^{2k}(2k)!^s\\ &\leqslant C'_{K,\varepsilon}\left\{N^{2p}(2p)!^s+N^{2p-2}(2p-2)!^s\,p\left(\frac{N+\varepsilon/2}{N}\right)^{2p-2}\right\}\\ &=C'_{K,\varepsilon}\left\{N^{2p}(2p)!^s+(2p)!^s\frac{p}{[(2p-1)2p]^s}(N+\varepsilon/2)^{2p-2}\right\}\\ &\leqslant C''_{K,\varepsilon}(N+\varepsilon)^{2p}(2p)!^s\end{aligned}$$

so that u is of Laplacian order ρ and type N. If $\rho=1$, that is $s=0$, we have as above

$$\begin{aligned}|\Delta^p u|&\leqslant N^{2p}C_K+N^{2(p-1)}C_{K,\varepsilon/2}\sum_{k=0}^{p-1}\left(\frac{N+\varepsilon/2}{N}\right)^{2k}\\ &\leqslant C'_{K,\varepsilon/2}\left\{N^{2p}+(N+\varepsilon)^{2p}\left(\frac{N+\varepsilon/2}{N+\varepsilon}\right)^{2p}p\right\}\\ &\leqslant C''_{K,\varepsilon}(N+\varepsilon)^{2p}.\end{aligned}$$

The argument for $\tau=\infty$ is similar. ||

Notice that if $(\Delta-\lambda)u=w$ then $\Delta^p w=\Delta^{p+1}u-\lambda\Delta^p u$, so that the Laplacian order of w is less than or equal to the Laplacian order of u.

Corollary 5.1. If $(\Delta-\lambda)u=w$ *and* w *is of least Laplacian order* $\rho\geqslant1$, *then* u *is of least Laplacian order* ρ.

Let $P(t)=t^m+a_1t^{m-1}+\ldots+a_{m-1}t+a_m$ so that the corresponding metaharmonic equation is

$$P(\Delta)u=\Delta^m u+a_1\Delta^{m-1}u+\ldots+a_{m-1}\Delta u+a_m u=0.$$

Factor $P(t)$ over the complex numbers and write

$$P(t)=\prod_{k=1}^{m}(t-\lambda_k)$$

$$P(\Delta)u=\prod_{k=1}^{m}(\Delta-\lambda_k)u.$$

Theorem 5.1. If $P(\Delta)u=0$ then u is polyharmonic of order $\rho=1$ and type

$$\tau=\max_{1\leq k\leq m}\{|\lambda_k|^{\frac{1}{2}}\}.$$

Furthermore, there is always a solution having least order $\rho=1$ and least type

$$\tau=\max_{1\leq k\leq m}\{|\lambda_k|^{\frac{1}{2}}\}.$$

Proof. Let

$$v=\prod_{k=1}^{m-1}(\Delta-\lambda_k)u.$$

Since $(\Delta-\lambda_m)v=0$, by Lemma 5.1 v has order $\rho=1$ and type $|\lambda_m|^{\frac{1}{2}}$. However,

$$v=(\Delta-\lambda_{m-1})\left\{\prod_{k=1}^{m-2}(\Delta-\lambda_k)u\right\},$$

so that

$$\prod_{k=1}^{m-2}(\Delta-\lambda_k)u$$

has order $\rho=1$ and type $\max\{|\lambda_m|^{\frac{1}{2}}, |\lambda_{m-1}|^{\frac{1}{2}}\}$. Continuing in this way, we see that u has order $\rho=1$ and type

$$\tau=\max_{1\leq k\leq m}\{|\lambda_k|^{\frac{1}{2}}\}.$$

If

$$|\lambda_1|^{\frac{1}{2}}=\max_{1\leq k\leq m}\{|\lambda_k|^{\frac{1}{2}}\},$$

the function $u(x_1,\ldots, x_n)=\exp(x_1\lambda_1^{\frac{1}{2}})$ is a solution of $P(\Delta)u=0$ having least order $\rho=1$ and least type $|\lambda_1|^{\frac{1}{2}}$. ‖

Remark 5.1. The problem of determining the type of a solution u in Theorem 5.1 from the coefficients $a_1,\ldots, a_m$ of $P(t)$ is a problem of estimating zeros of a polynomial in terms of its coefficients.

Theorem 5.2. Let A be a linear partial differential operator with constant coefficients and order $q\leq 2m-1$. If $\Delta^m u+Au=0$ on an open set $D\subset\mathbb{R}^n$, then u is polyharmonic of order $\rho=2m/(2m-q)$ and some type $\tau\leq\infty$.

Proof. Every solution of $\Delta^m u+Au=0$ in D is analytic there (see Friedman [34, p. 205]). Therefore u can be continued as a single-valued analytic function $\tilde{u}$ on an open set $D'\subset\mathbb{C}^n$ such that $D'\cap\mathbb{R}^n=D$. Let K be a compact subset of D, let K_1 be a compact subset of $\mathbb{C}^n$ such that

$K \subset K_1^0 \subset K_1 \subset D'$, let d_0 be the minimum distance from points of K to points of ∂K_1, and let $d = \min\{1, d_0\}$. If α is a multi-index and v is any analytic function on D', the Cauchy bounds give

$$\|D^\alpha v\|_K \leqslant |\alpha|!\, d^{-|\alpha|}\, \|v\|_{K_1}. \tag{5.1}$$

Let

$$A = \sum_{|\alpha| \leqslant q} c_\alpha D^\alpha,$$

let

$$L_0 = \sum_{|\alpha| \leqslant q} |c_\alpha|,$$

and let $L = \max\{1, L_0\}$. If k is any positive integer and if v is analytic on D', from (5.1) we obtain

$$\|A^k v\|_K \leqslant n^{q/2} L^k (kq)!\, d^{-kq}\, \|v\|_{K_1}. \tag{5.2}$$

Suppose now that u satisfies the equation $\Delta^m u + Au = 0$ on D' and set

$$M = \max_{0 \leqslant l \leqslant m-1} \{\|\Delta^l u\|_{K_1}\}.$$

If p is any positive integer we can write $p = km + l$, $0 \leqslant l \leqslant m-1$, $k \geqslant 0$. Applying (5.2) we obtain

$$\|\Delta^p u\|_K = \|\Delta^{km+l} u\|_K = \|A^k(\Delta^l u)\|_K \leqslant n^{q/2} L^k (kq)!\, d^{-kq} M. \tag{5.3}$$

Since

$$kq = \frac{p-l}{m} q \leqslant 2p\left(\frac{q}{2m}\right),$$

the factor $(kq)!$ can be majorized by $C^{2p}(2p)!^{q/2m}$ with C independent of p, and the exponents k and kq can be majorized by $2p$. With these substitutions, inequality (5.3) shows that u is of Laplacian order $\rho = 2m/(2m-q)$ with type $\tau \leqslant \infty$ on D. ‖

The order $\rho = 2m/(2m-q)$ in Theorem 5.2 is, in general, the best possible result. The following example shows that there always exist solutions of certain equations of the form $\Delta^m u + Au = 0$ which are exactly of order $\rho = 2m/(2m-q)$.

Example 5.1. Let $(x, y) \in \mathbb{C}^2$ and make the change of variables $z_1 = x + iy$, $z_2 = x - iy$. In the new variables $\Delta = \partial^2/\partial z_1\, \partial z_2$. If A is linear with constant coefficients and order $q \leqslant 2m-1$, the equation $\Delta^m u - Au = 0$ becomes

$$\frac{\partial^{2m} u}{\partial z_1^m\, \partial z_2^m} - A_1 u = 0$$

in the variables (z_1, z_2) where A_1 is also linear with constant coefficients and order

q. Since every solution is analytic,

$$u(z_1, z_2) = \sum_{k,l \geq 0} a_{k,l} \frac{z_1^k}{k!} \frac{z_2^l}{l!}$$

where the condition

$$|a_{k,l}| < \frac{1}{R^{k+l}} k!\, l!$$

for some $R > 0$ is necessary and sufficient for local convergence of the series about (0, 0).

Let $A_1 = \partial^q/\partial z_2^q$. Substituting the series expansion for u in the equation we obtain the relations $a_{k,l} = a_{k-m,l+q-m}$. These relations are necessary and sufficient for u to satisfy

$$\frac{\partial^{2m}}{\partial z_1^m \partial z_2^m} u - \frac{\partial^q}{\partial z_2^q} u = 0.$$

Since these relations connect only the $a_{k,l}$ with k in the same residue class modulo m, we can assume that all the coefficients vanish except those of the form $a_{km,l}$, $k, l = 0, 1, 2, \ldots$. The relations now take the form

$$a_{km,l} = a_{(k-1)m,l+q-m}.$$

Let $j = q - m$ and note that the terms

$$\frac{z_1^{km}}{(km)!} \frac{z_2^l}{l!}$$

and

$$\frac{z_1^{(k-1)m}}{\{(k-1)m\}!} \frac{z_2^{l+j}}{(l+j)!}$$

have equal coefficients. To group all such terms together, set $s = l + kj$, define

$$f_s = \sum_{\substack{k \geq 0 \\ l \geq 0 \\ l = s - kj}} \frac{z_1^{km}}{(km)!} \frac{z_2^l}{l!},$$

and let $b_s = a_{km,l}$ with $s = l + kj$. The expansion of u becomes $u = \sum b_s f_s$, and the domain of the index s will be determined by the following two cases: (1) $-m < j < 0$, so that $-\infty < s < +\infty$, and (2) $0 \leq j < m$, so that $s \geq 0$. Define

$$b_s = \begin{Bmatrix} s! & \text{if} \quad s \geq 0 \\ 1/|s|! & \text{if} \quad s < 0 \end{Bmatrix}.$$

In case (1), if $s \geq 0$ then

$$a_{km,l} = s! = (l + kj)! \leq (l - kj)! \leq 2^{l-kj} l!\, (-kj)! \leq 2^{l+km} l!\, (km)!.$$

If $s < 0$,

$$a_{km,l} = \frac{1}{|s|!} = \frac{1}{|l + kj|!} \leq 1 \leq 2^{l+km} l!\, (km)!.$$

In case (2)

$$a_{km,l} = s! = (l+kj)! \leqslant 2^{l+km} l!\,(km)!.$$

Therefore in both cases the series expansion for u is convergent in a neighbourhood about $(0, 0)$.

Since

$$\Delta^{pm} = \frac{\partial^{2pm}}{\partial z_1^{pm}\, \partial z_2^{pm}},$$

it follows that $\Delta^{pm} f_s = f_{s-pq}$ where in case (2) $f_{s-pq} \equiv 0$ if $s - pq < 0$. Therefore in case (1)

$$\Delta^{pm} u = \sum_{s=-m}^{\infty} b_{s+pq} f_s,$$

and in case (2)

$$\Delta^{pm} u = \sum_{s=0}^{\infty} b_{s+pq} f_s.$$

We now show that in both cases the least Laplacian order of u is $2m/(2m-q)$ and with this order the type is strictly positive.

Case (1). If $s < 0$ and $s + pq \geqslant 0$, then

$$b_{s+pq} = (s+pq)! = \frac{(pq-|s|)!\,|s|!}{(pq)!} \frac{(pq)!}{|s|!} \geqslant \frac{(pq)!}{2^{pq}}\, b_s.$$

If $s < 0$ and $s + pq < 0$, then

$$b_{s+pq} = \frac{1}{|s+pq|!} = \frac{|s|!}{(|s|-pq)!\,(pq)!} \frac{(pq)!}{|s|!} \geqslant \frac{(pq)!}{2^{pq}}\, b_s.$$

If $s \geqslant 0$ then

$$b_{s+pq} = (s+pq)! \geqslant s!\,(pq)! = (pq)!\, b_s \geqslant \frac{(pq)!}{2^{pq}}\, b_s.$$

For $z_1, z_2 > 0$ we now obtain

$$|\Delta^{pm} u| \geqslant \frac{(pq)!}{2^{pq}} |u|.$$

From the definition of order and type, the desired conclusion follows by Stirling's formula.

Case (2). In this case $s \geqslant 0$, so that as above we obtain

$$b_{s+pq} \geqslant (pq)!\, b_s$$

and the desired conclusion follows.

Notes

The book by Vekua [90] contains a detailed study of the solutions of the meta-harmonic equation, including expansions of the solutions. Friedman

[33] discusses bounded solutions (see also §3), as does Svešnikov [83]. Ghermanescu [41, 42] and Nicolescu [72] obtain representations of the solutions in terms of means and prove theorems of the Harnack and Liouville type. These results are generalized in Chapter VII of the paper by Hervé [48].

6. Boundary value problems for polyharmonic functions

In the class of functions polyharmonic of a fixed finite degree p, boundary value problems in various domains $D \subset \mathbb{R}^n$ have been studied for a long time. We can think of the class of functions of Laplacian order ρ and type τ in a domain $D \subset \mathbb{R}^n$ as a class of functions satisfying a differential equation of infinite order, and boundary value problems can be posed for such a class if an infinite number of boundary conditions are prescribed. The boundary conditions we shall consider will be of the form $\check{B}_k u = g_k$ where the $\check{B}_k$ are boundary differential operators (of finite order) and the functions g_k are continuous on ∂D. Since the polyharmonic functions are analytic in D but, in general, not even continuous on $\bar{D}$, a meaning must be given to the boundary values $\check{B}_k u$.

A boundary value problem is said to be properly posed for a given class $\mathscr{F}$ of solutions and a given class $\mathscr{G}$ of boundary data if for every $u \in \mathscr{F}$ the boundary values $\check{B}_k u$ exist and $\{\check{B}_k u\} \in \mathscr{G}$, and if for every system of boundary values $\{g_k\} \in \mathscr{G}$ there exists a unique function $u \in \mathscr{F}$ with $\check{B}_k u = g_k$. If $\mathscr{F}$ is the class of analytic functions in D, that is the class $\mathscr{P}^{\infty,\infty}(D)$, such a boundary value problem cannot be properly posed. Indeed, uniqueness must always fail since there exist analytic functions in D which converge to zero, together with all derivatives, at ∂D. (Choose a sequence of points $\{\alpha_j\}$ dense in ∂D and set

$$f(x) = -\sum_j \varepsilon_j \, |x - \alpha_j|^{-1}.$$

If ε_j approaches zero rapidly enough, then $f(x)$ is analytic in D and $\exp\{f(x)\}$ converges to zero, together with all its derivatives, at ∂D.)

In this section we show that for certain systems of boundary operators $\{\check{B}_k\}$, classes of polyharmonic functions $\mathscr{F}$, and classes of data $\mathscr{G}$, the boundary value problem is properly posed. In order to establish the results we shall need some specific properties of Green's function of a boundary value problem of finite order. The first part of our discussion concerns elliptic boundary value problems of finite order, and in the second part we apply the results to boundary value problems in the class $\mathscr{P}^1$.

The domain $D \subset \mathbb{R}^n$ will be bounded with $\mathscr{C}^\infty$ boundary of dimension $n-1$. A boundary differential operator $\check{B}$ will be the restriction to ∂D of a differential operator B with coefficients in $\mathscr{C}^\infty(\bar{D})$. The operator B is called an *extension* of $\check{B}$. Extensions are not unique, but a unique representation of the boundary operator can be obtained in terms of normal and tangential derivatives on ∂D. To see this, note that every point x in a sufficiently small neighbourhood of ∂D can be written $x = x' + t\nu_{x'}$ where x' is the normal projection of x on ∂D, $\nu_{x'}$ is the unit interior normal at x', and t is the normal distance from x to ∂D. This establishes a $\mathscr{C}^\infty$ diffeomorphism between $\partial D \times \{t : |t| < \omega\}$ and a neighbourhood of ∂D, provided that $\omega > 0$ is sufficiently small. Using this diffeomorphism we can write

$$\check{B} = \sum_{i=0}^{k} \Gamma_i \frac{\partial^i}{\partial t^i}$$

where the Γ_i are uniquely determined differential operators with $\mathscr{C}^\infty$ coefficients on ∂D and Γ_k is not identically zero. The *normal order* of $\check{B}$ is k, as distinguished from its total order.

When the normal order equals the total order (in which case Γ_k is of order zero, that is a function) and $\Gamma_k \neq 0$ everywhere on ∂D, $\check{B}$ is called a *normal* boundary operator. A sequence $\{\check{B}_k\}$ is called a *normal sequence* of boundary operators if each $\check{B}_k$ is normal and if the orders are strictly increasing.

Since we shall consider functions u in the class $\mathscr{C}^\infty(D)$ but not necessarily in $\mathscr{C}^\infty(\bar{D})$, we must give a meaning to boundary values $\check{B}u$. To do this, let B be an extension of $\check{B}$. If $\check{B}u$ has a continuous extension to $\bar{D}$ then we say that *u has boundary value $\check{B}u$ relative to the extension B*. The restriction of this extension to ∂D is then, by definition, $\check{B}u$. It is clear that the existence and values of $\check{B}u$ generally depend on the extension B. It will always be assumed that the order of an extension does not exceed the total order of $\check{B}$ on ∂D.

The above definition of boundary value is equivalent to the following one. The boundary value $\check{B}u$, relative to an extension B, exists and equals $v(x')$, $x' \in \partial D$, if the functions $Bu(x' + t\nu_{x'})$, as functions of x', converge uniformly to $v(x')$ as $t \to 0$.

We are now ready to discuss an elliptic differential problem given by the differential system $\{\Delta^m, \{\check{B}_k\}_{k=1,\ldots,m}\}$ where m is a fixed integer not less than unity and $\{\check{B}_k\}$ is a (finite) normal sequence of boundary operators with orders not exceeding $2m-1$. Three assumptions will be made. The first one avoids some slight complications but is not essential to our later treatment of problems of infinite order. (For references concerning the three conditions given below and the three general statements which follow, see, for example, Agmon, Douglis and Nirenberg [1], Friedman [34], or Morrey [66].)

(1) *m is even and $m > n/2$. The coefficients of the $\check{B}_k$ are real.*

(2) *The $\check{B}_k$ satisfy the 'complementing condition' on ∂D relative to Δ^m.*

In our case condition (2) can be expressed as follows. If x is any point of ∂D, consider the characteristic polynomials $\check{B}_k^0(x, \xi)$ of the principal parts $\check{B}_k^0$ of $\check{B}_k$. For any tangent vector $\eta \neq 0$ at x we obtain a system of polynomials $\{\check{B}_k^0(x, \eta + \zeta\nu_x)\}$ in the complex variable ζ. The condition means that for any fixed x and η, no linear combination $\Sigma_k c_k \check{B}_k^0(x, \eta + \zeta\nu_x)$ with constant coefficients is divisible by $(\zeta - |\eta| i)^m$ unless all the c_k are zero.

(3) *The system* $\{\Delta^m, \{\check{B}_k\}\}$ *is self-adjoint and positive definite.*

Condition (3) means that, for every u and v in $\mathscr{C}^\infty(\bar{D})$ satisfying all the boundary conditions $\check{B}_k u = \check{B}_k v = 0$ on ∂D,

$$\int_D (\Delta^m u)\bar{v}\, dx = \int_D u(\Delta^m \bar{v})\, dx \tag{6.1}$$

and

(6.2)

there exists a constant $C > 0$ such that $\displaystyle\int_D (\Delta^m u)\bar{u}\, dx \geqslant C \int_D |u|^2\, dx.$

While there are no known algebraic criteria for the system to satisfy condition (3), it is easy to establish large classes of boundary operators $\check{B}_k$ for which conditions (2) and (3) hold.

The following facts pertaining to systems satisfying the above conditions are known.

(A) The boundary value problem $\Delta^m u = f$ in D, $\check{B}_k u = g_k$ on ∂D, $k = 1, 2, \ldots, m$, with $f \in \mathscr{C}^\infty(\bar{D})$ and $g_k \in \mathscr{C}^\infty(\partial D)$, has a unique solution $u \in \mathscr{C}^\infty(\bar{D})$.

(B) The system has a sequence $\{\psi_l\}$ of eigenfunctions corresponding to eigenvalues $\{\lambda_l\}$, $l = 1, 2, \ldots$, which satisfy $\Delta^m \psi_l = \lambda_l \psi_l$ in D, $\check{B}_k \psi_l = 0$ on ∂D, for all k and l. After normalization and suitable indexing, the sequence $\{\psi_l\}$ is orthonormal and complete in $\mathscr{L}^2(D)$ and the eigenvalues satisfy

$$0 < \lambda_1 \leqslant \lambda_2 \leqslant \ldots, \qquad \lim_{l \to \infty} \lambda_l = \infty.$$

Moreover, all ψ_l are in $\mathscr{C}^\infty(\bar{D})$. Since all the operators $\check{B}_k$ have real coefficients, we can assume that all the eigenfunctions are real.

(C) Let $\{A^{(j)}, \{\check{B}_k^{(j)}\}, k = 1, \ldots, m\}$ be a sequence of differential systems with each $A^{(j)}$ a differential operator of order $2m$ with coefficients in $\mathscr{C}^\infty(\bar{D})$, and each $\check{B}_k^{(j)}$ of the same order as $\check{B}_k$. Suppose that the coefficients of $A^{(j)}$ and $\check{B}_k^{(j)}$ converge in the Fréchet topology of $\mathscr{C}^\infty(\bar{D})$ to the coefficients of Δ^m and $\check{B}_k$, and that $\{f^{(j)}\}$ and $\{g_k^{(j)}\}$ are sequences of functions in $\mathscr{C}^\infty(\bar{D})$ and $\mathscr{C}^\infty(\partial D)$ respectively which converge in the Fréchet topologies of $\mathscr{C}^\infty(\bar{D})$ and $\mathscr{C}^\infty(\partial D)$ respectively to functions f and g_k. For all j sufficiently large the boundary value problems $A^{(j)}u^{(j)}=f^{(j)}$ in D, $\check{B}_k^{(j)}u^{(j)}=g_k^{(j)}$ on ∂D have unique solutions $u^{(j)}$ in $\mathscr{C}^\infty(\bar{D})$, and these solutions converge uniformly, together with all their derivatives, to the solution of the problem $\Delta^m u=f$ in D, $\check{B}_k u=g_k$ on ∂D.

The systems $\{A^{(j)},\{\check{B}_k^{(j)}\}\}$ are not assumed to satisfy all the conditions imposed upon $\{\Delta^m,\{\check{B}_k\}\}$, but for j sufficiently large each $A^{(j)}$ is elliptic and the complementing condition holds.

Next we define Green's function $G(x, y)$ for the system $\{\Delta^m,\{\check{B}_k\}\}$ in D. Let $F(x, y)$ be a fundamental solution of Δ^m in $\mathbb{R}^n$, for instance $F(x, y)=\imath_m(x-y)$ as in §2, Chapter I. For each $y\in D$ let $v_y(x)$ be the solution of the boundary value problem $\Delta^m v_y=0$ in D, $\check{B}_k v_y(x)=\check{B}_k(x)F(x, y)$ on ∂D. (The notation $B(x)$ means that the differential operator is applied to the variable x.) Define $G(x, y)=F(x, y)-v_y(x)$. It is known that $G(x, y)=G(y, x)$ for $(x, y)\in D$ and, by statement (A), $G(x, y)$ is of class $\mathscr{C}^\infty$ on $(\bar{D}\times D)\cup(D\times\bar{D})$ except at points $x=y$ where it has the same singularity as the fundamental solution. Because Green's function is a reproducing kernel (Aronszajn [9]), we obtain the formula

$$G(x, y)=\sum_{l=1}^{\infty}\lambda_l^{-1}\psi_l(x)\psi_l(y). \tag{6.3}$$

Before discussing additional boundary value problems associated with the system $\{\Delta^m,\{\check{B}_k\}\}$, we need further information about the boundary operators (see Aronszajn and Milgram [13]). To the system $\{\Delta^m,\{\check{B}_k\}\}$ there corresponds a system of linear differential boundary operators $\check{\Phi}_k$, $k=1, 2,\dots, m$, not unique, with (order $\check{\Phi}_k$) $=(2m-1)-$(order $\check{B}_k$) such that for all u and v in $\mathscr{C}^\infty(\bar{D})$

$$\int_D(\Delta^m u)\bar{v}\,dx=\int_D u(\Delta^m\bar{v})\,dx+\int_{\partial D}\sum_{k=1}^{m}\{\check{\Phi}_k u(x)\overline{\check{B}_k v(x)}-\check{B}_k u(x)\overline{\check{\Phi}_k v(x)}\}\,d\sigma(x) \tag{6.4}$$

where $d\sigma$ is the $(n-1)$-dimensional area measure on ∂D. Such a system $\{\check{\Phi}_k\}$ will be called a *complementary system.* Since the coefficients of the $\check{B}_k$ are real we can assume that the same is true of the coefficients of the $\check{\Phi}_k$. By a standard argument (§2, Chapter I, for example), for all $v\in\mathscr{C}^\infty(\bar{D})$ and all $y\in D$ we have

$$v(y)=\int_D G(x, y)\Delta^m v(x)\,dx+\int_{\partial D}\sum_{k=1}^{m}\{\check{\Phi}_k(x)G(x, y)\}\check{B}_k v(x)\,d\sigma(x). \tag{6.5}$$

Formula (6.5) shows that, even though the operators $\check{\Phi}_k$ are not uniquely determined, the functions $\check{\Phi}_k(x)G(x, y)$, $x\in\partial D$ and $y\in D$, are independent of the choice of $\{\check{\Phi}_k\}$.

The next lemma shows that (6.5) is valid for v in a much larger class than $\mathscr{C}^\infty(\bar{D})$.

Lemma 6.1. If B_k, $k=1,2,\ldots,m$, are extensions of the boundary operators $\check{B}_k$, and if v is a function in $\mathscr{C}^\infty(D)$ such that $\Delta^m v \in \mathscr{L}^1(D)$ and such that the boundary values $\check{B}_k v$ exist relative to B_k, then (6.5) *holds.*

Proof. We begin with a preliminary construction. Let $\omega > 0$ be chosen so that $\partial D \times \{t : |t| < \omega\}$ is diffeomorphic to a neighbourhood of ∂D, and let $\phi \in \mathscr{C}^\infty(\mathbb{R}^1)$ be a function which is non-increasing and such that $\phi(t) = 1$ for $t \leq 0$, $\phi(t) = 0$ for $t \geq \omega/3$. Define mappings T_δ from $\mathbb{R}^1$ onto $\mathbb{R}^1$ by $t' = T_\delta(t) = t - \delta\phi(t-\delta)$. There is a number $\delta_0 < 0$ such that for $\delta > \delta_0$ the mapping T_δ is a $\mathscr{C}^\infty$ diffeomorphism. Note that T_δ is the identity when $\delta = 0$, and for $0 \leq \delta \leq \omega/3$ the mapping T_δ transforms the half-line $t \geq \delta$ onto the half-line $t' \geq 0$ and keeps the points $t \geq 2\omega/3$ invariant.

For each $\delta \in [0, \omega]$ denote by D_δ the open set $\{x \in D : d(x, \partial D) > \delta\}$. For each $\delta \in [0, \omega/3]$ define a diffeomorphism $\tilde{T}_\delta$ of $\bar{D}_\delta$ onto $\bar{D}$ by

$$\begin{aligned} \tilde{T}_\delta x &= x \qquad &&\text{for} \quad x \in \bar{D}_\omega \\ \tilde{T}_\delta x &= x' + T_\delta(t)\nu_{x'} \qquad &&\text{for} \quad x = x' + t\nu_{x'},\ \delta \leq t < \omega. \end{aligned}$$

The mapping $\tilde{T}_\delta$ is actually defined by the second part of the formula for $t \in [0, \delta)$, so that it can be considered to be defined on all of $\bar{D}$. As δ approaches zero, $\tilde{T}_\delta$ converges in the Fréchet topology of $\mathscr{C}^\infty(\bar{D})$ to the identity on $\bar{D}$.

Next we extend formula (6.4) to a more general case (Aronszajn and Milgram [13]). We consider a differential system $\{A, \{\check{C}_k\}\}$ in a domain D with compact closure in a $\mathscr{C}^\infty$ manifold $\mathscr{M}$ of dimension n, with ∂D a $\mathscr{C}^\infty$ submanifold of dimension $n-1$. All the coefficients of all the operators A and $\check{C}_k$ are assumed to be real $\mathscr{C}^\infty$ functions, A is assumed to be elliptic of order $2m$, and $\{\check{C}_k\}$ is a normal system of m boundary operators with orders no greater than $2m-1$. On the manifold $\mathscr{M}$ we take a Riemannian metric and assume that it and the operators A and $\check{C}_k$ have $\mathscr{C}^\infty$ dependence on a real parameter δ. Denote by $\mathrm{d}x$ and $\mathrm{d}\sigma$ the measures on D and ∂D, respectively, induced by the metric, and denote by A^* the adjoint of A. There exist systems of boundary operators $\{\check{C}'_k\}$, $\{\check{\Psi}_k\}$, and $\{\check{\Psi}'_k\}$ such that for every $u, v \in \mathscr{C}^\infty(\bar{D})$

$$(6.6) \quad \int_D (Au)\bar{v}\,\mathrm{d}x = \int_D u(A^*\bar{v})\,\mathrm{d}x + \int_{\partial D} \sum_{k=1}^m \{\check{\Psi}'_k u(x)\check{C}'_k\bar{v}(x) - \check{C}_k u(x)\check{\Psi}\bar{v}(x)\}\,\mathrm{d}\sigma(x).$$

The system $\{\check{C}'_k\}$ is called an *adjoint system* to $\{\check{C}_k\}$, and $\{\check{\Psi}_k\}$ and $\{\check{\Psi}'_k\}$ are corresponding complementary systems. The systems $\{\check{C}'_k\}$, $\{\check{\Psi}_k\}$, and $\{\check{\Psi}'_k\}$

are not unique, but they can be chosen to have real $\mathscr{C}^\infty$ coefficients and to have $\mathscr{C}^\infty$ dependence on the parameter δ with any suitable choice of the systems for $\delta = 0$.

Now consider the region D_δ and the differential system $\{\Delta^m, \{\check{B}_{k,\delta}\}\}$ where $\check{B}_{k,\delta} = B_k|_{\partial D_\delta}$. By using the diffeomorphism $\tilde{T}_\delta$ we transform this system to $\{A_\delta, \{\check{C}_{k,\delta}\}\}$ and transform the Euclidean metric into a Riemannian metric on a neighbourhood of $\bar{D}$. (This neighbourhood is the manifold $\mathscr{M}$.) The operator A_δ is self-adjoint and we obtain corresponding systems $\{\check{C}'_{k,\delta}\}$, $\{\check{\Psi}_{k,\delta}\}$, and $\{\check{\Psi}'_{k,\delta}\}$ as described above. For $\delta = 0$ we have $A_\delta = \Delta^m$, $\check{C}_{k,\delta} = \check{B}_k$, $\check{C}'_{k,\delta} = \check{B}_k$, $\check{\Psi}_{k,\delta} = \check{\Phi}_k$, and $\check{\Psi}'_{k,\delta} = \check{\Phi}_k$. By statement (C), for all sufficiently small δ there exist unique $\mathscr{C}^\infty$ solutions $\tilde{v}_{y,\delta}(x)$ of the boundary value problem

$$A_\delta \tilde{v}_{y,\delta} = 0 \quad \text{in} \quad D, \qquad \check{C}'_{k,\delta}\tilde{v}_{y,\delta}(x) = F(\tilde{T}_\delta^{-1}x, y) \quad \text{on} \quad \partial D,$$

the solutions $\tilde{v}_{y,\delta}$ converging in the Fréchet topology of $\mathscr{C}^\infty(\bar{D})$ to the solution v_y which determines Green's function $G(x, y)$.

We now transform back by $\tilde{T}_\delta^{-1}$ and return to $\{\Delta^m, \{\check{B}_{k,\delta}\}\}$ with corresponding adjoint system $\{\check{B}'_{k,\delta}\}$ and complementary systems $\{\check{\Phi}_{k,\delta}\}$ and $\{\check{\Phi}'_{k,\delta}\}$. The function $\tilde{v}_{y,\delta}$ becomes $v_{y,\delta}$ satisfying $\Delta^m v_{y,\delta} = 0$ in D_δ and $\check{B}'_{k,\delta}v_{y,\delta}(x) = F(x, y)$ on ∂D. Setting $G_\delta(x, y) = F(x, y) - v_{y,\delta}$ we obtain a fundamental solution for Δ^m, and by the usual procedure, using (6.6) with $v = G_\delta(x, y)$ and assuming δ is small enough so that $y \in D_\delta$, we obtain

$$u(y) = \int_{D_\delta} \Delta^m u(x) G_\delta(x, y)\,dx + \int_{\partial D} \sum_{k=1}^{m} \{\check{\Phi}_{k,\delta}(x) G_\delta(x, y)\} \check{B}_{k,\delta} u(x)\,d\sigma(x).$$

Writing $x = x' + \delta\nu_{x'}$, $x \in \partial D_\delta$, as $\delta \to 0$ we obtain

$$G_\delta(x, y) \to G(x', y),$$

$$\check{\Phi}_{k,\delta}(x) G_\delta(x, y) \to \check{\Phi}_k(x') G(x', y),$$

and

$$\check{B}_{k,\delta} u(x) = B_k u(x)|_{\partial D_\delta} \to \check{B}_k u(x'),$$

each convergence being uniform. Since $\Delta^m u \in \mathscr{L}^1(D)$, the limit gives the desired formula (6.5). ‖

Corollary 6.1 (Uniqueness for $\{\Delta^m, \{\check{B}_k\}\}$). If u' and u'' are solutions of the boundary value problem

$$\Delta^m u = f \quad \text{in} \quad D, \qquad \check{B}_k u = g_k \quad \text{on} \quad \partial D, \qquad k = 1, 2, \ldots, m,$$

with $f \in \mathscr{C}^\infty(D) \cap \mathscr{L}^1(D)$, $g_k \in \mathscr{C}^0(\partial D)$, the boundary values $\check{B}_k u'$ and $\check{B}_k u''$ being taken relative to extensions $\{B'_k\}$ and $\{B''_k\}$ of $\{\check{B}_k\}$, then $u' = u''$.

Proof. By Lemma 6.1, both solutions are represented by the formula

$$\int_D f(x) G(x, y)\,dx + \int_{\partial D} \sum_{k=1}^{m} g_k(x) \{\check{\Phi}_k(x) G(x, y)\}\,d\sigma(x). \quad \|$$

We now turn to the question of existence for the system $\{\Delta^m, \{\check{B}_k\}\}$. The problem which will be considered is specific to our later purposes.

Existence problem. Find a function $u \in \mathscr{C}^\infty(D) \cap \mathscr{L}^\infty(D)$ which satisfies $\Delta^m u = f$ in D and $\check{B}_k u = g_k$ on ∂D where $f \in \mathscr{C}^\infty(D) \cap \mathscr{C}^0(\bar{D})$, $g_k \in \mathscr{C}^0(\partial D)$, and the boundary values $\check{B}_k u$ exist relative to given extensions B_k of $\check{B}_k$.

From the formulation of the problem, the existence of a solution might depend upon the given extensions of the boundary operators. The following example shows that in fact it does.

Example 6.1. Let D be the unit disc in $\mathbb{R}^2$ and consider the system $\{\Delta^2, \{I|_{\partial D}, \Delta|_{\partial D}\}\}$. If we use the extensions I and Δ, there always exists a solution of the boundary value problem. Indeed, extend f to be zero outside $\bar{D}$ and define

$$w(y) = \int_{\mathbb{R}^2} f(x)(-1/2\pi)\log|x-y|\,dx.$$

The function w is in $\mathscr{C}^0(\mathbb{R}^2)$, in $\mathscr{C}^\infty(D)$, and satisfies $\Delta w = f$ in D. Let u_1 and u_2 be the solutions of the Dirichlet problems $\Delta u_1 = 0$ in D, $u_1|_{\partial D} = g_1$, and $\Delta u_2 = 0$ in D, $u_2|_{\partial D} = g_2 - w|_{\partial D}$. The desired solution u is

$$u(y) = u_1(y) + \int_D \{w(x) + u_2(x)\} G_0(x, y)\,dx$$

where $G_0(x, y)$ is the usual Green function for D. However, suppose we use the extensions I and $\Delta + r^2(1-r^2)(\partial^2/\partial r^2)$ of the boundary operators. If the given data are $f = 0$, g_1 arbitrary, and $g_2 = 0$, then with the previous extensions the solution would be $u(y) = u_1(y)$. By uniqueness, Corollary 6.1, the solution for the new extensions would also be u_1. However, we can choose a continuous function g_1 such that $(1-r^2)(\partial^2 u_1/\partial r^2)$ is unbounded in D. Thus there is no solution.

We call a system $\{B_k\}$ of extensions *distinguished* if the existence problem is solvable for any choice of data.

Open question. Does there exist a distinguished system of extensions for every system $\{\Delta^m, \{\check{B}_k\}\}$?

Next we shall express the distinguished character of the extensions $\{B_k\}$ in terms of the Green function $G(x, y)$ of the system $\{\Delta^m, \{\check{B}_k\}\}$. First we define a system of properties (P-0),..., (P-7). Our differential system is assumed to satisfy the conditions (1), (2), and (3) stated before, and $\{\check{\Phi}_k\}$ denotes a fixed complementary system.

Properties P. There exists a constant c depending only on the domain D and the system $\{\Delta^m, \{\check{B}_k\}\}$ such that

(P-0) $|G(x, y)| \le c$ *for* $x, y \in D$.

(P-1) $$\begin{cases} \displaystyle\int_D |G(x, y)|\,dx \le c \quad \textit{for} \quad y \in D \\ \displaystyle\int_{\partial D} |\check{\Phi}_k(x) G(x, y)|\,d\sigma(x) \le c \quad \textit{for} \quad y \in D, \qquad k = 1, \dots, m \\ \textit{except for } k = 1 \textit{ in the case when } \check{B}_1 \textit{ is of order } 0. \end{cases}$$

(P-2) *For every* $v \in \mathscr{C}^\infty(D) \cap \mathscr{C}^0(\bar{D})$, *the integral* $\int_D v(x)G(x, y)\,dx$ *is in* $\mathscr{C}^0(\bar{D})$.

(P-3) *For every* $g \in \mathscr{C}^0(\partial D)$, *and* $k = 1, \ldots, m$ *(except* $k = 1$ *if* $\check{B}_1$ *is of order* 0), *the integral* $\int_{\partial D} g(x)\{\check{\Phi}_k(x)G(x, y)\}\,d\sigma(x)$ *is in* $\mathscr{C}^0(\bar{D})$.

(P-4) $$\begin{cases} \int_D |B_l(y)G(x, y)|\,dx \le c \quad \textit{for} \quad y \in D, \qquad l = 1, 2, \ldots, m, \\ \int_{\partial D} |\check{B}_l(y)\check{\Phi}_k(x)G(x, y)|\,d\sigma(x) \le c \quad \textit{for} \quad k, l = 1, \ldots, m, \\ \textit{and all } y \in D. \end{cases}$$

(P-5) *For* $v \in \mathscr{C}^\infty(D) \cap \mathscr{C}^0(\bar{D})$ *and* $l = 1, \ldots, m$, *the integral* $\int_D v(x)\{B_l(y)G(x, y)\}\,dx$ *is in* $\mathscr{C}^0(\bar{D})$ *and vanishes on* ∂D.

(P-6) *For* $g \in \mathscr{C}^0(\partial D)$ *and* $k \ne l$, *the integral* $\int_{\partial D} g(x)\{\check{B}_l(y)\check{\Phi}_k(x)G(x, y)\}\,d\sigma(x)$ *is in* $\mathscr{C}^0(\bar{D})$ *and vanishes on* ∂D.

(P-7) *For* $g \in \mathscr{C}^0(\partial D)$ *and every* k, *the integral* $\int_{\partial D} g(x)\{\check{B}_k(y)\check{\Phi}_k(x)G(x, y)\}\,d\sigma(x)$ *is in* $\mathscr{C}^0(\bar{D})$ *and takes the values* $g(y)$ *on* ∂D.

Remark 6.1. There are relations between the different properties, the obvious one being that (P-0) implies the first part of (P-1). Other relations will be dealt with in Lemma 6.2. In the second part of property (P-1) we excluded the case $k = 1$ with $\check{B}_1$ of order 0. In this case $\check{B}_1$ is the operation of multiplication by a $\mathscr{C}^\infty$ function non-vanishing on ∂D, but if this case arises all inequalities of property (P-1) are actually included in property (P-4) for $l = 1$.

Krasovskiĭ [58] gives estimates for Green's function and its derivatives, including the following.

Krasovskiĭ's inequalities

(a) $|G(x, y)| \le M(|x - y|^{2m-n} + M_0 + 1)$ for $2m - n \ne 0$ and $|G(x, y)| \le M(|\log|x - y|| + M_0 + 1)$ for $2m - n = 0$, with M and M_0 positive constants, $x, y \in D$.

(b) If m_k is the order of $\check{B}_k$ then $|\check{\Phi}_k(x)G(x, y)| \le M(|x - y)|^{m_k - n + 1} + M_0 + 1)$ for $m_k - n + 1 \ne 0$ and $|\check{\Phi}_k(x)G(x, y)| \le M(|\log|x - y|| + M_0 + 1)$ for $m_k - n + 1 = 0$, with M and M_0 positive constants, $x \in \partial D$, $y \in D$.

Using these inequalities we immediately obtain the following.

Proposition 6.1. Property (P-0) holds if and only if $m > n/2$. Property (P-1) holds always.†

Lemma 6.2. The properties (P-2) and (P-3) follow from (P-1). Properties (P-5), (P-6), and (P-7) follow from property (P-4).

Proof. Take $v \in \mathscr{C}^\infty(D) \cap \mathscr{C}^0(\bar{D})$ and $g \in \mathscr{C}^0(\partial D)$. There exists a sequence of functions $v^{(j)} \in \mathscr{C}^\infty(\bar{D})$ and a sequence of functions $g^{(j)} \in \mathscr{C}^\infty(\partial D)$ such that $v^{(j)}$ converges uniformly in $\bar{D}$ to v and $g^{(j)}$ converges uniformly in ∂D to g. Define

$$u_0^{(j)}(y) = \int_D v^{(j)}(x) G(x, y)\, dx$$

$$u_k^{(j)}(y) = \int_{\partial D} g^{(j)}(x) \{\check{\Phi}_k(x) G(x, y)\}\, d\sigma(x), \qquad k = 1, \ldots, m.$$

By formula (6.5), $u_0^{(j)}$ is the solution in $\mathscr{C}^\infty(\bar{D})$ of the problem $\Delta^m u = v^{(j)}$ in D, $\check{B}_k u = 0$ on ∂D, $k = 1, \ldots, m$. Further, $u_k^{(j)}$ is a solution in $\mathscr{C}^\infty(\bar{D})$ of the problem $\Delta^m u = 0$ in D, $\check{B}_l u = 0$ on ∂D for $l \neq k$, $\check{B}_k u = g^{(j)}$ on ∂D. If any integral in (P-1) or (P-4) is uniformly bounded in D, then the corresponding functions $B_l(y) u_k^{(j)}(y)$ converge uniformly in D. For instance, if

$$\int |B_l(y) \check{\Phi}_k(x) G(x, y)|\, d\sigma(x) \leq c,$$

then the functions converge uniformly to

$$\int g(x) \{B_l(y) \check{\Phi}_k(x) G(x, y)\}\, d\sigma(x).$$

Since the functions $B_l u_k^{(j)}$ are continuous in $\bar{D}$, the limit function is also continuous in $\bar{D}$ and takes on ∂D the value zero or $g(x)$ depending on whether $l \neq k$ or $l = k$. This finishes the proof of the lemma. ‖

Lemma 6.3. A system of extensions B_k of the boundary operators $\check{B}_k$ is distinguished if and only if the property (P-4) holds.

Proof. Consider the class of functions $\mathscr{F}$ defined by $u \in \mathscr{F}$ if and only if $u \in \mathscr{C}^\infty(D) \cap \mathscr{L}^\infty(D)$, $\Delta^m u \in \mathscr{C}^\infty(D) \cap \mathscr{C}^0(\bar{D})$, and $B_k u \in \mathscr{C}^0(\bar{D})$, $k = 1, \ldots, m$. $\mathscr{F}$ is a Fréchet space with Fréchet metric defined by adding to the Fréchet metric of $\mathscr{C}^\infty(D)$ the norms

$$\|u\|_{\mathscr{L}^\infty} + \|\Delta^m u\|_{\mathscr{C}^0} + \sum_{k=1}^{m} \|B_k u\|_{\mathscr{C}^0}.$$

† The reason for making an exception in property (P-1) is that Krasovskii's inequalities do not imply boundedness of the corresponding integral.

Let $\mathcal{K}$ be the Fréchet space composed of systems $\{v, \{g_k\}\}$ where $v \in \mathscr{C}^\infty(D) \cap \mathscr{C}^0(\bar{D})$ and $g_k \in \mathscr{C}^0(\partial D)$, $k = 1, \ldots, m$. Define the mapping $T: \mathcal{F} \to \mathcal{K}$ by setting $Tu = \{\Delta^m u, \{B_k u|_{\partial D}\}\}$. This mapping is continuous, and by Lemma 6.1 it is one-to-one.

Clearly the distinguished character of the extensions B_k is equivalent to the fact that T maps $\mathcal{F}$ onto $\mathcal{K}$.

If property (P-4) holds, by Lemma 6.2 and Proposition 6.1 all the properties (P-1),..., (P-7) hold. It follows then that for any given data $\{v, \{g_k\}\} \in \mathcal{K}$ the function

$$u(y) = \int_D v(x)G(x, y)\,dx + \sum_{k=1}^{m} \int_{\partial D} g_k(x)\{\check{\Phi}_k(x)G(x, y)\}\,d\sigma(x) \tag{6.7}$$

is a corresponding solution in $\mathcal{F}$.

Conversely, if the mapping T is onto $\mathcal{K}$, by the closed graph theorem the inverse mapping $T^{-1}: \mathcal{K} \to \mathcal{F}$ is continuous. By Lemma 6.1 the inverse mapping $T^{-1}\{v, \{g_k\}\} = u(y)$ is given by formula (6.7), from which we obtain

$$\begin{aligned} B_l u(y) = {} & \int_D v(x)\{B_l(y)G(x, y)\}\,dx \\ & + \sum_{k=1}^{m} \int_{\partial D} g_k(x)\{B_l(y)\check{\Phi}_k(x)G(x, y)\}\,d\sigma(x). \end{aligned} \tag{6.8}$$

By the continuity of T^{-1} we see that

$$v \to \int_D v(x)\{B_l(y)G(x, y)\}\,dx$$

is a continuous mapping of $\mathscr{C}^\infty(D) \cap \mathscr{C}^0(\bar{D})$ into $\mathscr{C}^0(\bar{D})$. Hence, by the uniform boundedness principle and a standard argument,

$$\int_D |B_l(y)G(x, y)|\,dx \in \mathscr{L}^\infty(D).$$

Furthermore,

$$g \to \int_{\partial D} g(x)\{B_l(y)\Phi_k(x)G(x, y)\}\,d\sigma(x)$$

is a continuous mapping of $\mathscr{C}^0(\partial D)$ into $\mathscr{C}^0(\bar{D})$, so in the same way as above

$$\int_{\partial D} |\check{B}_l(y)\Phi_k(x)G(x, y)|\,d\sigma(x) \in \mathscr{L}^\infty(D),$$

which finishes the proof of (P-4). ||

Lemma 6.3 allows us to prove an existence theorem that is more general than the existence problem stated above.

Corollary 6.2 (*Existence Theorem*). *If* (*P*-4) *is satisfied and* $v \in \mathscr{C}^\infty(D)$ *is such that there exists a function* $u_0 \in \mathscr{C}^\infty(D) \cap \mathscr{C}^{2m-1}(\bar{D})$ *with* $\Delta^m u_0 = v$ *in* D, *then the boundary value problem* $\Delta^m u = v$, $B_k u|_{\partial D} = g_k$, $k = 1, \ldots, m$ *has a solution for any given* g_k *in* $\mathscr{C}^0(\partial D)$. *In fact this is true when* $v \in \mathscr{C}^\infty(D) \cap \mathscr{L}^p(D)$ *for some* p, $n < p \leq \infty$.

Proof. By Lemma 6.3 there is a solution u' for the data $v' = 0$ and $g'_k = g_k - B_k u_0|_{\partial D}$. The solution u of the problem is given by $u = u_0 + u'$. If $v \in \mathscr{C}^\infty(D) \cap \mathscr{L}^p(D)$, $n < p \leq \infty$, we extend v to be zero outside of D and form the convolution with the fundamental solution $\mathscr{e}_m$ corresponding to Δ^m. The resulting function u_0, when restricted to D, has the desired properties since for every fixed y the derivatives of $F(x, y) = \mathscr{e}_m(x - y)$ of order up through $2m - 1$ are in $\mathscr{L}^q_{\text{loc}}(\mathbb{R}^n)$ for $1 \leq q < n/(n-1)$. ∥

Remark 6.2. We did not need property (P-0) in the above considerations. It follows that our results about uniqueness and existence for the system $\{\Delta^m, \{\check{B}_k\}\}$ are valid without the restriction $m > n/2$, which means that condition (1) is not needed for these results. However, property (P-0) will be essential when we consider boundary value problems of infinite order.

Remark 6.3. The operator Δ^m in D is obviously chosen because we wish to consider boundary value problems for polyharmonic functions. However, our preceding developments concerning uniqueness and existence for a system can be automatically extended to systems $\{A, \{\check{B}_k\}\}$ with an arbitrary elliptic operator A with coefficients in $\mathscr{C}^\infty(\bar{D})$ satisfying (2) and (3).

We now introduce the iterated systems

$$\left\{\Delta^{mp}, \{\check{B}_k^{(i)}|_{\partial D}\}_{\substack{k=1,\ldots,m \\ i=0,\ldots,p-1}}\right\}$$

for each positive integer p, where $\check{B}_k^{(i)} = B_k \Delta^{mi}|_{\partial D}$ for fixed extensions B_k of boundary operators $\check{B}_k$. (Clearly the boundary operators $\check{B}_k^{(i)}$ do not depend on the choice of the extensions B_k.)† For $p = 1$ we obtain the original system $\{\Delta^m, \{\check{B}_k\}\}$.

It is immediately checked that if the original system satisfies (1), (2), and (3), the same is true for the pth iterated system. We can therefore define the corresponding Green function $G_p(x, y)$, with $G_1(x, y) = G(x, y)$, and speak about the properties (P-0), (P-1), and (P-4). In (P-4) we shall consider the extensions of the boundary operators $\check{B}_k^{(i)}$ given by $B_k \Delta^{mi}$.

† The boundary operators $\check{B}_k^{(i)}$ form a normal system if ordered by their orders, that is $\check{B}_k^{(i)}$ precedes $\check{B}_{k'}^{(i')}$ if $i < i'$ or $i = i'$ and $k < k'$.

The following facts are well known and immediately verified.

(6.9) $$G_{p+q}(x, y)=\int_D G_p(x, z)G_q(z, y)\,dz \quad \textit{for positive integers } p \textit{ and } q$$

(6.10) $$\Delta^{qm}(x)G_p(x, y)=\Delta^{qm}(y)G_p(x, y)=G_{p-q}(x, y)$$
$$\textit{for}\quad q<p,\ (x, y)\in(D\times\bar{D})\cup(\bar{D}\times D)$$

(6.10′) $$\Delta^{qm}(y)G_p(x, y)=0 \quad \textit{for}\quad q\geq p,\ x\neq y.$$

The eigenfunctions ψ_l of our original system are eigenfunctions of the pth iterated system with corresponding eigenvalues λ_l^p.
Thus

(6.11) $$G_p(x, y)=\sum_{l=1}^{\infty}\lambda_l^{-p}\psi_l(x)\psi_l(y).$$

In the pth iterated system a complementary system to $\{\check{B}_k^{(i)}\}$ is given by $\{\check{\Phi}_k^{(i)}\}$ with

(6.12) $$\check{\Phi}_k^{(i)}=\check{\Phi}_k\Delta^{m(p-i-1)}|_{\partial D}.\dagger$$

Formulae (6.10) and (6.12) give

(6.13) $$\check{\Phi}_k^{(i)}G_p(x, y)=\check{\Phi}_k(x)G_{i+1}(x, y).$$

Also,

(6.14) $$\begin{cases} B_l\Delta^{mj}(y)G_p(x, y)=B_l(y)G_{p-j}(x, y) \quad \text{for}\quad 0\leq j\leq p-1, \\ \qquad\qquad\qquad\qquad\qquad\qquad\qquad\qquad y\in D,\ x\neq y \\ B_l\Delta^{mj}(y)\check{\Phi}_k^{(i)}(x)G_p(x, y)=B_l(y)\check{\Phi}_k(x)G_{i-j+1}(x, y) \\ \qquad\qquad\qquad\qquad\qquad \text{for}\quad i-j\geq 0, \qquad x\in\partial D,\ y\in D. \end{cases}$$

By an application of Lemma 6.1 to the system $\{\Delta^{mp}, \{\check{B}_k^{(i)}\}\}$ we obtain the following.

Lemma 6.4. If $u\in\mathscr{C}^\infty(D)$, $\Delta^{mp}u\in\mathscr{L}^1(D)$, *and for some extensions* $B_k^{(i)}$ *of* $\check{B}_k^{(i)}$ *the boundary values* $g_k^{(i)}=\check{B}_k^{(i)}u$ *exist, then*

(6.15) $$u(y)=\int_D \Delta^{mp}u(x)G_p(x, y)\,dx$$
$$+\sum_{i=0}^{p-1}\sum_{k=1}^{m}\int_{\partial D} g_k^{(i)}(x)\{\check{\Phi}_k(x)G_{i+1}(x, y)\}\,d\sigma(x)$$

for $y\in D$.

† For a boundary operator $\check{\Phi}$ and an operator A with coefficients in $\mathscr{C}^\infty(\bar{D})$ we use the notation $\check{\Phi}A|_{\partial D}$ for the boundary operator $\Phi A|_{\partial D}$ where Φ is any extension of $\check{\Phi}$. This boundary operator does not depend on the extension Φ. Thus $\check{B}_k^{(i)}=\check{B}_k\Delta^{mi}|_{\partial D}$. It should be noted that there is no meaning in general for $\check{\Phi}\check{B}$ or for $A\check{\Phi}|_{\partial D}$.

Lemma 6.5. Let the original system satisfy (1), (2), *and* (3).

(a) *The following inequality holds, where* λ_1 *is the first eigenvalue of the original system:*

$$|G_p(x, y)| \leq c\lambda_1^{-p+1} \quad \textit{for} \quad p > 1. \tag{6.16}$$

(b) *If* (*P*-4) *holds with extensions* B_k *we have the following inequalities, with* $|D|$ *being the volume of* D:

$$\int_D |G_p(x, y)|\, dx \leq c\lambda_1^{-p+1} |D| \quad \textit{for} \quad p \geq 1 \tag{6.17}$$

$$\int_{\partial D} |\check{\Phi}_k(x) G_p(x, y)|\, d\sigma(x) \leq c^2 \lambda_1^{-p+2} |D| \quad \textit{for} \quad p \geq 2 \quad \textit{and} \quad k = 1, \ldots, m \tag{6.18}$$

$$\int_{\partial D} |\check{B}_l(y) G_p(x, y)|\, dx \leq c^2 \lambda_1^{-p+2} |D| \quad \textit{for} \quad p \geq 2, \quad l = 1, \ldots, m \tag{6.19}$$

$$\int_{\partial D} |\check{B}_l(y) \check{\Phi}_k(x) G_p(x, y)|\, d\sigma(x) \leq c^3 \lambda_1^{-p+3} |D| \quad \textit{for} \quad p \geq 3,\ 1 \leq k,\ l \leq m. \tag{6.20}$$

Remark 6.4. In (6.18) and (6.19) the evaluation for $p = 1$ is directly given as c by (P-1) and (P-4) respectively. In (6.20) for $p = 1$ or $p = 2$ the evaluations are given as c and $c^2 |D|$ respectively by (P-1) and (P-4).

Proof of Lemma 6.5.

(a) By (6.3) and (P-0) we have that

$$\sum_{l=1}^{\infty} \lambda_l^{-1} |\psi_l(x)|^2 = G(x, x) \leq c \quad \text{for} \quad x \in D.$$

Therefore by (6.11)

$$|G_p(x, y)| \leq \left(\sum_{l=1}^{\infty} \lambda_l^{-p} |\psi_l(x)|^2\right)^{\frac{1}{2}} \left(\sum_{l=1}^{\infty} \lambda_l^{-p} |\psi_l(y)|^2\right)^{\frac{1}{2}} \leq \lambda_1^{-p+1} c$$

which gives (6.16).

(b) Inequality (6.17) follows immediately from (6.16). We obtain (6.18) if we write

$$G_p(x, y) = \int_D G_1(x, z) G_{p-1}(z, y)\, dz$$

and use (P-1) and (6.16),† and (6.19) follows if we write

$$G_p(x, y) = \int_D G_{p-1}(x, z)G_1(z, y)\,dz$$

and use (P-4) and (6.16).

Finally, (6.20) is obtained by writing

$$G_p(x, y) = \int_D \int_D G_1(x, z_1)G_{p-2}(z_1, z_2)G_1(z_2, y)\,dz_1\,dz_2$$

and using (P-1), (P-4), and (6.16). ||

Corollary 6.3. If the operators B_k, $k = 1,\ldots, m$, *are distinguished extensions of boundary operators for the original differential system, then* $B_k\Delta^{mi}$, $i = 0,\ldots, p-1$, *are distinguished extensions for the* pth *iterated system.*

Proof. This is an immediate consequence of Lemma 6.5 in view of (6.13) and (6.14). ||

We now consider boundary value problems of infinite order. They will be of a special type which we shall call *periodic relative to* Δ *of period* m. Starting with the original system $\{\Delta^m, \{\check{B}_k\}\}$ we form the infinite sequence of boundary operators $\{\check{B}_k^{(i)}\}$, $k = 1,\ldots, m$, $i = 0, 1,\ldots$ from all the iterated systems. We look for solutions $u \in \mathscr{C}^\infty(D)$ satisfying the infinite system of boundary conditions $\check{B}_k^{(i)}u = g_k^{(i)}$ relative to some fixed extensions $B_k^{(i)}$ of the boundary operators. We shall determine classes $\mathscr{F}$ of admissible solutions and corresponding classes $\mathscr{G}$ of admissible data so that the problem will be well posed. To this effect we introduce certain notations.

Recall that polyharmonic functions of Laplacian order ρ and type τ in D are functions u such that for every $\varepsilon > 0$ and every compact $K \subset D$ there exists a constant $C_{K,\varepsilon}$ such that

$$\max_{x \in K} |\Delta^p u(x)| \leq C_{K,\varepsilon}(\tau + \varepsilon)^{2p}(2p!)^{1-1/\rho}$$

for $p = 0, 1,\ldots$. We may consider other inequalities independent of the choice of the compact K with some other expressions involving $\Delta^p u(x)$ on the left-hand side and different bounds on the right-hand side. They will be called *uniform* inequalities. These inequalities may hold only for those exponents which are multiples mj of a fixed integer m. We call such inequalities *periodically uniform with period* m.

† In the exceptional case of (P-1) when $\check{B}_1$ is of order 0 we can always assume that $\check{B}_1$ is the identity on ∂D and B_1 is the identity on $\bar{D}$. The missing inequality in (P-1) will be given by the corresponding inequality in (P-4) for $l = 1$.

Lemma 6.6. Let $\{M_j\}$ be a sequence of positive constants such that for every $\varepsilon > 0$ there exists a constant $C_\varepsilon > 0$ with

$$\left(\frac{M_j}{(2mj)!^{\varepsilon}}\right)^{\frac{1}{2mj}} \leq C_\varepsilon$$

for all j. If $u \in \mathscr{C}^\infty(D)$ and if $\|\Delta^{mj}u\|_D \leq M_j$ for all j then u has Laplacian order $\rho = 1$.

Proof. By the theorem on elliptic iterates (Lions and Magenes [65, Chapter 8]) u is analytic on D. Continue u as a holomorphic function on a domain $D' \subset \mathbb{C}^n$ with $D = D' \cap \mathbb{R}^n$. Let K be any compact subset of D and let K_1 be the closure of an open set in $\mathbb{R}^n$ such that $K \subset K_1^0 \subset K_1 \subset D$. Finally, choose a set G open in $\mathbb{C}^n$, with compact closure in D', such that $K_1 \subset G$. By definition of the capacity function, §A1, Appendix, $\alpha(K_1, G; z) = 1$ for all $z \in K_1$. By Corollary A4.3, Appendix, $\alpha(K_1, G; z)$ is continuous at every point of K_1^0.

Since u is holomorphic on D', there exists a constant C_G such that $\|\Delta^p u\|_G \leq C_G^{2p}(2p)!$ for all p. In particular, $\|\Delta^{mj}u\|_G \leq C_G^{2mj}(2mj)!$, $j = 0, 1, \ldots$. If η is any number with $0 < \eta < 1$, we can cover K with a finite union U of polydiscs such that $U \subset G$ and $\alpha(K_1, G; Z) \geq 1 - \eta$ for all $z \in U$. Assuming, without loss of generality, that each $M_j \geq 1$, we obtain

$$\|\Delta^{mj}u\|_U \leq M_j((2mj)!\, C_G^{2mj})^{1-\eta}.$$

Let U' be a subdomain of U such that $K \subset U'$ and let $R = d(U', \partial U) > 0$. For every p, $0 < p < m$, the Cauchy bounds yield

$$\|\Delta^{mj+p}u\|_K \leq (2p)!\, R^{-2p} n^p M_j((2mj)!\, C_G^{2mj})^{1-\eta}.$$

Let C be the maximum of $(2p)!\, R^{-2p} n^p$, $0 < p < m$. For any $\varepsilon > 0$ we have

$$\|\Delta^{mj+p}u\|_K \leq C C_\varepsilon^{2mj}(C_G^{2mj})^{1-\eta}(2mj)!^{1-\eta+\varepsilon}.$$

Since η can be chosen arbitrarily close to unity and $\varepsilon > 0$ is arbitrary, this last inequality implies that u has Laplacian order $\rho = 1$. ‖

For the existence problem we consider the class of functions $u \in \mathscr{C}^\infty(D)$ satisfying the periodically uniform condition

(U_∞, m, τ): *There exist constants $C_j > 0$ with*

$$\lim_{j\to\infty} C_j = 0 \quad \textit{such that} \quad \|\Delta^{mj}u\|_D \leq C_j \tau^{2mj}, \quad j = 0, 1, \ldots.$$

By Lemma 6.6, such functions are polyharmonic of order $\rho = 1$.

Theorem 6.1 (Existence). Assume that the original system satisfies (1), (2), (3), *and (P-4) with extensions B_k. Consider any system $\{g_k^{(i)}\}$ with $g_k^{(i)} \in$*

$\mathscr{C}^0(\partial D)$ satisfying

$$(6.21)\qquad \sum_{i=0}^{\infty} \lambda_1^{-i} \sum_{k=1}^{m} \|g_k^{(i)}\|_{\partial D} < \infty.$$

There exists a solution u of the boundary value problem $B_k^{(i)}u = g_k^{(i)}$, the boundary values being taken relative to the extensions $B_k\Delta^{mi}$, with u in the class of functions satisfying $(U_\infty, m, \lambda_1^{1/2m})$. The solution (which is unique by Theorem 6.2 below) is given by

$$(6.22)\quad u(y) = \sum_{i=0}^{\infty} \sum_{k=1}^{m} \int_{\partial D} g_k^{(i)}(x)\{\check{\Phi}_k(x)G_{i+1}(x, y)\}\, d\sigma(x), \qquad y \in D,$$

and the constants C_j in condition $(U_\infty, m, \lambda_1^{1/2m})$ are

$$(6.23)\qquad C_j = \sum_{i=j}^{\infty} c'\lambda_1^{-i} \sum_{k=1}^{m} \|g_k^{(i)}\|_{\partial D}, \qquad c' = \max\{c, c^2\lambda_1\}.$$

Proof. By (6.18) and (6.21) the double series (6.22) is uniformly convergent in D. Since each term is in $\mathscr{C}^0(\bar{D})$, by Lemma 6.2 applied to consecutive iterated problems, the function u is in $\mathscr{C}^0(\bar{D})$. Applying Δ^{mj} term by term to the series (6.22), we obtain by formulae (6.10) and (6.10′) the series

$$\sum_{i=j}^{\infty} \sum_{k=1}^{m} \int_{\partial D} g_k^{(i)}(x)\{\check{\Phi}_k(x)G_{i-j+1}(x, y)\}\, d\sigma(x).$$

By (6.18) and Remark 6.4 this series is term by term majorized by

$$\lambda_1^j \sum_{i=j}^{\infty} c'\lambda_1^{-i} \sum_{k=1}^{m} \|g_k^{(i)}\|_{\partial D}.$$

Therefore $|\Delta^{mj}u(y)| \leq C_j\lambda_1^j$ with C_j given by (6.23), and

$$\lim_{j\to\infty} C_j = 0$$

by (6.21). Hence u satisfies condition $(U_\infty, m, \lambda_1^{1/(2m)})$.

For arbitrarily large j, the application of Δ^{mj} term by term to the series (6.22) yields a uniformly convergent series in D. It follows that any differential operator can be applied term by term to (6.22), the resulting series being uniformly convergent on compacts. For the operator $B_l\Delta^{mj}$ we obtain

$$B_l\Delta^{mj}u(y) = \sum_{i=j}^{\infty} \sum_{k=1}^{m} \int_{\partial D} g_k^{(i)}(x)\{\check{B}_l(y)\check{\Phi}_k(x)G_{i-j+1}(x, y)\}\, d\sigma(x).$$

By (6.20) and (6.21) this series is uniformly convergent in D, by Lemma 6.2 each term is in $\mathscr{C}^0(\bar{D})$, and hence $B_l\Delta^{mj}u$ is in $\mathscr{C}^0(\bar{D})$. Applying (P-6) and (P-7) to the iterated systems we obtain $B_l\Delta^{mj}u|_{\partial D}=g_l^{(j)}$. Thus u is the desired solution. ‖

A periodically uniform condition weaker than $(U_\infty, m, \lambda_1^{1/2m})$ will guarantee uniqueness of the boundary value problem. The condition is

(U_1, m, τ): *There exist constants* $C_j>0$ *with*

$$\lim_{j\to\infty} C_j=0$$

such that

$$\int_D |\Delta^{mj}u(x)|\,dx\le C_j\tau^{2mj}, \qquad j=0,1,\ldots.$$

Theorem 6.2 (Uniqueness). Let the original system satisfy (1), (2), and (3). If for a given system $\{g_k^{(i)}\}$ *of continuous functions on* ∂D *there exist two functions* u_1 *and* u_2 *for which* $(U_1, m, \lambda_1^{1/2m})$ *holds and such that* $\check{B}_k^{(i)}u_1=\check{B}_k^{(i)}u_2=g_k^{(i)}$, *the boundary values being taken relative to two possibly different systems of extensions of the boundary operators* $\check{B}_k^{(i)}$, *then* $u_1=u_2$ *in* D.

Proof. By Lemma 6.1 applied to the pth iterated system we have

$$u_j(y)=\int_D \Delta^{mp}u_j(x)G_p(x,y)\,dx+\sum_{i=0}^{p-1}\sum_{k=1}^{m}\int_{\partial D} g_k^{(i)}(x)\{\check{\Phi}_k(x)G_{i+1}(x,y)\}\,d\sigma(x),$$

$j=1,2$, $y\in D$. By using inequality (6.16) and the condition $(U_1, m, \lambda_1^{1/2m})$ with constants $C_j^{(1)}$ and $C_j^{(2)}$ for the functions u_1 and u_2 respectively, we obtain

$$\begin{aligned}|u_1(y)-u_2(y)|&\le\int_D \{|\Delta^{mp}u_1(x)|+|\Delta^{mp}u_2(x)|\}\,|G_p(x,y)|\,dx\\&\le (C_p^{(1)}+C_p^{(2)})\lambda_1^p c\lambda_1^{-p+1}.\end{aligned}$$

Since the last member of the inequality has limit zero as p approaches infinity, $u_1(y)=u_2(y)$. ‖

Remark 6.5. Since the first eigenfunction ψ_1 is in $\mathscr{C}^\infty(\bar{D})$, satisfies all the homogeneous boundary conditions $\check{B}_k^{(i)}\psi_1=0$, and satisfies $\Delta^{mp}\psi_1=\lambda_1^p\psi_1$, the uniqueness theorem is not valid for any linear class of functions which contains ψ_1. In particular, if in Theorem 6.2 the condition $(U_1, m, \lambda_1^{1/2m})$ is replaced by (U_1, m, τ) or (U_∞, m, τ) with $\tau>\lambda_1^{1/2m}$, uniqueness does not hold.

Remark 6.6. Property (P-0) was essential in obtaining the inequalities (6.16)–(6.20) and this required the restriction $m>n/2$. However, if p_0 is chosen so large that $p_0m>n/2$, then inequalities (6.16)–(6.20) hold with

$p > p_0$ and p replaced by $p - p_0 + 1$. The constant c appearing in (P-0) may need to be increased. Theorems 6.1 and 6.2 still hold with these slight changes, and so condition (1) is not needed for these two results.

Theorems 6.1 and 6.2 give rise to an infinite number of well-posed boundary value problems of infinite order. The largest class of solutions obtained by these theorems is the class of functions satisfying $(U_\infty, m, \lambda_1^{1/2m})$ and having boundary values $\check{B}_k^{(i)}u$ relative to $B_k\Delta^{mi}$ with

$$\sum_{i=0}^{\infty} \lambda_1^{-i} \sum_{k=1}^{m} \|B_k^{(i)}u\|_{\partial D} < \infty.$$

The corresponding class of admissible boundary data $\{g_k^{(i)}\}$ is determined by (6.21).

The boundary value problems discussed above can also be studied in the class of functions $u \in \mathscr{C}^\infty(D)$ which satisfy the condition

(U, m, τ): *For every* $\varepsilon > 0$ *there exists a constant* $C_\varepsilon > 0$ *such that*

$$\|\Delta^{mj}u\|_D \leq C_\varepsilon(\tau + \varepsilon)^{2mj}, \qquad j = 0, 1, \ldots.$$

Uniqueness holds in this class if and only if $\tau < \lambda_1^{1/2m}$. For existence we require that the class of admissible solutions consists of functions u which satisfy (U, m, τ), $\tau < \lambda_1^{1/2m}$ and which have boundary values $\check{B}_k^{(i)}u$ as above such that for every $\varepsilon > 0$ there exists a constant C_ε with $\|\check{B}_k^{(i)}u\|_{\partial D} \leq C_\varepsilon(\tau + \varepsilon)^{2mi}$. The corresponding class of admissible boundary data $\{g_k^{(i)}\}$ is determined by the condition that for every $\varepsilon > 0$ there exists a constant C_ε with $\|g_k^{(i)}\|_{\partial D} \leq C_\varepsilon(\tau + \varepsilon)^{2mi}$. In this class the boundary value problem is well posed.

7. Analytic solutions of the heat equation

In this section we examine solutions $u(x, t)$ of the heat equation $\partial u/\partial t = \Delta u$ in $\mathbb{R}^n \times \mathbb{R}^1$. Our discussion will concern solutions which are analytic in (x, t), will develop connections with polyharmonic functions, and will ultimately focus on analytic continuation of the solutions as a function of t. This material, along with the general theory of polyharmonic functions, comprises the basic detail referred to by Aronszajn [10, Chapter III, §1], particularly Theorems III and IV† of that paper. A theory of traces was developed in the succeeding paper [11]. See Baouendi [6] for an extension of this work.

Recall that

$$E(x, t) = (4\pi t)^{-n/2} \exp\left(\frac{-|x|^2}{4t}\right)$$

† Because of a misprint, both of these were called Theorem III.

is the fundamental solution for the Cauchy problem. If we replace $|x|^2$ by $\sum_{j=1}^n x_j^2$, let x range throughout $\mathbb{C}^n$, and let t range throughout $\mathbb{C}^1\backslash\{0\}$, then $E(x, t)$ is analytic in x and, for each branch of $t^{n/2}$ if n is odd, is analytic in t.

If $v(x)$ has suitable growth as a function of real x, then the function

$$u(y, t)=\{E(x, t-t_0)*v(x)\}(y, t)=\int_{\mathbb{R}^n} E(y-x, t-t_0)v(x)\,dx$$

satisfies the heat equation and has initial data $u(y, t_0)=v(y)$. For this it is sufficient that $v(x)\exp(-\alpha|x|^2)\in\mathscr{L}^1(\mathbb{R}^n)$ for some $\alpha>0$ and $0<t-t_0<(4\alpha)^{-1}$. Actually the convolution will be valid for complex x and t provided that $0<\mathscr{Re}(t-t_0)<(4\alpha)^{-1}$.

Remark 7.1. The function E can be used to construct solutions of the heat equation which are $\mathscr{C}^\infty$ but not analytic in t. Choose constants a_k so that the series $\sum_{k=1}^\infty a_k E(x, t-i/k)$, together with its derivatives, converges on the whole real t axis. This can be done so that convergence is obtained outside a small half-disc $|t|\leq\beta$, $\mathscr{Im}\, t>0$, together with a finite number of points. The series will represent a solution of the heat equation which is $\mathscr{C}^\infty$ in t on the real axis, analytic in the lower half-plane, and not analytic at $t=0$. By choosing a dense set of points on the real t axis, constructing solutions by the method above which fail to be analytic at these points, and adding all the solutions, we obtain a solution which is $\mathscr{C}^\infty$ but nowhere analytic on the real axis.

Definition 7.1. Let $u(x, t)$ be an analytic solution of the heat equation in a domain $D\subset\mathbb{R}^n\times\mathbb{R}^1$ (or $\mathbb{C}^n\times\mathbb{C}^1$). For fixed t_0 the function $u(x, t_0)$ is called a (*space*) *section* of u at t_0.

A section is defined on an open set in $\mathbb{R}^n$ (or $\mathbb{C}^n$) which is not necessarily connected.

Let $u(x, t)$ be an analytic solution of the heat equation in $D\subset\mathbb{R}^n\times\mathbb{R}^1$, and let $D_{t_0}=\{x\in\mathbb{R}^n:(x, t_0)\in D\}$. For any compact $K\subset D_{t_0}$ there is a disc $\mathscr{B}_R(t_0)$ in $\mathbb{C}^1$ and a domain $D'\subset\mathbb{C}^n\times\mathbb{C}^1$ such that $K\times\overline{\mathscr{B}_R(t_0)}\subset D'$ and u is analytically extendable to D'. There is a constant $C_{K,R}$ such that $|u(x, t)|\leq C_{K,R}$ for all $(x, t)\in K\times\mathscr{B}_R(t_0)$, and for each $x\in K$ the Cauchy bounds give

$$\left|\frac{\partial^k u}{\partial t^k}(x, t)\right|\leq C_{K,R}\frac{k!}{R^k}.$$

Since u is a solution of the heat equation we have

$$\frac{\partial^k u}{\partial t^k}(x, t)=\Delta^k u(x, t),$$

so that

$$|\Delta^k u(x,t)| \leq C_{K,R} \frac{k!}{R^k}. \tag{7.1}$$

An application of Stirling's formula shows that the section $u(x, t_0)$ is polyharmonic of order $\rho=2$ and some type $\tau \leq \infty$ in D_{t_0}.

Conversely, suppose that $v(x)$ is polyharmonic of order $\rho=2$ and type $\tau \leq \infty$ in an open set $D \subset \mathbb{R}^n$. Setting

$$u(x,t) = \sum_{k=0}^{\infty} \frac{(t-t_0)^k}{k!} \Delta^k v(x),$$

we can easily see that u is an analytic solution of the heat equation for t near t_0 and that $u(x, t_0) = v(x)$. Therefore we have proved the following.

Theorem 7.1. For a function $v(x)$ to be a section of an analytic solution of the heat equation it is necessary and sufficient that v be polyharmonic of order $\rho = 2$ and some type $\tau \leq \infty$.

Corollary 7.1. A function $v(x)$ defined on a domain $U \subset \mathbb{R}^n$ (or $\mathbb{C}^n$) is the section of an analytic solution u of the heat equation in $U \times \mathcal{B}_R(t_0)$ at time t_0 if and only if v is polyharmonic in U of order 2 and type $(2R)^{-\frac{1}{2}}$. In this case

$$u(x,t) = \sum_{k=0}^{\infty} \frac{(t-t_0)^k}{k!} \Delta^k v(x). \tag{7.2}$$

If $U \subset \mathbb{R}^n$, then u has analytic continuation to $\hat{U} \times \mathcal{B}_R(t_0)$.

The proof is the same as that of Theorem 7.1 because inequality (7.1) holds for $K \times \mathcal{B}_{R'}(t_0)$ with $K \subset U$ and any $R' < R$. Corollary 7.1 is the same as that given by Aronszajn [10, Theorem III, Chapter III, §1].

By looking at the series (7.2) we see that if $v(x)$ has order $\rho = 2$ and type 0 in U, the solution $u(x, t)$ is entire in t for each $x \in U$ and is therefore a solution on $U \times \mathbb{C}^1$. If U is a domain in $\mathbb{R}^n$ (rather than $\mathbb{C}^n$), the solution $u(x, t)$ can be continued to $\tilde{U} \times \mathbb{C}^1$.

If v has Almansi expansion

$$v(x) = \sum_{k=0}^{\infty} r^{2k} h_k(x)$$

with

$$h_k(x) = \left(\tau + \frac{1}{k}\right)^{2k} (2k)!^{-\frac{1}{2}},$$

then v has order 2 and type τ on all of $\mathbb{R}^n$ and no lower type (see Ex. 3.4, Chapter V). Thus the solution given by (7.2) is valid on $\mathbb{R}^n \times \mathcal{B}_R(t_0)$, where $R = 1/(2\tau^2)$, and no larger disc in $\mathbb{C}^1$ can be used. The solution can be continued to $\tilde{\mathbb{R}}^n \times \mathcal{B}_R(t_0) = \mathbb{C}^n \times \mathcal{B}_R(t_0)$. With examples of functions $v(x)$ with order 2 and variable type (§3, Chapter V), the series (7.2) also enables us to construct solutions on domains $U(R) \times \mathcal{B}_R(t_0)$ whose section at time t_0 is v restricted to $U(R)$. Notice that the domains $U(R)$ expand as the radius R decreases.

As a result of Theorem 7.1, if $u(x, t)$ is an analytic solution of the heat equation in $D \subset \mathbb{R}^n \times \mathbb{R}^1$, then for each fixed t_0 the function $u(x, t_0)$ can be continued to the harmonicity hull $\tilde{D}_{t_0}$ of D_{t_0}. We now ask for a domain in $\mathbb{C}^1$ into which $u(x_0, t)$ can be continued as a function of t for fixed x_0. We have seen that if $u(x, t_0)$ has order 2 and type 0 the function $u(x_0, t)$ is entire in t. Therefore the remaining case to consider is that of order 2 and type τ, $0 < \tau \leq \infty$. The technique used will involve making a suitable interpretation of the formula

$$u(y, t) = \int_{\mathbb{R}^n} E(y - x, t - t_0) u(x, t_0)\, dx$$

when $u(x, t_0)$ is not defined on all of $\mathbb{R}^n$.† In order to do so we shall need some tools from complex function theory. The required material will be sketched and the reader is referred to Boas [18, Chapter 5], Levin [64, Chapter 1], and Titchmarsh [86, Chapters 4 and 5] for details.

Hadamard multiplication theorem (Titchmarsh [86, §4.6]). *Let*

$$f(z) = \sum_{k=0}^{\infty} a_k z^k, \qquad |z| < R,$$

and

$$g(z) = \sum_{k=0}^{\infty} b_k z^k, \qquad |z| < R',\ z \in \mathbb{C}^1.$$

The function

$$F(z) = \sum_{k=0}^{\infty} a_k b_k z^k$$

is holomorphic near the origin and the singularities of F are to be found among the products of the singularities of f with the singularities of g. In particular, if f is holomorphic in a star domain about the origin and g has its singularities contained in the half-line $1 \leq z < \infty$, then F is holomorphic at least on the domain of f.

Let $f(z)$ be a function of one complex variable which is holomorphic in an angle with vertex at the origin. The notions of order and type (§4, Chapter I) can be carried over to such functions if $M(f; R)$ is taken to mean the maximum of $|f(z)|$ for $|z| \leq R$ and z in the angle.

† This is the idea behind the 'symbolic integrals' in Aronszajn [10, Chapter I, §5].

Let $f(z)$ be of order ρ and finite type in the angle $|\theta| \leq \alpha$. We define the *Phragmén–Lindelöf indicator function of f*, or simply the *indicator* of f, to be

$$h(\theta) = \limsup_{r \to \infty} \frac{\log |f(re^{i\theta})|}{r^\rho}, \qquad |\theta| \leq \alpha.$$

This function is bounded above, and if f is entire it is also bounded below. For the function $f(z) = \exp\{(a - ib)z^\rho\}$ the indicator is $a \cos \rho\theta + b \sin \rho\theta \equiv h(\theta)$. That function $h(\theta)$ which takes the values h_k at θ_k, $k = 1, 2$, is needed below. It is denoted

$$H(\theta) = \frac{h_1 \sin \rho(\theta_2 - \theta) + h_2 \sin \rho(\theta - \theta_1)}{\sin \rho(\theta_2 - \theta_1)}.$$

Lemma 7.1. Let f be holomorphic of order 1 *and finite type in the angle* $|\theta| \leq \alpha$, *let* $h(\theta)$ *be its indicator, let* $0 < \theta_2 - \theta_1 < \pi$, $|\theta_1| \leq \alpha$, $|\theta_2| \leq \alpha$, *and suppose that* $h(\theta_1) \leq h_1$, $h(\theta_2) \leq h_2$. *For all* θ *with* $\theta_1 < \theta < \theta_2$ *we have* $h(\theta) \leq H(\theta)$.

The property appearing in Lemma 7.1 is called *trigonometric convexity*. This lemma can be restated as follows.

Lemma 7.2. Let f be holomorphic of order 1 *and finite type in the angle* $|\theta| \leq \alpha$, *let* $h(\theta)$ *be its indicator, and let* $0 < \theta_3 - \theta_1 < \pi$, $|\theta_1| \leq \alpha$, $|\theta_3| \leq \alpha$. *If* $\theta_1 < \theta_2 < \theta_3$, *then*

$$h(\theta_1)\sin(\theta_2 - \theta_3) + h(\theta_2)\sin(\theta_3 - \theta_1) + h(\theta_3)\sin(\theta_1 - \theta_2) \leq 0. \tag{7.3}$$

Moreover, h is continuous.

Let C be a bounded closed convex set in the plane. The *support function* of C is

$$k(\theta) = \sup_{(x, y) \in C} (x \cos \theta + y \sin \theta).$$

The line $l_\theta : x \cos \theta + y \sin \theta - k(\theta) = 0$ is a *supporting line* of C. Thus $k(\theta)$ is the (signed) distance from the supporting line l_θ to the origin.

Lemma 7.3. A necessary and sufficient condition that a function $k(\theta)$ *be the support function of a bounded closed convex set C is that*

(a) $k(\theta + 2\pi) = k(\theta)$, *and*

(b) *if* $\theta_1 \leq \theta_2 \leq \theta_3$, $\theta_3 - \theta_2 < \pi$, *and* $\theta_2 - \theta_1 < \pi$, *then*

$$k(\theta_1)\sin(\theta_2 - \theta_3) + k(\theta_2)\sin(\theta_3 - \theta_1) + k(\theta_3)\sin(\theta_1 - \theta_2) \leq 0.$$

Moreover, any such k is continuous.

A function f which is holomorphic in an angle $|\theta| \leq \alpha$, and of order 1 and finite type or of order less than 1 is sometimes called a function of *exponential type* or of *exponential growth* (see Boas [18]). For such a function f, which is entire if $\alpha = 2\pi$, the function

$$\limsup_{r\to\infty} \frac{\log|f(re^{i\theta})|}{r}, \qquad |\theta| \leq \alpha,$$

has all of the announced properties of the indicator function $h(\theta)$. If f has order less than 1 then the above limit supremum is clearly zero, so the above function of θ is the support function of the convex set consisting only of the origin. The properties which we shall need depend only on this fact and therefore *we shall now adopt the convention, following Boas* [18], *that the indicator function* $h(\theta)$ *is defined to be*

$$h(\theta) = \limsup_{r\to\infty} \frac{\log|f(re^{i\theta})|}{r}$$

where f is a function of exponential growth.

From Lemma 7.3 and inequality (7.3) it follows that the indicator of an entire function of exponential growth is the support function of a bounded closed convex set, the *indicator diagram* of f, denoted I_f. Lemma 7.3 also shows that if $k(\theta)$ is an indicator then so is $h(\theta) = \max(0, k(\theta))$. The set determined by $h(\theta)$ is the convex hull of the set determined by $k(\theta)$ and the origin.

To each function

$$f(z) = \sum_{k=0}^{\infty} \frac{a_k}{k!} z^k,$$

entire of exponential growth, there corresponds a function

$$\varphi(z) = \sum_{k=0}^{\infty} \frac{a_k}{z^{k+1}}$$

called the *Borel transform* of f. From Theorem 4.1, Chapter I, with a_k replaced by $a_k/k!$, it follows that if f is of order 1 and finite type τ then

$$\tau = \limsup_{k\to\infty} |a_k|^{1/k}$$

and therefore φ is holomorphic for $|z| > \tau$. If f has order less than 1, φ is holomorphic for $|z| > 0$. The smallest convex set $\bar{I}_f$ containing all of the singularities of φ is called the *conjugate diagram* of f. Certainly $\bar{I}_f$ is contained in the disc $|z| \leq \tau$.

Pólya's Theorem. The conjugate diagram of an entire function of exponential growth is the reflection in the real axis of its indicator diagram.

A consequence of Pólya's Theorem is that, given any support function $h(\theta)$ or equivalently given any bounded closed convex set I, there is a function f entire of exponential growth such that $h(\theta)$ is the indicator and I is the diagram of f. Indeed, reflect I in the real axis, obtaining a bounded closed convex set $\bar{I}$, and then choose a countable set of points $\{\lambda_k\}$ dense in the boundary of $\bar{I}$. If a sequence $\{a_k\}$ is chosen appropriately, the function

$$\varphi(z)=\sum_{k=1}^{\infty}\frac{a_k}{z-\lambda_k}$$

will be holomorphic outside $\bar{I}$ and all the points λ_k will be singular points. Then we note that $\varphi(z)$ is the Borel transform of

$$f(z)=\sum_{k=1}^{\infty} a_k \exp(\lambda_k z).$$

If I consists only of the origin, the construction is clear.

The relation between $f(z)$ and $\varphi(z)$ can be used to provide *Borel's method of continuation.* It can be verified that

$$\varphi(z)=\int_0^{\infty} e^{-z\zeta} f(\zeta)\,d\zeta$$

where the integration is performed along the ray $\zeta=\rho e^{-i\theta}$, $\rho>0$. Moreover, this integral converges uniformly and absolutely for $\mathcal{Re}(ze^{-i\theta})>h(-\theta)+\varepsilon$ where h is the indicator of f. Setting $z=1/w$ we obtain

$$\varphi\left(\frac{1}{w}\right)=\sum_{k=0}^{\infty} a_k w^{k+1}=\int_0^{\infty} e^{-\zeta/w} f(\zeta)\,d\zeta,$$

and if the integration is performed along the ray $\zeta=\rho w$, $\rho>0$, it follows that

$$\varphi\left(\frac{1}{w}\right)=\sum_{k=0}^{\infty} a_k w^{k+1}=\int_0^{\infty} e^{-\rho} f(\rho w) w\,d\rho$$

and

(7.4) $$\sum_{k=0}^{\infty} a_k w^k=\int_0^{\infty} e^{-\rho} f(\rho w)\,d\rho.$$

We can apply these results in the reverse order as follows. Given a holomorphic function $g(w)$ in a disc about the origin such that the Taylor series $\sum_{k=0}^{\infty} a_k w^k$ for g has a finite radius of convergence, the function

$$f(z)=\sum_{k=0}^{\infty}\frac{a_k}{k!}z^k$$

is entire of order 1 and positive finite type and the integral relations above hold. The integral representation (7.4) of g provides analytic continuation across any arc of the circle of convergence of the series for g where g is analytic. If the indicator of f is everywhere non-negative, then the continuation is valid in a star domain about the origin which is the complement of the conjugate diagram transformed by the mapping $w = 1/z$.

The integral formula providing analytic continuation can be extended to include functions which are holomorphic of order 1 and positive finite type in an angle $|\theta| \leq \alpha \leq \pi/2$. If f is such a function, its indicator $h(\theta)$ is the support function of an unbounded convex set. In this case

$$\varphi(z) = \int_0^\infty e^{-z\rho} f(\rho)\, d\rho$$

defines a function which is holomorphic outside the unbounded convex set whose support function is $h(-\theta)$, $|\theta| \leq \alpha$. As before, we set $w = 1/z$ and obtain

$$\varphi\left(\frac{1}{w}\right) = w \int_0^\infty e^{-\rho} f(\rho w)\, d\rho,$$

so that the set on which $\int_0^\infty e^{-\rho} f(\rho w)\, d\rho$ is holomorphic is the set described above transformed by the mapping $w = 1/z$.

This completes the discussion of the tools from complex function theory. Before returning to the heat equation, we make a final observation. Let v be holomorphic on a neighbourhood of x, and write the Almansi expansion in the form

$$v(x+y) = \sum_{k=0}^{\infty} |y|^{2k} h_k(x; y) \tag{7.5}$$

for y sufficiently near zero. Calculating the successive Laplacians and evaluating at $y = 0$, we obtain

$$\Delta^k v(x) = \frac{4^k k!\, \Gamma(k + n/2)}{\Gamma(n/2)} h_k(x; 0). \tag{7.6}$$

Now let $u(x, t)$ be an analytic solution of the heat equation. We saw that for each fixed t, $u(x, t)$ has Laplacian order $\rho = 2$ and type $\tau \leq \infty$. To interpret the symbolic expression

$$\int_{\mathbb{R}^n} \{4\pi(t - t_0)\}^{-n/2} \exp\left\{-\frac{|x-y|^2}{4(t-t_0)}\right\} u(y, t_0)\, dy$$

we proceed as follows. For sufficiently small $R > 0$ we set $x - y = R\theta$,

$\theta \in S_1^{n-1}$, and obtain

$$(7.7)\quad R^{n-1}\int_{S_1^{n-1}} \{4\pi(t-t_0)\}^{-n/2} \exp\left\{-\frac{|R\theta|^2}{4(t-t_0)}\right\} u(x-R\theta, t_0)\, d\theta$$

$$= R^{n-1}\{4\pi(t-t_0)\}^{-n/2}\omega_n \sum_{k=0}^{\infty} R^{2k} \exp\left\{-\frac{R^2}{4(t-t_0)}\right\} h_k(t_0; x; 0)$$

$$= R^{n-1}\{4\pi(t-t_0)\}^{-n/2}\omega_n \sum_{k=0}^{\infty} \exp\left\{-\frac{R^2}{4(t-t_0)}\right\} R^{2k} \frac{\Gamma(n/2)}{4^k k!\,\Gamma(k+n/2)} \Delta^k u(x, t_0)$$

which follows from (7.5) and (7.6) applied to $v(z) \equiv u(z, t_0)$. Since $u(x, t_0)$ has Laplacian order $\rho = 2$ and type $\tau \leq \infty$, the last series in (7.7) defines a function which is entire in R^2 of exponential growth. Thus the last expression in (7.7) is defined for all complex R. Integrating (7.7) we obtain

$$(7.8)\quad \int_0^{\infty} R^{n-1}\int_{S_1^{n-1}} \{4\pi(t-t_0)\}^{-n/2} \exp\left\{-\frac{|x-y|^2}{4(t-t_0)}\right\} u(y, t_0)\, d\theta\, dR$$

$$= \int_0^{\infty} R^{n-1}\{4\pi(t-t_0)\}^{-n/2}\omega_n \sum_{k=0}^{\infty} \exp\left\{-\frac{R^2}{4(t-t_0)}\right\}$$

$$\times R^{2k} \frac{\Gamma(n/2)}{4^k k!\,\Gamma(k+n/2)} \Delta^k u(x, t_0)\, dR.$$

For $t > t_0$ the substitution $\rho = R^2/4(t-t_0)$ in the right-hand side of (7.8) yields

$$(7.9)\qquad \tfrac{1}{2}\omega_n \pi^{-n/2} \int_0^{\infty} \rho^{n/2-1} e^{-\rho} \sum_{k=0}^{\infty} \rho^k \frac{\Gamma(n/2)(t-t_0)^k}{k!\,\Gamma(k+n/2)} \Delta^k u(x, t_0)\, d\rho.$$

Let

$$g(w) = \sum_{k=0}^{\infty} \frac{\Gamma(n/2)\Delta^k u(x, t_0)}{\Gamma(k+n/2)} w^k = \sum_{k=0}^{\infty} a_k w^k$$

and let

$$f(z) = \sum_{k=0}^{\infty} \frac{\Gamma(n/2)\Delta^k u(x, t_0)}{k!\,\Gamma(k+n/2)} z^k = \sum_{k=0}^{\infty} \frac{a_k}{k!} z^k.$$

First we note that g is the Hadamard product of the series

$$\sum_{k=0}^{\infty} \frac{\Gamma(n/2)k!}{\Gamma(k+n/2)} z^k$$

and

$$\sum_{k=0}^{\infty} \frac{\Delta^k u(x, t_0)}{k!} z^k.$$

With $z=t-t_0$, the second series is the Taylor expansion of $u(x, t_0)$. To study the first series, observe that for $n>2$

$$\frac{\Gamma(n/2)k!}{\Gamma(k+n/2)}=\left(\frac{n}{2}-1\right)B\left(k+1, \frac{n}{2}-1\right)=\left(\frac{n}{2}-1\right)\int_0^1 \xi^k(1-\xi)^{n/2-2}\,\mathrm{d}\xi$$

where B is the beta function. Therefore

$$\sum_{k=0}^{\infty}\frac{\Gamma(n/2)k!}{\Gamma(k+n/2)}z^k=\left(\frac{n}{2}-1\right)\sum_{k=0}^{\infty}\left\{\int_0^1 \xi^k(1-\xi)^{n/2-2}\,\mathrm{d}\xi\right\}z^k$$
$$=\left(\frac{n}{2}-1\right)\int_0^1\frac{1}{1-\xi z}(1-\xi)^{n/2-2}\,\mathrm{d}\xi, \tag{7.10}$$

and from the last expression in (7.10) we see that the function is holomorphic in z except on the ray $[1, \infty)$ of the real axis. If $n=2$ the series in (7.10) becomes

$$\sum_{k=0}^{\infty} z^k=(1-z)^{-1}$$

which is holomorphic if $|z|<1$. By the Hadamard multiplication theorem, the Hadamard product of the two series, which is g, is therefore holomorphic on any star domain about the origin on which the function

$$\sum_{k=0}^{\infty}\frac{\Delta^k u(x, t_0)}{k!}z^k$$

is holomorphic. (As remarked earlier, if $u(x, t_0)$ has Laplacian order 2 and type 0, this function is entire.) If h is the indicator function of the entire function f, which clearly has exponential growth, and if $k(\theta)=\max(0, h(\theta))$, then the largest such star domain is the complement of the image under the map $z=1/w$ of the convex set determined by $k(\theta)$.

If n is even, say $n/2-1=m\geq 0$, in the integral (7.9), we take ρ^m inside the summation, the constant $\frac{1}{2}\omega_n\pi^{-n/2}$ inside the integral, and multiply by $(t-t_0)^m$. The result has the form

$$\int_0^{\infty} \mathrm{e}^{-\rho}F(\rho(t-t_0))\,\mathrm{d}\rho \tag{7.11}$$

where

$$F(z)=\tfrac{1}{2}\omega_n\pi^{-n/2}\sum_{k=0}^{\infty}\frac{a_k}{k!}z^{k+m}=\tfrac{1}{2}\omega_n\pi^{-n/2}z^m f(z),$$

the a_k being the same as in the definition of $f(z)$. Denoting by $G(w)$ the corresponding transformed function, we obtain

$$G(t-t_0)=\int_0^{\infty}\mathrm{e}^{-\rho}F(\rho(t-t_0))\,\mathrm{d}\rho \tag{7.12}$$

and the integral expression (7.12) provides analytic continuation of G into the star domain described before. Performing the integration in (7.12), which is the expression (7.9) multiplied by $(t-t_0)^m$, we obtain

$$\tfrac{1}{2}\omega_n\pi^{-n/2}\Gamma\left(\frac{n}{2}\right)(t-t_0)^m\sum_{k=0}^{\infty}\frac{(t-t_0)^k}{k!}\Delta^k u(x,t_0)=(t-t_0)^m u(x,t).$$

Thus $G(t-t_0)=(t-t_0)^m u(x,t)$, and $u(x,t)$ can be continued into the domain in the complex t-space corresponding to G.

If n is odd, say $n/2-1=m+\frac{1}{2}\geqslant 0$, instead of (7.11) we obtain

$$\text{(7.13)}\qquad \int_0^{\infty} e^{-\rho}\rho^{\frac{1}{2}}F(\rho(t-t_0))\,d\rho=(t-t_0)^{-\frac{1}{2}}\int_0^{\infty} e^{-\rho}\{\rho(t-t_0)\}^{\frac{1}{2}}F(\rho(t-t_0))\,d\rho.$$

Set $F_1(z)=z^{\frac{1}{2}}F(z)$ and choose a branch for $z^{\frac{1}{2}}$ by cutting the plane along the negative real axis. It follows that $F_1(z)$ is holomorphic in the angle $|\theta|\leqslant\pi/2$, and since $F(z)$ is entire of exponential growth, $F_1(z)$ is holomorphic of exponential growth in the angle $|\theta|\leqslant\pi/2$. Thus the integral

$$\text{(7.14)}\qquad G(t-t_0)=(t-t_0)^{-\frac{1}{2}}\int_0^{\infty} e^{-\rho}\{\rho(t-t_0)\}^{\frac{1}{2}}F(\rho(t-t_0))\,d\rho$$

provides analytic continuation as discussed before. Once again, by performing the integration we obtain the analytic continuation of $u(x,t)$.

APPENDIX

HOLOMORPHIC CAPACITY

The use of what is essentially holomorphic capacity appears in the study of polyharmonic functions by Aronszajn [8]. The theory was developed and the name established in the 1970s. However, the results in this Appendix are applicable to the whole class of holomorphic functions and their development here is therefore independent of the concept of polyharmonicity.

A1. Basic properties

Definition A1.1. If G is a domain in $\mathbb{C}^n$ and A is any subset of G, the *holomorphic capacity of A on G* is the function defined at each $z \in G$ by

$$\alpha(z) = \alpha(A, G; z) = \inf \frac{\log |\phi(z)|}{\log \|\phi\|_A}$$

where the infimum is taken over the class $\Phi(A, G)$ of functions ϕ holomorphic on G, less than 1 in modulus there, such that $\|\phi\|_A = \sup\{|\phi(x)| : x \in A\} > 0$. Of course, $\alpha(z)$ is also the largest number α such that

$$|f(z)| \leqslant \|f\|_A^{\alpha} \|f\|_G^{1-\alpha}$$

for all f holomorphic on G such that $\|f\|_A > 0$.

Clearly $\alpha(z) \geqslant 0$, and since $\phi \equiv \frac{1}{2}$ lies in the class it follows that $\alpha(z) \leqslant 1$. Such basic inequalities are set out below in quick succession. Illustrative examples are collected in §A2.

Proposition A1.1. If G is any domain in $\mathbb{C}^n$ and A any subset of G, then $1 \geqslant \alpha(A, G; z) \geqslant 0$ and

$$\alpha(\bar{A} \cap G, G; z) = \alpha(A, G; z) \geqslant \chi_A(z)$$

where χ_A is the characteristic function of A, taking the value 1 *on A and zero on the complement of A.*

Proposition A1.2 (*Montononicity*). *If $A \subset A' \subset G \subset G'$, then*

$$\alpha(A, G; z) \leqslant \alpha(A', G', z) \quad \textit{for} \quad z \in G.$$

Proof. For any $\phi' \in \Phi(A', G') \cap \Phi(A, G)$ and $z \in G$

$$\frac{\log |\phi'(z)|}{\log \|\phi'\|_{A'}} \geqslant \frac{\log |\phi'(z)|}{\log \|\phi'\|_A} \geqslant \alpha(A, G; z).$$

If $\phi' \in \Phi(A', G') \backslash \Phi(A, G)$ then $\|\phi'\|_A = 0$. In this case replace ϕ' by $(\phi' + \varepsilon)/(1+\varepsilon)$, and then take the limit as ε approaches zero. ‖

Remark A1.1. In the preceding proof, the replacement of ϕ' by $(\phi' + \varepsilon)/(1+\varepsilon)$ resolves the problem posed by $\|\phi'\|_A = 0$. It is also possible to avoid any problems posed by $\log \|\phi\|_A = 0$ or $\|\phi\|_A = 1$, replacing ϕ by $\phi/(1+\varepsilon)$.

Proposition A1.3 (Subadditivity). If $A_0 = A_1 \cup A_2 \subset G$ and $\alpha_k(z) = \alpha(A_k, G; z)$, $k = 0, 1, 2$, then

$$\max\{\alpha_1(z), \alpha_2(z)\} \leq \alpha_0(z) \leq \alpha_1(z) + \alpha_2(z).$$

Proof. The left-hand inequality follows from monotonicity (Proposition A1.2) and the right-hand inequality from considering powers of two functions $\phi_k \in \Phi(A_k, G)$, as follows:

$$\alpha_0(z) \leq \frac{\log |\phi_1^p(z)\phi_2^q(z)|}{\log \|\phi_1^p \phi_2^q\|_{A_0}} = \frac{p \log |\phi_1(z)| + q \log |\phi_2(z)|}{\log \|\phi_1^p \phi_2^q\|_{A_0}}$$

for all positive integers p and q. In the denominator

$$\begin{aligned}\log \|\phi_1^p \phi_2^q\|_{A_0} &= \max\{\log \|\phi_1^p \phi_2^q\|_{A_1}, \log \|\phi_1^p \phi_2^q\|_{A_2}\} \\ &\leq \max\{p \log \|\phi_1\|_{A_1}, q \log \|\phi_2\|_{A_2}\} < 0\end{aligned}$$

since $|\phi_1|$ and $|\phi_2|$ are less than 1. It follows from Kronecker's theorem† that for any $\varepsilon > 0$ there exist arbitrarily large positive integers p and q such that

$$|p \log \|\phi_1\|_{A_1} - q \log \|\phi_2\|_{A_2}| < \epsilon.$$

Thus

$$\log \|\phi_1^p \phi_2^q\|_{A_0} \leq \min\{p \log \|\phi_1\|_{A_1} + \varepsilon, q \log \|\phi_2\|_{A_2} + \varepsilon\} < 0$$

and

$$\begin{aligned}\alpha_0(z) &\leq \frac{p \log |\phi_1(z)| + q \log |\phi_2(z)|}{\min\{p \log \|\phi_1\|_{A_1} + \varepsilon, q \log \|\phi_2\|_{A_2} + \varepsilon\}} \\ &\leq \frac{\log |\phi_1(z)|}{\log \|\phi_1\|_{A_1} + \varepsilon/p} + \frac{\log |\phi_2(z)|}{\log \|\phi_2\|_{A_2} + \varepsilon/q}.\end{aligned}$$

Now the infimum is taken, first on ε, then over ϕ_1 and ϕ_2. ‖

Proposition A1.3 generalizes, being valid for finite unions, but Example A2.6 will show that there is no countable subadditivity.

Proposition A1.4 (Supermultiplicativity). If $A \cup K \subset G$ then

$$\alpha(A, G; z) \geq \alpha(K, G; z) \inf\{\alpha(A, G; x) : x \in K\}.$$

† If θ is irrational then the set of points $(n\theta) = \{n\theta - [n\theta] : n$ an integer$\}$ is dense in the interval $(0, 1)$ (Hardy and Wright [46, p. 376]). (For the present application use $\theta = \log \|\phi_1\|_{A_1} / \log \|\phi_2\|_{A_2}$.)

Proof. If $\phi \in \Phi(A, G) \cap \Phi(K, G)$ and $\log \|\phi\|_K < 0$, then

$$\begin{aligned}\frac{\log|\phi(z)|}{\log\|\phi\|_A} &= \frac{\log|\phi(z)|}{\log\|\phi\|_K}\frac{\log\|\phi\|_K}{\log\|\phi\|_A}\\ &\geqslant \alpha(K, G; z)\frac{\log\|\phi\|_K}{\log\|\phi\|_A}\\ &= \alpha(K, G; z)\frac{\inf\{\log|\phi(x)| : x \in K\}}{\log\|\phi\|_A}\\ &\geqslant \alpha(K, G; z)\inf\{\alpha(A, G; x) : x \in K\}.\end{aligned}$$

If $\phi \in \Phi(A, G)\backslash\Phi(K, G)$ consider $\phi_\varepsilon = (\phi + \varepsilon)/(1 + \varepsilon)$ instead, as in the proof of Proposition A1.2. Taking the infimum over ϕ (and ϕ_ε) gives the result (see Remark A1.1). ‖

Every property of infima of continuous positive functions must also hold, the following one in particular.

Proposition A1.5. *$\alpha(A, G; z)$ is upper semi-continuous in that*

$$\alpha(A, G; z) \geqslant \limsup_{x \to z} \alpha(A, G; x).$$

Proposition A1.6. (*Invariance*). *If T: $G \to \mathbb{C}^n$ is a holomorphic mapping of a domain $G \subset \mathbb{C}^m$ into a domain $G' \subset \mathbb{C}^n$ and $A \subset G$, then $\alpha(A, G; z) \leqslant \alpha(T[A], G'; T(z))$ for all $z \in G$. If T is a homeomorphism and $T[G] = G'$, then $\alpha(A, G; z) = \alpha(T[A], T[G]; T(z))$.*

The proof consists only of the observation that if $\phi \in \Phi(T[A], G')$, then $\phi \circ T \in \Phi(A, G)$. If T is a homeomorphism the converse is also true.

As examples of holomorphic homeomorphisms on $\mathbb{C}^n$ we point out the affine transformations

$$T(z) = a + Az$$

where $a \in \mathbb{C}^n$ and A is a non-singular $n \times n$ complex matrix. On $\mathbb{C}^2$ the function

$$T(z_1, z_2) = (z_1, z_2 + f(z_1))$$

is a holomorphic homeomorphism if f is an entire function of one variable.

A2. Some examples

The simplest example occurs if $\bar{A} \supset G$ or if $G = \mathbb{C}^n$. In this case $\alpha(A, G; z) \equiv 1$. However, there is no general formula for evaluating a holomorphic capacity function and consequently it is of considerable importance to have a collection of known examples. Some of the most important for applications stem from the Hadamard three-circle theorem.

Example A2.1. Let $G = \mathscr{B}_R(0)$ and $A = \mathscr{B}_r(0)$ be concentric discs in $\mathbb{C}^1$. If $0<r<R$ then

$$\alpha(\mathscr{B}_r(0), \mathscr{B}_R(0); z) = \min\left\{1, \frac{\log|z/R|}{\log(r/R)}\right\}.$$

If $z \in \overline{\mathscr{B}_r(0)}$ then the minimum is 1, as expected. If $r<|z|<R$ then the Hadamard three-circle theorem shows that

$$\text{(A2.1)} \qquad \log|\phi(z)| \leq \frac{\log|z/R|}{\log(r/R)} \log\|\phi\|_{\mathscr{B}_r(0)} + \frac{1-\log|z/R|}{\log(r/R)} \log\|\phi\|_{\mathscr{B}_R(0)}$$

for all ϕ holomorphic on $G\backslash\bar{A}$, so that

$$\alpha(\mathscr{B}_r(0), \mathscr{B}_R(0); z) \geq \frac{\log|z/R|}{\log(r/R)}$$

there, but equality holds in (A2.1) for $\phi(z) \equiv z/R$. Hence the infimum α is given by $\log|z/R|/\log(r/R)$.

Example A2.2. Let f be a non-constant holomorphic function on a bounded region $G \subset \mathbb{C}^1$ and continuous on $\bar{G}$ taking values with $|f(z)|=1$ for $z \in \partial G$. If $A = \{z \in G : |f(z)| < r < 1\}$, then

$$\text{(A2.2)} \qquad \alpha(A, G; z) \leq \min\left\{1, \frac{\log|f(z)|}{\log r}\right\}$$

since $f \in \Phi(A, G)$. Since $(-\log|f|)/(-\log r)$ is harmonic and positive on $G\backslash\bar{A}$, taking the value zero on ∂G and the value 1 on ∂A, it follows that

$$\frac{-\log|\phi|}{-\log\|\phi\|_A} \geq \frac{-\log|f|}{-\log r} = \frac{-\log|f|}{-\log\|f\|_A} \text{ on } G\backslash A$$

for each $\phi \in \Phi(A, G)$. This gives the inequality which reverses (A2.2) and shows that

$$\alpha(A, G; z) = \min\left\{1, \frac{\log|f(z)|}{\log r}\right\}.$$

If, instead, $|f|$ takes the constant value R on ∂G and r on ∂A then

$$\alpha(A, G; z) = \min\left\{1, \frac{\log|f(z)/R|}{\log(r/R)}\right\},$$

generalizing Example A2.1.

The next four examples and two lemmas concern domains in $\mathbb{C}^n$ with $n \geq 2$. Examples A2.1 and A2.2, together with Proposition A1.6, give lower bounds for α.

Example A2.3. Let f be a non-constant polynomial in n variables. If G is a component of $\{z \in \mathbb{C}^n ; |f(z)| < R\}$ and $A = \{z \in G : |f(z)| < r\}$ then

$$\alpha(A, G; z) = \min\left\{1, \frac{\log|f(z)/R|}{\log(r/R)}\right\}.$$

To see this, consider a point $z \in G \backslash \bar{A}$ and some $z_0 \in \bar{A}$. Define $T: \mathbb{C} \to \mathbb{C}^n$ by

$$T(\zeta) = (1-\zeta)z + \zeta z_0,$$

let G' be the component of $T^{-1}[G]$ containing $\zeta = 0$ and let $A' = G' \cap T^{-1}[A]$. Since $f \circ T$ is a polynomial in ζ and $|f(T(0)| = |f(z)| > r \geq |f(z_0)| = |f(T(1))|$, it follows that $f \circ T$ is not constant and so G' is bounded. Thus, the previous example applies, giving the inequality

$$\alpha(A, G; z) \geq \alpha(A', G'; 0) = \frac{\log |f(T(0))/R|}{\log (r/R)}$$

$$= \frac{\log |f(z)/R|}{\log (r/R)}.$$

On the other hand, since $f \in \Phi(A, G)$ the opposite inequality also holds.

Lemma A2.1. Let G' and G'' be domains in $\mathbb{C}^n$ and $\mathbb{C}^m$ respectively. Let $A' \subset G'$, $A'' \subset G''$, and define $\alpha'(z') = \alpha(A', G'; z')$ and $\alpha''(z'') = \alpha(A'', G''; z'')$. For the cartesian products

$$A' \times A'' \subset G' \times G'' \subset \mathbb{C}^n \times \mathbb{C}^m$$

the following inequalities hold:

$$\alpha'(z')\alpha''(z'') \leq \alpha(A' \times A'', G' \times G''; (z'; z'')) \leq \min\{\alpha'(z'), \alpha''(z'')\}.$$

Proof. If $\phi' \in \Phi(A', G')$, then $\psi(z', z'') = \phi'(z')$ defines a function in $\Phi(A' \times A'', G' \times G'')$. Thus

$$\frac{\log |\phi'(z')|}{\log \|\phi'\|_A} = \frac{\log |\psi(z', z'')|}{\log \|\psi\|_{A' \times A''}} \geq \inf \frac{\log |\phi(z', z'')|}{\log \|\phi\|_{A' \times A''}},$$

the infimum being taken over $\phi \in \Phi(A' \times A'', G' \times G'')$. Therefore

$$\alpha'(z') \geq \alpha(A' \times A'', G' \times G''; (z'; z'')) = \alpha(z', z'').$$

Since the same is true of α'', it follows that

$$\alpha(z', z'') \leq \min\{\alpha'(z'), \alpha''(z'')\}.$$

Now consider instead the other stated inequality and let $\phi \in \Phi(A' \times A'', G' \times G'')$. For a given point (z', z'') and a given $\varepsilon > 0$ we can find $\zeta' \in A'$ such that

$$\frac{\log |\phi(z', z'')|}{\log \|\phi\|_{A' \times A''}} = \frac{\log |\phi(z', z'')|}{\log \|\phi\|_{A' \times \{z''\}}} \left(\frac{-\log \|\phi\|_{A' \times \{z''\}}}{-\log \|\phi\|_{A' \times A''}} \right)$$

$$\geq \alpha'(z') \left\{ \frac{-\log |\phi(\zeta', z'')| - \varepsilon}{-\log \|\phi\|_{A' \times A''}} \right\}$$

$$\geq \alpha'(z') \left\{ \frac{\log |\phi(\zeta', z'')|}{\log \|\phi\|_{\{\zeta'\} \times A''}} - \frac{\varepsilon}{-\log \|\phi\|_{A' \times A''}} \right\}$$

$$\geq \alpha'(z') \left\{ \alpha''(z'') + \frac{\varepsilon}{\log \|\phi\|_{A' \times A''}} \right\}$$

where $\zeta' \in A'$ is chosen so that

$$0 > \log|\phi(\zeta', z'')| + \varepsilon > \log\|\phi\|_{A'\times\{z''\}}.$$

Taking the infimum on ε and then on ϕ gives the result. $\|$

The next example illustrates the possibility of equality between α and $\min\{\alpha', \alpha''\}$.

Example A2.4. For two concentric polydiscs in $\mathbb{C}^n$,

$$G = \mathscr{B}_{R_1}(0)\times\ldots\times\mathscr{B}_{R_n}(0) \supset A = \mathscr{B}_{r_1}(0)\times\ldots\times\mathscr{B}_{r_2}(0)$$

with $0 < r_k < R_k$, $k = 1,\ldots, n$, the holomorphic capacity function is

$$\alpha(A, G; (z_1,\ldots, z_n)) = \min\{\alpha_1(z_1),\ldots, \alpha_n(z_n)\}$$

where

$$\alpha_k(z_k) = \alpha(\mathscr{B}_{r_k}(0), \mathscr{B}_{R_k}(0); z_k) = \min\left\{1, \frac{\log|z_k/R_k|}{\log(r_k/R_k)}\right\}.$$

Define $T: \mathscr{B}_1(0) \to G$ by

$$T(\zeta) = (R_1\lambda_1\zeta^{p_1},\ldots, R_n\lambda_n\zeta^{p_n})$$

where $\lambda_k \in \mathbb{C}$ with $|\lambda_k| = 1$, and p_k is a non-negative integer, $k = 1,\ldots, n$. Let r be the least of the numbers $(r_k/R_k)^{1/p_k}$, so that

$$T[\mathscr{B}_r(0)] \subset A.$$

If $\phi \in \Phi(A, G)$ then $\phi\circ T \subset \Phi(\mathscr{B}_r(0), \mathscr{B}_1(0))$, and if $r < |\zeta| < 1$ then

$$\begin{aligned}\log|\phi\circ T(\zeta)| &\leq \log\|\phi\circ T\|_{\mathscr{B}_r(0)}\alpha(\mathscr{B}_r(0), \mathscr{B}_1(0); \zeta)\\ &\quad + \log\|\phi\circ T\|_{\mathscr{B}_1(0)}\{1 - \alpha(\mathscr{B}_r(0), \mathscr{B}_1(0); \zeta)\}\\ &\leq \log\|\phi\|_A \frac{\log|\zeta|}{\log r} < 0,\end{aligned}$$

that is

$$\begin{aligned}\frac{\log|\phi\circ T(\zeta)|}{\log\|\phi\|_A} &\geq \frac{\log|\zeta|}{\log r}\\ &= \min\left\{\frac{\log|\zeta|}{(1/p_1)\log(r_1/R_1)},\ldots, \frac{\log|\zeta|}{(1/p_n)\log(r_n/R_n)}\right\}\\ &= \min\left\{\frac{\log|z_1/R_1|}{\log(r_1/R_1)},\ldots, \frac{\log|z_n/R_n|}{\log(r_n/R_n)}\right\}\end{aligned}$$

if $z = T(\zeta) \notin \bar{A}$. Thus

$$\frac{\log|\phi(z)|}{\log\|\phi\|_A} \geq \min\{\alpha_1(z_1),\ldots, \alpha_n(z_n)\}$$

for all z on the dense set

$$\{(R_1\lambda_1\zeta^{p_1},\ldots, R_n\lambda_n\zeta^{p_n}) : \lambda_k \in \mathbb{C}, |\lambda_k| = 1, \text{ and } p_k \text{ is a non-negative integer, all } k\},$$

and also for every other point of G by the continuity of ϕ. The reverse inequality is shown by Lemma A2.1. Therefore

$$\alpha(A, G; (z_1,\dots, z_n)) = \min\{\alpha_1(z_1),\dots, \alpha_n(z_n)\}.$$

Example A2.5. The *Lie Ball* $\tilde{B}$ is one of the six classical symmetric domains of E. Cartan (see Hua [51, Chapter IV]). It also arises in the study of harmonic and polyharmonic functions, where there is need for a generalization: the Lie ball with centre x and radius R is the set

$$\tilde{B}_R(x) = \{z \in \mathbb{C}^n : q(z-x) < R\}$$

where

(A2.3) $$q(\xi + i\eta) = [|\xi|^2 + |\eta|^2 + 2\sqrt{\{|\xi|^2\,|\eta|^2 - (\xi, \eta)^2\}}]^{\frac{1}{2}}$$

if $\xi, n \in \mathbb{R}^n$, $(\cdot,\cdot)$ being the inner product in $\mathbb{R}^n$. It will be shown that if $0<r<R$ then the holomorphic capacity for concentric Lie balls is

(A2.4) $$\alpha(\tilde{B}_r(0), \tilde{B}_R(0); z) = \min\left\{1, \frac{\log\{q(z)/R\}}{\log(r/R)}\right\}.$$

In terms of co-ordinates, $\xi_j + i\eta_j$, the formula for q is

(A2.5) $$q(\xi_1 + i\eta_1,\dots, \xi_n + i\eta_n) = \left[\sum_{j=1}^{n} \xi_j^2 + \sum \eta_j^2 + 2\sqrt{\left\{\sum_{j<k} (\xi_j\eta_k - \xi_k\eta_j)^2\right\}}\right]^{\frac{1}{2}}.$$

Consequently, if $\zeta \in \mathbb{C}$ then

(A2.6) $$q(\zeta z) = |\zeta|\, q(z)$$

so that q is homogeneous. If z is any fixed point of $\tilde{B}_R(0)$ except the origin, then

$$T(\zeta) = \zeta z/q(z)$$

defines a holomorphic mapping from $\mathcal{B}_R(0) \subset \mathbb{C}$ into $\mathbb{C}^n$ such that $q(T(\zeta)) = |\zeta|$. By Propositions A1.6 and A1.2, it follows that

$$\alpha(\tilde{B}_r(0), \tilde{B}_R(0); T(\zeta)) \geq \alpha(\mathcal{B}_r(0), \mathcal{B}_R(0), \zeta) = \min\left\{1, \frac{\log|\zeta/R|}{\log(r/R)}\right\}.$$

Since $T(q(z)) = z$, taking $\zeta = q(z)$ shows that

(A2.7) $$\alpha(\tilde{B}_r(0), \tilde{B}_R(0); z) \geq \min\left\{1, \frac{\log\{q(z)/R\}}{\log(r/R)}\right\}.$$

There remains the task of reversing this inequality. For each $z \in \tilde{B}_R(0)\backslash\tilde{B}_r(0)$ a holomorphic function ϕ will be found such that

$$\frac{\log\{q(z)/R\}}{\log(r/R)} = \frac{\log|\phi(z)|}{\log\|\phi\|_{\tilde{B}_r(0)}}.$$

Certainly q itself is not eligible, for it is real and not holomorphic.

It is known (Hua [51], Siciak [81]) that in $\mathbb{C}^2$

$$q(z) = q(z_1, z_2) = \max\{|z_1 + iz_2|, |z_1 - iz|\}.$$

In fact, if $z_1 = x_1 + iy_1$, $z_2 = x_2 + iy_2$, $\xi = (x_1, x_2)$ and $\eta = (y_1, y_2)$, then

$$|z_1 + iz_2|^2 = |\xi|^2 + |\eta|^2 + 2(x_2y_1 - x_1y_2)$$
$$|z_1 - iz_2| = |\xi|^2 + |\eta|^2 - 2(x_2y_1 - x_1y_2)$$

and

$$\{q(z_1, z_2)\}^2 = |\xi|^2 + |\eta|^2 + 2\,|x_2y_1 - x_1y_2|.$$

If $\phi(z_1, z_2) = (z_1 + iz_2)/R$ and $(z_1, z_2) \in \tilde{B}_R(0) \backslash \tilde{B}_r(0)$, then

$$\log|\phi(z_1, z_2)| = \log\{q(z_1, z_2)/R\} < 0 \quad \text{if} \quad x_2y_1 \geqslant x_1y_2$$
$$\log\|\phi\|_A = \log\|q(z_1, z_2)/R\|_A = \log(r/R) < 0$$

where $A = \tilde{B}_r(0)$, and so

$$\frac{\log|\phi(z_1, z_2)|}{\log\|\phi\|_A} = \frac{\log\{q(z_1, z_2)/R\}}{\log(r/R)} \quad \text{if} \quad x_2y_1 \geqslant x_1y_2.$$

Naturally, if $x_2y_1 \leqslant x_1y_2$ then

$$\frac{\log|z_1 - iz_2|}{\log\|z_1 - iz_2\|_A} = \frac{\log\{q(z_1, z_2)/R\}}{\log(r/R)}.$$

Thus, the infimum in the definition of α is attained for $n = 2$ by a constant multiple of one of the two functions $z_1 \pm iz_2$.

For $n > 2$, the calculation is reduced to that of the previous case by a transformation. From (A2.5) it is clear that

$$q(z_1, z_2, z_3, \ldots, z_n) \geqslant q(z_1, z_2, 0, \ldots, 0)$$

and that

$$q(z_1, \ldots, z_n) \geqslant \max\{|z_j + iz_k|, |z_j - iz_k|;\ 1 \leqslant j, k \leqslant n\},$$

where $z_j = \xi_j + i\eta_j$ in terms of the co-ordinates on $\mathbb{R}^n$. Let z be any point of $\tilde{B}_R(0)$, and consider all points

$$\begin{aligned}\zeta z &= (\cos\theta + i\sin\theta)(\xi + i\eta)\\ &= \xi\cos\theta + \eta\sin\theta + i(\xi\sin\theta + \eta\sin\theta)\\ &= \xi' + i\eta'\end{aligned}$$

where $\xi, \eta, \xi', \eta' \in \mathbb{R}^n$ and ζ is a complex number with unit modulus. By the homogeneity of q,

$$q(\zeta z) = q(z).$$

Also, if $z = (z_1, \ldots, z_n)$ and $\zeta z = (z_1', \ldots, z_n')$, then $|z_1' + iz_2'| = |\zeta z_1 + i\zeta z_2| = |z_1 + iz_2|$. By choosing θ correctly it can be arranged that ξ' and η' are orthogonal. Construct a new orthonormal basis for $\mathbb{R}^n$ beginning with multiples of those of ξ' and η' which are not zero. In terms of the new basis, let

$$\phi_+(z_1', \ldots, z_n') = z_1' + iz_2'$$
$$\phi_-(z_1', \ldots, z_n') = z_1' - iz_2'$$

and note that

$$q(y) \geqslant \max\{|\phi_+(y)|, |\phi_-(y)|\}, \qquad y \in \mathbb{C}^n$$

but

$$q(z) = q(\zeta z) = \max\{|\phi_+(z')|, |\phi_-(z')|\} \quad \text{where} \quad z' = \zeta z.$$

Thus the calculation reduces to the case of $n = 2$ and the conclusion holds:

$$\alpha(\tilde{B}_r(0), \tilde{B}_R(0); z) = \min\left\{1, \frac{\log(q(z)/R)}{\log(r/R)}\right\}.$$

Example A2.6. If G is any bounded domain in $\mathbb{C}^n$ and $A = \{x\}$ any single point, for instance the origin, then $\alpha(A, G; z) = \chi_A(z) = \chi_{\{x\}}(z)$, the characteristic function, taking the value 1 on A and zero elsewhere. To see this, let G' be any polydisc about x large enough to contain G, apply Ex. A2.4 using for A' a polydisc in G' about x, and take the limit as $r_1, \ldots, r_n$ decrease to zero.

Of course $\chi_A(z) \leqslant \alpha(A, G; z)$ for any A.

If A is a finite set then subadditivity shows that

$$\chi_A(z) \leqslant \alpha(A, G; z) \leqslant \sum_{x \in A} \chi_{\{x\}}(z) = \chi_A(z).$$

If A is a sequence of distinct points with a single limit point x_0, then

$$\chi_{\bar{A}}(z) \leqslant \alpha(\bar{A}, G; z) = \alpha(A, G; z) \leqslant \alpha(P, G; z) + \chi_A(z)$$

where P is a polydisc about x_0. Taking the limit as the radius of P decreases shows that

$$\chi_{\bar{A}}(z) = \alpha(\bar{A}, G; z) = \chi_{\{x_0\}}(z) + \chi_A(z)$$

which is strictly greater than $\chi_A(z)$ if $z = x_0$. That is, $\alpha(A, G; x_0) \neq \sum_{x \in A} \chi_{\{x\}}(x_0) = \sum_{x \in A} \alpha(\{x\}, G; x_0)$.

This inequality shows that there is no property of countable subadditivity.

For another example, let G and P be concentric polydiscs in $\mathbb{C}^n$, and let A be a countable subset of $\bar{P}$ which is dense in $\bar{P}$.

$$\alpha(A, G; z) = \alpha(\bar{P}, G; z)$$

which is positive at every point of G, while

$$\chi_A(z) = \sum_{x \in A} \chi_{\{x\}}(z)$$

takes value zero outside A, for instance on $G \backslash \bar{P}$.

The point used above, that

$$\chi_{\{0\}}(z) = \lim_{r \to 0} \alpha(\mathcal{B}_r(0), \mathcal{B}_R(0); z)$$

suggests many examples if A and G are cartesian products. Holomorphic capacity is not only finitely additive when applied to characteristic functions, it is also finitely multiplicative. The next Lemma illustrates the possibilities.

Lemma A2.2. If $A' \subset G' \subset \mathbb{C}^m$ *and* $A'' \subset G'' \subset \mathbb{C}^n$, *and if* $\alpha(A', G'; z') = \chi_{A'}(z')$, *then*

$$\alpha(A' \times A'', G' \times G''; (z'; z'')) = \min\{\chi_{A'}(z'), \alpha(A'', G''; z'')\}.$$

If $G' = G'' = G$ *and* $\alpha(A', G; z) = \chi_{A'}(z)$, *then*

$$\alpha(A' \cap A'', G; z) = \min\{\chi_{A'}(z), \alpha(A'', G; z)\} = \chi_{A' \cap A''}(z)$$

and

$$\alpha(A' \cup A'', G; z) = \max\{\chi_{A'}(z), \alpha(A'', G; z)\}.$$

The first assertion follows from Lemma A2.1, and the last two from monotonicity and subadditivity.

Applications of holomorphic capacity usually depend on showing that the capacity functions of interest are not characteristic functions but are positive on $G \backslash A$.

The remaining examples of this section concern bounded plane domains. Interest settles finally on sets A of real numbers, which are the subject of a very important positivity lemma in the next section.

Example A2.7. If $\mathcal{B}_r(x)$ is a subdisc of $\mathcal{B}_1(0) \subset \mathbb{C}^1$ with compact closure in it, then a linear fractional transformation

$$T(z) = \frac{z - \lambda}{1 - \bar{\lambda} z}$$

can be found which carries $\mathcal{B}_r(x)$ to a disc $\mathcal{B}_\rho(0)$ centred at the origin. If $p = |x| - r$ and $q = |x| + r$ with $x \neq 0$, then

$$\lambda = \frac{1 + pq - \sqrt{\{1 - p^2)(1 - q^2)\}}}{pq} \frac{x}{|x|}$$

$$\rho = \frac{1 - pq - \sqrt{\{(1 - p^2)(1 - q^2)\}}}{q - p}$$

and

$$\alpha(\mathcal{B}_r(x), \mathcal{B}_1(0); z) = \min\left\{1, \frac{\log |T(z)|}{\log \rho}\right\}.$$

The limiting case is that in which the two discs have mutually tangent boundaries. It will follow the next example, which generates several more examples as well.

Example A2.8. Let G be the half-plane $\{z : \mathcal{R}e\, z < 1\}$ and $A = \{z : \mathcal{R}e\, z < c\}$. If $f(z) = e^{z-1}$ then $|f(z)| = \exp(\mathcal{R}e\, z - 1)$ and

$$G = \{z \in \mathbb{C} : |f| < 1\}, \qquad A = \{z \in G : |f| < e^{c-1}\}.$$

Ex. A.2. does not apply directly because G is not a bounded set. However, a linear fractional transformation T can be used to map G conformally onto a bounded domain, and so

$$\alpha(A, G; z) = \alpha(T[A], T[G]; T(z)) = \frac{\log |f \circ T^{-1}(T(z))|}{\log \|f \circ T^{-1}\|_{T[A]}}$$

$$= \frac{\log |f(z)|}{\log \|f\|_A} = \frac{1 - \mathcal{R}e\, z}{1 - c} \quad \text{if} \quad z \in G \backslash \bar{A}.$$

Thus,

$$\alpha(A, G; z) = \min\left\{1, \frac{1 - \mathcal{R}e\, z}{1 - c}\right\}.$$

Example A2.9. Let $G = \mathcal{B}_1(0)$ and let A be a subdisc with boundary tangent to ∂G. This is the case not covered by Ex. A2.7.

Specifically, let $A = \mathcal{B}_r(x)$ where $x \in \mathcal{B}_1(0)$ and $r = 1 - |x|$. The transformation

$$\zeta = T(z) = \frac{z \exp(-i \arg x) + 1}{z \exp(-i \arg x) - 1} + 1 = \frac{2z \exp(-i \arg x)}{z \exp(-i \arg x) - 1}$$

carries G and A conformally onto the left half-planes

$$T[G] = \{\zeta \in \mathbb{C} : \mathcal{R}e\, \zeta < 1\} \quad \text{and} \quad T[A] = \left\{\zeta \in \mathbb{C} : \mathcal{R}e\, \zeta < \frac{1 - 2|x|}{1 - |x|}\right\}.$$

By the previous example

$$\alpha(A, G; z) = \alpha(T[A], T[G]; T(z))$$

$$= \min\left\{1, \frac{(1 - |x|)\{1 - \mathcal{R}e\, T(z)\}}{|x|}\right\}$$

$$= \min\left\{1, \frac{1 - |x|}{|x|} \mathcal{R}e\left\{\frac{1 + z \exp(-i \arg x)}{1 - z \exp(-i \arg x)}\right\}\right\}.$$

Exx. A2.10 and A2.11 lead to the holomorphic capacity for the real diameter of a disc.

Example A2.10. Let $G = \{z \in \mathbb{C} : |\mathcal{R}e\, z| < 1\}$ and let A be the imaginary axis. If h is any function harmonic on the right half of the strip, $\{z \in \mathbb{C}; 0 < \mathcal{R}e\, z < 1\}$, positive there, and with limit infimum 1 at points of the imaginary axis, then $h(z) \geq 1 - \mathcal{R}e\, z$. Similar treatment of the left half of the strip shows that

$$\alpha(A, G; z) \geq 1 - |\mathcal{R}e\, z|.$$

However, since e^{z-1} and e^{-z-1} are both in $\Phi(A, G)$, it follows that

$$\alpha(A, G; z) = 1 - |\mathcal{R}e\, z|.$$

Example A2.11. Let G be the complex unit disc $\mathcal{B}_1(0)$, and let A be its real diameter. The mapping $w = (1 + z)/(1 - z)$ carries G onto the right half-plane and $\zeta = \log w$ then carries it to the strip $|\mathcal{I}m\, \zeta| < \pi/2$, each mapping the real axis onto itself. Thus

$$T(z) = i \frac{2}{\pi} \log \frac{1 + z}{1 - z}$$

carries $\mathscr{B}_1(0)$ conformally to the strip $|\mathscr{R}e\, z|<1$, as in Ex. A2.10. It follows that

$$\alpha(A, G; z) = 1 - \left|\mathscr{I}m\left(\frac{2}{\pi}\log\frac{1+z}{1-z}\right)\right|.$$

Example A2.12. The two-to-one function

$$z = \frac{r}{2}\left(\zeta + \frac{1}{\zeta}\right), \qquad \zeta \in \mathbb{C}$$

can be provided with an inverse

$$\zeta = T(z) = \frac{z + \sqrt{(z^2 - r^2)}}{r}$$

which has well-defined branches on the region between the ellipse

$$x^2 + \frac{y^2}{1-r^2} = 1$$

and the segment $[-r, r]$. The inverse takes values with modulus $\{1+\sqrt{(1-r^2)}\}/r$ on the ellipse and modulus 1 on the segment. In between, $\log|T(x+iy)|$ is a harmonic function of x and y. If G_r is the region enclosed by the ellipse and $A = [-r, r]$, then

$$\frac{\log|\phi(z)|}{\log\|\phi\|_A} \geqslant \frac{\log|rT(z)/\{1+\sqrt{(1-r^2)}\}|}{\log[r/\{1+\sqrt{(1-r^2)}\}]}$$
$$= \frac{\log\{|z+\sqrt{(z^2-r^2)}|/(1+\sqrt{(1-r^2)}\}}{\log[r/\{1+\sqrt{(1-r^2)}\}]}, \qquad z \in G_r \backslash A$$

provided that $\phi \in \Phi(A, G_r)$. The same is also true if $\phi \in \Phi(A, \mathscr{B}_1(0))$, and so

$$\min\left\{1, \frac{\log|z|}{\log r}\right\} = \alpha(\bar{\mathscr{B}}_r(0), \mathscr{B}_1(0); z)$$
$$\geqslant \alpha([-r, r], \mathscr{B}_1(0); z) \geqslant \min\left\{1, \frac{\log|\{z+\sqrt{(z^2-r^2)}\}/\{1+\sqrt{(1-r^2)}\}|}{\log[r/\{1+\sqrt{(1-r^2)}\}]}\right\},$$

the lower bound being positive inside the ellipse

$$x^2 + \frac{y^2}{1-r^2} = 1.$$

Example A2.13. If A is a compact subset of $\mathbb{R}$, a closed subset of the bounded interval $[a, b]$ for instance, then it is clear that the Lebesgue measure of A can be obtained as

$$\lim_{k\to\infty} \frac{N(b-a)}{k} = \sigma(A)$$

where N is the number of intervals intersecting A in the partition of $[a, b]$ into k intervals of length $(b-a)/k$. Thus, A will have measure zero if

$$\lim_{k\to\infty} N/k = 0.$$

We shall show that if A belongs to the restricted class for which in addition

$$\lim_{k\to\infty} N/\log k = 0,$$

then

$$\alpha(A, G; z) = \chi_A(z)$$

for each bounded domain $G \subset \mathbb{C}^n$ containing A.

It is no real restriction to assume that $A \subset G \subset \mathcal{B}_1(0)$. If $z \notin A$ and if the k intervals of the kth partition are

$$\left[x_j - \frac{b-a}{2k}, x_j + \frac{b-a}{2k}\right], \qquad j = 1, \ldots, k,$$

then monotonicity and subadditivity give

$$\begin{aligned}
\alpha(A, G; z) &\leq \sum_j \alpha(\mathcal{B}_{(b-a)/2k}(x_j), G; z) \\
&\leq \sum_j \alpha(\mathcal{B}_{(b-a)/2k}(x_j), \mathcal{B}_2(x_j); z) \\
&= \sum_j \frac{\log\{|z - x_j|/2\}}{\log\{(b-a)/4k\}} \\
&\leq \log(\min\{|z - x|/2 : x \in A\}) \sum_j \frac{1}{\log\{(b-a)/4k\}}.
\end{aligned}$$

Since

$$\sum_j \frac{1}{\log\{(b-a)/4k\}} = \frac{N}{\log\{(b-a)/4\} - \log k}$$

and since

$$\lim_{k\to\infty} N/\log k = 0$$

it follows that the upper bound for $\alpha(A, G; z)$ approaches zero as k increases.

A3. Polar and non-polar sets

The usefulness of holomorphic capacity in applications depends on giving the capacity function a positive lower bound on sets of interest. In this section some conditions are established which are sufficient to ensure the existence of positive lower bounds. The first lemma is fundamental.

Lemma A3.1 (Positivity lemma). Let G be a domain in $\mathbb{C}^n$ and let A be any subset of G.

(a) *Let $D = \prod_{j=1}^n D_j$ be a product of simply connected plane domains and let D have compact closure in G. If each D_j has a boundary consisting of a piecewise analytic Jordan curve ∂D_j, if ν_j is the inward normal for ∂D_j, and* *if g_j is the Green function for D_j, then for each $z=(z_1,\ldots, z_n)$ in D*

$$\text{(A3.1)}\quad \alpha(A, G; z) \geqslant \int_{\partial D_1} \cdots \int_{\partial D_n} \left\{\prod_{j=1}^{n} \frac{\partial g_j}{\partial \nu_j}(z_j, \zeta_j)\right\} \alpha(A, G; (\zeta_1,\ldots, \zeta_n))\, d\sigma(\zeta_n) \ldots d\sigma(\zeta_1).$$

(b) *The integral on the right-hand side of* (A3.1) *is a function of z which is harmonic on D with respect to the $2n$ underlying real variables. It is positive there if $\alpha(A, G; \zeta)$ is positive on a subset of $\prod_{j=1}^{n} \partial D_j$ having positive n-dimensional measure.*

Proof. Let $\phi \in \Phi(A, G)$. For any fixed point

$$(z_1,\ldots, z_{n-1}) \in \prod_{j=1}^{n-1} D_j$$

the function $\phi(z_1,\ldots, z_{n-1}, \zeta_n)$ depending on ζ_n alone is holomorphic on a neighbourhood of $\bar{D}_n$ and has in D_n only finitely many zeros, $a_1,\ldots, a_m$. If F_k is a conformal mapping of D_n onto $\mathscr{B}_1(0)$ with its zero at a_k, then

$$\check{\phi}(\zeta_n) = \frac{\phi(z_1,\ldots, z_{n-1}, \zeta_n)}{\prod_{k=1}^{m} F_k(\zeta_n)}$$

is holomorphic and without zeros on D_n, and

$$|\check{\phi}(\zeta_n)| = |\phi(z_1,\ldots, z_{n-1}, \zeta_n)| \quad \text{if} \quad \zeta_n \in \partial D_n.$$

Naturally, $\log |\check{\phi}|$ is harmonic, and so for each $z_n \in D_n$

$$\begin{aligned}\log |\phi(z_1,\ldots, z_{n-1}, z_n)| &\leqslant \log |\check{\phi}(z_n)| \\ &= \int_{\partial D_n} \frac{\partial g_n}{\partial \nu_n}(z_n, \zeta_n) \log |\check{\phi}(\zeta_n)|\, d\sigma(\zeta_n) \\ &= \int_{\partial D_n} \frac{\partial g_n}{\partial \nu_n}(z_n, \zeta_n) \log |\phi(z_1,\ldots, z_{n-1}, \zeta_n)|\, d\sigma(\zeta_n).\end{aligned}$$

Iteration is followed by division by the negative number $\log \|\phi\|_A$. (See Remark A1.1 if $\log \|\phi\|_A = 0$.) Thus,

$$\frac{\log |\phi(z)|}{\log \|\phi\|_A} \geqslant \int_{\partial D_1} \cdots \int_{\partial D_n} \left\{\prod_{j=1}^{n} \frac{\partial g_j}{\partial \nu_j}(z_j, \zeta_j)\right\} \frac{\log |\phi(\zeta)|}{\log \|\phi\|_A}\, d\sigma(\zeta_n) \ldots d\sigma(\zeta_1).$$

Taking the infima over ϕ, first on the right and then on the left, shows the desired inequality for part (a).

Since the Poisson kernels $\partial g_j/\partial \nu_j$ are harmonic and strictly positive, part (b) follows immediately. ||

The general thrust of the lemma is that if $\alpha(A, G; z)$ is positive on a subset of ∂D that is large enough, then it is positive on all of D. To make 'large enough' precise in more general cases, we introduce definitions.

Definition A3.1. Let G be a domain in $\mathbb{C}^n$. A subset M is an n-dimensional *submanifold of real type* if there exists a holomorphic homeomorphism of a subdomain G' of G onto a domain $G'' \subset \mathbb{C}^n$ such that the image of M is $G'' \cap \mathbb{R}^n$. As the n-dimensional measure on the submanifold of real type we take the measure transferred from $\mathbb{R}^n$ by any such mapping. (The differences among these equivalent measures will not affect the results obtained below.)

Definition A3.2. A set A is *polar* in a domain $G \subset \mathbb{C}^n$ if $A \subset G$ and $\alpha(A, G; z) = 0$ for some $z \in G$. If $\alpha(A, G; z) > 0$ for every $z \in G$, then A is *non-polar* in G.

Theorem A3.1. Let G be a domain in $\mathbb{C}^n$ and let A be an arbitrary subset of G.

(a) *A is non-polar if and only if $\alpha(A, G; z) > 0$ for every z in some open subset of G, that is, A is polar if and only if $\alpha(A, G; z) = 0$ for every z in some dense subset of G.*

(b) *A is non-polar if and only if $\alpha(A, G; z) > 0$ for every z in some subset having positive n-dimensional measure in some n-dimensional submanifold of real type in G, that is, A is polar if and only if $\alpha(A, G; z)$ vanishes almost everywhere on n-dimensional submanifolds of real type.*

Proof

(a) Let G_0 be the interior of the set $\{z \in G: \alpha(A, G; z) > 0\}$. It must be shown that either $G_0 = \phi$ or $G_0 = G$. Any point of $\bar{G}_0 \cap G$ has a polydisc neighbourhood D in G whose boundary intersects G_0. The intersection of G_0 with the distinguished boundary of D has positive n-dimensional measure, and so the positivity lemma shows that α is positive on D. That is, every point of $\bar{G}_0 \cap G$ lies in G_0 and the connectivity of G must imply that $G_0 = G$.

(b) It must be shown that if F is a set having positive n-dimensional measure on an n-dimensional real submanifold and if $\alpha(A, G; z)$ is positive at each point of F, then α is positive on an open subset of G, so that part (a) can be applied.

Because real submanifolds are defined by a transformation, it can be assumed that $F \subset \mathbb{R}^n$ and that some compact real cube $Q = Q_1 \times \ldots \times Q_n$ intersects F in a set of positive measure. If r is the distance from Q to ∂G, then let D_j be a rectangle in $\mathbb{C}$ which has Q_j as one side and $r/2$ for its other dimension. Thus, $D = \prod_{j=1}^{n} D_j$ satisfies the hypotheses of Lemma A3.1 and $F \cap Q$ is a subset of ∂D with positive n-dimensional measure. By the lemma, then, $\alpha(A, G; z) > 0$ for all z in D, which is open. $\|$

Corollary A3.1. If G is a domain in $\mathbb{C}^n$ then a set A in G is non-polar if it contains either an open subset of G or a subset with positive n-dimensional measure of an n-dimensional submanifold of real type.

Theorem A3.2. If A is non-polar in a domain G in $\mathbb{C}^n$, then $\alpha(A, G; \cdot)$ has positive infimum on each compact subset of G.

Proof. A compact K can be covered by finitely many polydiscs with compact closure in G. In fact, it can be assumed that they all have the same radius R and that the closed concentric polydiscs of radius $2R$ also lie in G. On the larger polydiscs $\alpha(A, G; \cdot)$ is bounded below by a positive harmonic function, according to Lemma A3.1. Its infimum on each of the smaller polydiscs is positive. The least of these finitely many infima is a positive lower bound for α on K. ‖

Proposition A3.1. If G is a domain in $\mathbb{C}^n$ and A is a subset of G define

$$A_\varepsilon = \{z \in G : \alpha(A, G; z) \geq \varepsilon\}, \qquad 0 < \varepsilon \leq 1.$$

The following relationships hold among the A_ε. For each ε,

$$\alpha(A_\varepsilon, G; z) = 0 \text{ if and only if } \alpha(A, G; z) = 0$$

so that A is polar if and only if A_ε is polar. Also,

$$\alpha(A_1, G; z) = \alpha(A, G; z)$$

and A_1 is the largest set with this property.

Proof. Naturally, $A_\varepsilon \supset A$, and so

$$\alpha(A_\varepsilon, G; z) \geq \alpha(A, G; z) \geq \varepsilon \alpha(A_\varepsilon, G; z)$$

by monotonicity and supermultiplicativity respectively from §A1. ‖

The set A_1 and its properties will be taken up again in §A6.

A4. Estimates and continuity properties

If A is not polar in G then α is positive everywhere. The estimates derived will be of the form $\alpha(A, G; z) \geq b(z)$ where b is a known function that is positive on some region of interest. Usually this region will be near A, but the first lemma gives a lower bound on the annular region between two concentric polydiscs in G.

Lemma A4.1. If A is non-polar in $G \subset \mathbb{C}^n$, if $\alpha(A, G; z) \geq t > 0$ on some polydisc $P = \mathcal{B}_{r_1}(a_1) \times \ldots \times \mathcal{B}_{r_n}(a_n)$ with compact closure in G, and if $P' = \mathcal{B}_{R_1}(a_1) \times \ldots \times \mathcal{B}_{R_n}(a_n)$ is any polydisc in G concentric with P and contain-

ing its closure, then

$$\alpha(A, G; z) \geqslant \min\left\{t\frac{\log(|z_j - a_j|/R_j)}{\log(r_j/R_j)} : j = 1, \ldots, n\right\}$$

$$\geqslant t \prod_{j=1}^{n} \frac{\log(|z_j - a_j|/R_j)}{\log(r_j/R_j)} > 0 \quad \textit{for} \quad z \in P' \backslash P.$$

Since $\alpha(A, G; z) \geqslant t\alpha(P, P'; z)$ the result follows at once from the example on concentric polydiscs and the lemma on cartesian products (§A2).

Hereafter A *will be assumed to be measurable.* Since the capacity function for A is the same as that for the measurable set $\bar{A}$, results can be stated using $\bar{A}$. The cases of particular interest allow A to be a subset of $\mathbb{R}^n$, but require it to have positive measure there.

The *cube* in $\mathbb{R}^n$ with centre $a \in \mathbb{R}^n$, sides parallel to the co-ordinate hyperplanes, and side length $2R$ will be denoted $Q_R(a)$. More generally, for $z \in \mathbb{C}^n$, $Q_R(z) = \{\zeta \in \mathbb{C}^n : \zeta - z \in \mathbb{R}^n$ and $|\zeta_j - z_j| < R,\ 1 \leqslant j \leqslant n\}$. The *relative density of* A *in* $Q_R(a)$ is

$$\frac{\sigma(Q_R(a) \cap A)}{\sigma(Q_R(a))} = (2R)^{-n}\sigma(Q_R(a) \cap A)$$

where σ is Lebesgue measure. The density of the complement is needed frequently and so a special notation is used:

$$\Theta_R(a, A) = (2R)^{-n}\sigma(Q_R(a) \backslash A) = 1 - (2R)^{-n}\sigma(Q_R(a) \cap A).$$

A point $a \in A$ is an *n-dimensional density point* of $A \cap \mathbb{R}^n$ if the relative density rises to 1 as R approaches zero, but the condition is more conveniently stated

$$\lim_{R \to 0} \Theta_R(a, A) = 0.$$

(The *density of* A *at* a *is*

$$1 - \lim_{R \to 0} \Theta_R(a, A).)$$

It is well known that almost all points of A are density points of A. The main result of this section is that at such points $\alpha(A, G; \cdot)$ is continuous.

The basic estimates for $\alpha(A, G; \cdot)$ are made on the complement of the central cube in the polydisc

$$P_R(a) = \prod_{j=1}^{n} \mathscr{B}_R(a)$$

whose closure is to lie entirely in G. This complement has 2^n open components, each of which can be called a 'poly-half-disc', a cartesian product of complex half-discs such as the upper half-disc

$$H_R(a)=\{z\in\mathcal{B}_R(a):\mathcal{I}m(z-a)>0\}.$$

The positivity lemma from the previous section is now applied. We take $D_j=H_R(a_j)$, that is $P_R(a)\subset G$,

$$D=\prod_{j=1}^{n} H_R(a_j),$$

and $A\subset\mathbb{R}^n$. For the integrand, the Green function of the half-disc is needed. Since the normalized Green function for the disc $\mathcal{B}_R(0)$ is

$$\frac{1}{2\pi}\log\left|\frac{R^2-\zeta\bar{z}}{R(\zeta-z)}\right|$$

(the real form is shown in §2, Chapter I), it readily follows that the Green function for the upper half-disc $H_R(0)$ is

$$g(z,\zeta)=\frac{1}{2\pi}\log\left|\frac{R^2-\zeta\bar{z}}{R(\zeta-z)}\right|-\frac{1}{2\pi}\log\left|\frac{R^2-\zeta z}{R(\zeta-\bar{z})}\right|$$

which is the real part of a branch of the logarithm of a holomorphic function. The derivative normal to $\mathbb{R}$ on $\mathbb{R}\cap\partial H_R(0)$ is therefore

$$\frac{\partial}{\partial\nu}=i\frac{\partial}{\partial i\eta}=i\frac{\mathrm{d}}{\mathrm{d}\zeta}\quad\text{if}\quad\zeta=\xi+i\eta$$

and so for $\zeta\in\mathbb{R}\cap\partial H_R(0)$

$$\begin{aligned}\frac{\partial g}{\partial\nu}(z,\zeta)&=\frac{1}{2\pi}\mathcal{R}e\left\{i\frac{\partial}{\partial\zeta}\left(\log\frac{R^2-\zeta\bar{z}}{R(\zeta-z)}-\log\frac{R^2-\zeta z}{R(\zeta-\bar{z})}\right)\right\}\\&=\frac{y}{\pi}\frac{(R^2-|z|^2)(R^2-\zeta^2)}{|\zeta-z|^2\,|R^2-\zeta z|^2}\end{aligned}$$

where $\zeta\in\mathbb{R}$ and $z=x+iy\in H_R(0)$. Replacing ζ and z by ζ_j and z_j and taking the product gives the required integrand.

The evaluation will be made at purely imaginary points $z=x+iy=i(y_1,\ldots,y_n)$, using as the region of integration a region as nearly symmetric about z as possible, namely $Q_R(0)$:

$$\begin{aligned}\alpha(A,G;0+iy)&\geq\int_{Q_R(0)}\cdots\int\left\{\prod_{j=1}^{n}\frac{\partial g}{\partial\nu_j}(iy_j,\zeta_j)\right\}\alpha(A,G;\zeta)\,\mathrm{d}\sigma(\zeta_1)\ldots\mathrm{d}\sigma(\zeta_n)\\&\geq\int_{Q_R(0)\cap\bar{A}}\cdots\int\prod_{j=1}^{n}\frac{\partial g}{\partial\nu_j}(iy_j,\zeta_j)\,\mathrm{d}\sigma(\zeta_1)\ldots\mathrm{d}\sigma(\zeta_n)\\&=\left(\prod_{j=1}^{n}\frac{|y_j|}{\pi}\right)\int_{Q_R(0)\cap A}\cdots\int\prod_{j=1}^{n}\frac{(R^2-y_j^2)(R^2-\zeta_j^2)}{(\zeta_j^2+y_j^2)(R^4+y_j^2\zeta_j^2)}\,\mathrm{d}\sigma(\zeta).\end{aligned}$$

Here we have used the fact that on A the function α takes the value 1.

For convenience the region of integration will be treated as the difference between $Q_R(0)$, where symmetry provides good evaluation, and $Q_R(0)\setminus A$, whose measure is $(2R)^n\Theta_R(0, \bar{A})$, a small number if 0 is a density point of A. For the first, each factor is of the form

$$\text{(A4.1)} \quad \frac{|y|}{\pi}\int_{-R}^{R}\frac{(R^2-y^2)(R^2-\zeta^2)}{(\zeta^2+y^2)(R^4+\zeta^2y^2)}\,\mathrm{d}\zeta=\frac{|y|}{\pi}\int_{-R}^{R}\left\{\frac{1}{\zeta^2+y^2}-\frac{R^2}{R^4+\zeta^2y^2}\right\}\mathrm{d}\zeta$$

$$=\frac{|y|}{\pi}\left(\frac{2}{|y|}\arctan\frac{R}{|y|}-\frac{2}{|y|}\arctan\frac{|y|}{R}\right)=1-\frac{4}{\pi}\arctan\frac{|y|}{R}$$

$$\geqslant 1-\frac{4\,|y|}{\pi R},$$

that is

$$\int_{Q_R(0)}\!\!\cdots\int\prod_{j=1}^{n}\frac{\partial g}{\partial \nu_j}(iy_j, \zeta_j)\,\mathrm{d}\sigma(\zeta)\geqslant\prod_{j=1}^{n}\left(1-\frac{4\,|y_j|}{\pi R}\right)$$

provided that $|y_j|\leqslant \pi R/4$.

The second integral is

$$\text{(A4.2)} \quad \frac{|y_1\ldots y_n|}{\pi^n}\int_{Q_R(0)\setminus A}\!\!\cdots\int\prod_{j=1}^{n}\frac{(R^2-y_j^2)(R^2-\zeta_j^2)}{(\zeta_j^2+y_j^2)(R^4+\zeta_j^2y_j^2)}\,\mathrm{d}\sigma(\zeta)$$

$$\leqslant\frac{|y_1\ldots y_n|}{\pi^n}\left(\prod_{j=1}^{n}\frac{R^4}{y_j^2R^4}\right)(2R)^n\Theta_R(0, \bar{A})=\frac{(2R)^n\Theta_R(0, \bar{A})}{\pi^n\,|y_1\ldots y_n|}.$$

Of course, instead of making the origin the centre of the cubes and polydiscs we could have taken any other point x in $\mathbb{R}^n\cap G$ or in G. Combining (A4.1) and (A4.2) now gives the following preliminary estimate.

Lemma A4.2. If A is a measurable subset of G, if $P_R(x)\subset G$, and if $(y_1,\ldots, y_n)\in\mathbb{R}^n$, then

$$\alpha(A, G; x+i(y_1,\ldots, y_n))\geqslant\prod_{j=1}^{n}\left(1-\frac{4\,|y_j|}{\pi R}\right)-\left(\frac{2R}{\pi}\right)^n\frac{\Theta_R(x, \bar{A})}{|y_1\ldots y_n|}$$

provided that $0<|y_j|<(\pi/4)R$, $j=1,\ldots, n$.

The continuity of $\Theta_R(x, \bar{A})$ as a function of R has allowed the weakening of the previous restriction that the closure of $P_R(x)$ also lie in G. Among the points $x\in\mathbb{R}^n$, $\Theta_R(x, \bar{A})$ also depends continuously on x. This suggests removing that dependence with a further estimate, provided that x lies far enough from the boundary of G. Specifically, if $a\in\mathbb{R}^n$, if

$P_{3R}(a) \subset G$, and $x \in Q_{2R}(a)$, then $\sigma(Q_R(x) \backslash \bar{A}) \leq \sigma(Q_{3R}(a) \backslash \bar{A})$ and so

$$\Theta_R(x, A) \leq \sigma(Q_{3R}(a) \backslash \bar{A})/(2R)^n = 3^n \Theta_{3R}(a, \bar{A}).$$

Combining with Lemma A4.2 leads to the conclusion that

$$\text{(A4.3)} \quad \alpha(A, G; x + i(y_1, \ldots, y_n)) \geq \prod_{j=1}^{n} \left(1 - \frac{4|y_j|}{\pi R}\right) - \left(\frac{6R}{\pi}\right)^n \frac{\Theta_{3R}(a, \bar{A})}{|y_1 \ldots y_n|}$$

provided that $x \in Q_{2R}(a)$ and $0 < |y_j| < (\pi/4)R$, $j = 1, \ldots, n$.

The difficulty with (A4.3) and Lemma A4.2 is that they fail to provide any positive estimate at all near some of the most interesting and important points, those that lie in $\mathbb{R}^n$ but not in A itself. This difficulty is now overcome by using the estimates in succession, capitalizing on the fact that their validity extends to open sets. The result is crucial.

Theorem A4.1 (Two-step estimate). Let G be a domain in $\mathbb{C}^n$ and let A be a measurable subset of G. If $a \in \mathbb{R}^n \cap G$ and $P_{3R}(a) \subset G$, then

$$\begin{aligned} \alpha(A, G; x) &\geq 1 - c_n(\Theta_{3R}(a, \bar{A}))^{1/(n+1)} \\ &= 1 - (n+1)\left(\frac{48}{\pi^2}\right)^{n/(n+1)} (\Theta_{3R}(a, \bar{A}))^{1/(n+1)} \\ &\geq 1 - 5(n+1)\{\Theta_{3R}(a, \bar{A})\}^{1/(n+1)} \end{aligned}$$

for each $x \in Q_R(a) = \mathbb{R}^n \cap P_R(a)$.

Proof. First consider points of the form $x + iy_0$, where $x \in Q_{2R}(a)$ and $y_0 = (y, \ldots, y) \in \mathbb{R}^n$, for some fixed y such that $0 < |y| < \pi R/4n$. By (A4.3),

$$\begin{aligned} \alpha(A, G; x + iy_0) &\geq \left(1 - \frac{4|y|}{\pi R}\right)^n - \left(\frac{6R}{\pi |y|}\right)^n \Theta_{3R}(a, \bar{A}) \\ &\geq 1 - \frac{4n|y|}{\pi R} - \left(\frac{6R}{\pi |y|}\right)^n \Theta_{3R}(a, \bar{A}). \end{aligned}$$

The set of such points $x + iy_0$ is the cube $A_0 = Q_{2R}(a + iy_0)$. On this set the inequality above gives a uniform lower bound for $\alpha(A, G; \cdot)$. By supermultiplicativity (Proposition A1.4) it follows that

$$\begin{aligned} \alpha(A, G; x) &\geq \alpha(A_0, G; x) \inf\{\alpha(A, G; z) : z \in A_0\} \\ &\geq \alpha(A_0, G; x)\left\{1 - \frac{4n|y|}{\pi R} - \left(\frac{6R}{\pi |y|}\right)^n \Theta_{3R}(a, \bar{A})\right\}. \end{aligned}$$

In order to use Lemma 2 to obtain an estimate at x it is necessary that $P_R(z)$ should lie in G, which it does, and that $\Theta_R(z, A_0)$ should be known. Since $Q_R(z) \subset Q_{2R}(a + iy_0) = A_0$, it actually follows that $\Theta_R(z, A_0) = 0$. Therefore Lemma A4.2 allows substitution in the previous inequality giving

$$\alpha(A, G; x) \geq \left(1 - \frac{4n\,|y|}{\pi R}\right)\left\{1 - \frac{4n\,|y|}{\pi R} - \left(\frac{6R}{\pi\,|y|}\right)^n \Theta_{3R}(a, \bar{A})\right\}$$

$$\geq 1 - \frac{8n\,|y|}{\pi R} - \left(\frac{6R}{\pi\,|y|}\right)^n \Theta_{3R}(a, \bar{A}).$$

Since α is non-negative, the last inequality holds whether or not $|y| < \pi R/4n$. Since the left-hand side is independent of y, all that remains is to find that y which maximizes the right-hand side. It is easily seen that it is

$$y_{\max} = \frac{6R}{\pi}\left(\frac{\pi^2}{48}\Theta_{3R}(a, \bar{A})\right)^{1/(n+1)}$$

and the corresponding maximum gives the stated inequality. ‖

The most important consequences are Corollary A4.3, whose proof is apparent from Corollaries A4.1 and A4.2, and Corollary A4.4, which is the key to applications in the next section and in Chapters IV and VI.

Corollary A4.1. If G is open in $\mathbb{C}^n$, if $A \subset G$, and if $P_R(a) \subset G$, then

$$\alpha(A, G; x + iy) \geq \{1 - c_n(\Theta_R(a, \bar{A}))^{1/(n+1)}\}\left(1 - \frac{24\,|y_0|}{\pi R}\right)^n$$

$$\geq [1 - 5(n+1)\{\Theta_R(a, \bar{A})\}^{1/(n+1)}]\left(1 - \frac{24n\,|y_0|}{\pi R}\right)$$

for each $x + iy \in P_{R/8}(a)$ and in fact for $x \in Q_{R/6}(a)$ and

$$|y_j| \leq |y_0| \leq \pi R/24, \qquad j = 1, \ldots, n.$$

Proof. This follows from supermultiplicativity with $K = Q_{R/3}(a)$, using $Q_{R/3}(a)$ in Theorem A4.1 and $P_{R/6}(a)$ in Lemma A4.2, for

$$\Theta_{R/6}(x, Q_{R/3}(a)) = 0 \quad \text{if} \quad x \in Q_{R/6}(a). \quad \|$$

Corollary A4.2. If G is open in $\mathbb{C}^n$, if $A \subset G$, and if the polydisc $P = \mathcal{B}_{R_1}(a_1) \times \ldots \times \mathcal{B}_{R_n}(a_n)$ is a subset of G, then

$$\alpha(A, G; z) \geq \{1 - c_n(\Theta_r(a, \bar{A}))^{1/(n+1)}\}$$

$$\times \left(1 - \frac{24}{\pi}\sigma\right)^n \min\left\{1, \frac{\log\{|z_j - a_j|/R_j\}}{\log\{\sigma r/R_j\}} : j = 1, \ldots, n\right\}$$

for $z \in P$, where $r \leq \min\{R_j; j = 1, \ldots, n\}$ and $\sigma < \pi/24$.

Proof. Corollary A4.1 is applied to $P_r(a)$ for $x + iy \in P_{\sigma r}(a)$, where $|y_j| \leq y_0 = \sigma r \leq \pi r/24$. Lemma A4.1 finishes the estimate. ‖

Corollary A4.3. If G is open in $\mathbb{C}^n$ and A is a subset of G, then $\alpha(A, G; \cdot)$

is continuous at every n-dimensional density point of the intersection of $\bar{A}$ *with any translate of* $\mathbb{R}^n$.

In the following corollary, the form in which Theorem A4.1 will be used in several applications, the continuity of α with respect to changes in A is used.

Corollary A4.4. Let G be a domain in $\mathbb{C}^n$ *and* $A \subset G \cap \mathbb{R}^n$. *If* $\{A_q\}$ *is a sequence of subsets of A and* $\alpha_q(x) = \alpha(A_q, G; x)$, *and if*

$$\lim_{q \to \infty} \sigma(A \backslash A_q) = 0,$$

then for each K compact in $G \cap \mathbb{R}^n$

$$\lim_{q \to \infty} \|1 - \alpha_q\|_K \leq 5(n+1)(1-\delta_K)^{1/(n+1)}$$

where δ_K *is the infimum of the density of A over points of K.*

Proof. If $t > 1$ and $x \in Q_{(t-1)3R}(z)$, then $Q_{3R}(x) \subset Q_{t3R}(z)$. Under this condition

$$\sigma(Q_{3R}(x) \backslash A_q) \leq \sigma(Q_{t3R}(z) \backslash A_q) \leq \sigma(Q_{t3R}(z) \backslash A) + \sigma(A \backslash A_q),$$

so that

$$\Theta_{3R}(x, A_q) \leq t^n \Theta_{t3R}(z, A) + (6R)^{-n} \sigma(A \backslash A_q).$$

For any $\varepsilon > 0$ fix t small enough that $P_{t3R}(z) \subset G$ for each $z \in K$, and choose $R(z)$ so small that

$$t^n \Theta_{t3R(z)}(z, A) < 1 - \delta_K + 2\varepsilon/3.$$

Thus, if $x \in Q_{(t-1)3R(z)}(z)$ then

$$\Theta_{3R(z)}(x, A_q) \leq 1 - \delta_K + 2\varepsilon/3 + \{6R(z)\}^{-n} \sigma(A \backslash A_q)$$

which, in turn, is less than $1 - \delta_K + \varepsilon$ if q is sufficiently large, say $q \geq q_z$.

By Theorem A4.1,

$$\alpha(A_q, G; x) > 1 - 5(n+1)(1 - \delta_K + \varepsilon)^{1/(n+1)}$$

for x in this neighbourhood of $z \in K$ and $q \geq q_z$. Since the compact set K can be covered by finitely many such neighbourhoods, the inequality holds for all $x \in K$ provided that q is sufficiently large. Thus the result follows. ‖

One of the curious properties of holomorphic capacity is that upon restriction to $\mathbb{R}^n$, continuity at a point x becomes indistinguishable from *approximate continuity at x*, which holds for a function f if for every $\varepsilon > 0$

$$\lim_{R \to 0} \frac{\sigma(A_{\varepsilon,R}(x))}{\sigma(Q_R)} = 0$$

where

$$A_{\varepsilon,R}(x)=\{z\in Q_R(x):|f(z)-f(x)|\geq\varepsilon\}.$$

Theorem A4.2. Let G be a domain in $\mathbb{C}^n$ *and A a subset. The restriction of* $\alpha(A, G;\cdot)$ *to* $\mathbb{R}^n$ *is approximately continuous at a point x if and only if* $\alpha(A, G;\cdot)$ *is actually continuous there as a function on G.*

Proof. It is obvious from the definition that continuity always implies approximate continuity. The opposite implication is built on a chain of inequalities, of which the first is upper semicontinuity (Proposition A1.5):

$$\begin{aligned}(\text{A4.4})\qquad \alpha(A, G; x) &\geq \limsup_{z\to x} \alpha(A, G; z)\\ &\geq \liminf_{z\to x} \alpha(A, G; z)\\ &\geq [\alpha(A, G; x)-\varepsilon]\liminf_{z\to x} \alpha(K_\varepsilon, G; z).\end{aligned}$$

The last is supermultiplicativity (Proposition A1.4), using the set

$$K_\varepsilon=\{z\in G\cap\mathbb{R}^n: \alpha(A, G; z)\geq\alpha(A, G; x)-\varepsilon\}.$$

The density of K_ε at x is

$$\begin{aligned}1-\lim_{R\to 0}\Theta_R(x, K_\varepsilon) &= 1-\lim_{R\to 0}\frac{\sigma(Q_R(x)\backslash K_\varepsilon)}{\sigma(Q_R(x))}\\ &= 1-\lim_{R\to 0}\frac{\sigma(A_{\varepsilon,R}(x))}{\sigma(Q_R(x))}=1\end{aligned}$$

by the approximate continuity of $\alpha(A, G;\cdot)$ in $\mathbb{R}^n$. By Corollary A4.3 $\alpha(K_\varepsilon, G;\cdot)$ is continuous at x, so that

$$\lim_{z\to x}\alpha(K_\varepsilon, G; z)=1$$

in the last inequality, that is for each $\varepsilon>0$, (A4.4) shows that

$$\begin{aligned}\alpha(A, G; x) &\geq \limsup_{z\to x}\alpha(A, G; z)\\ &\geq \liminf_{z\to x}\alpha(A, G; z)\geq\alpha(A, G; x)-\varepsilon.\end{aligned}$$

Therefore α is continuous at x. ‖

A5. Applications to sequences of holomorphic functions

Among the sequences of interest are the coefficients $\{f_p\}$ of a power series

$$f(x; y)=\sum_{p=0}^{\infty} f_p(x)y^p$$

where f is holomorphic in x and y, and also the sequence of successive Laplacians $\{\Delta^p u\}$ necessary in the study of a polyharmonic function u. A general technique of proving uniform boundedness properties by means of holomorphic capacity functions will be illustrated by results on exponential growth rates for functions of n variables, lemmas of Hartogs type on the uniform convergence of multiple power series, and a property of upper semi-continuous envelopes.

Definition A5.1. A sequence $\{g_p\}_{p=1}^{\infty}$ of real functions is said to be *eventually uniformly bounded by M on a set A* if there exists p_M such that if $p \geq p_M$ then $g_p(x) \leq M$ for all $x \in A$.

It should be noted that this uniform condition is substantially stronger than the pointwise condition

$$\limsup_{p\to\infty} g_p(x) \leq M \quad \text{for all } x \in A.$$

Part of the relationship between the two is brought out as follows.

Lemma A5.1. Let A be a measurable subset of $\mathbb{R}^n$ with finite non-zero n-dimensional measure $\sigma(A)$, and let $\{g_p\}_{p=1}^{\infty}$ be a sequence of measurable real functions on A such that

$$\limsup_{p\to\infty} g_p(x) \leq m < \infty \quad \textit{for all} \quad x \in A.$$

(The possibility $m = -\infty$ is specifically allowed.) For every $t > 0$ there exists a measurable subset $A_t \subset A$ with the properties

(a) $\sigma(A \backslash A_t) < t\sigma(A)$

(b) *For each $M > m$ the sequence is eventually uniformly bounded by M on A_t, that is*

$$\limsup_{p\to\infty}\left(\sup_{A_t} g_p\right) \leq m.$$

The proof is an application of Egoroff's theorem to the sequence

$$\left\{\sup_{q\geq p} g_q\right\}_{p=1}^{\infty},$$

which converges to

$$\limsup_{p\to\infty} g_p$$

at each point of A.

Theorem A5.1. Let G be a domain in $\mathbb{C}^n$. Consider a sequence $\{\log|f_p|^{\nu_p}\}$ where each f_p is holomorphic on G and $\{\nu_p\}$ is a sequence of positive real *numbers with limit zero such that*

$$\mu = \limsup_{p\to\infty}\left(\sup_G \log|f_p|^{\nu_p}\right)$$

is finite. If there exists m strictly less than μ *such that*

$$\limsup_{p\to\infty} \log|f_p(x)|^{\nu_p} \leqslant m$$

for each x in some measurable $A \subset \mathbb{R}^n \cap G$ *having positive n-dimensional measure, no matter how small, then the following hold.*

(a) *For each compact* $K \subset G$ *there exists* μ_K *strictly less than* μ *such that*

$$\limsup_{p\to\infty}\left(\sup_K \log|f_p|^{\nu_p}\right) \leqslant \mu_K < \mu.$$

(b) *If* $m = -\infty$ *then* $\mu_K = -\infty$ *for each compact K.*

(c) *If m is finite and*

$$\limsup_{p\to\infty}\left(\sup_A \log|f_p|^{\nu_p}\right) \leqslant m,$$

then

$$\mu_K = \mu - (\mu - m)\inf\{\alpha(A, G; z) : z \in K\}.$$

Proof. Lemma A5.1 shows that there exists A_t such that $\sigma(A_t) > 0$ and $\{\log|f_p|^{\nu_p}\}_{p=1}^{\infty}$ is eventually bounded on A_t by each $m' > m$. The fact that A_t is non-polar is essentially a consequence of the positivity lemma, and it follows that for each compact $K \subset G$ the number $c_K = \inf\{\alpha(A_t, G; z) : z \in K\}$ is positive (see Theorems A3.1 and A3.2).

Let $\mu' > \mu$ and $m' = m + (\mu' - \mu)$. There exists p' such that for each $x \in A_t$

$$\log|f_p(x)|^{\nu_p} \leqslant m' \quad \text{for all } p \geqslant p'.$$

Furthermore, p' may be taken so large that if z is any point of G then

$$\log|f_p(z)|^{\nu_p} \leqslant \mu' \quad \text{if} \quad p \geqslant p'.$$

Thus

$$\begin{aligned}\log|f_p(z)|^{\nu_p} &\leqslant \alpha(A_t, G; z)m' + \{1 - \alpha(A_t, G; z)\}\mu' \\ &= \mu' - \alpha(A_t, G; z)(\mu' - m') \\ &= \mu' - \alpha(A_t, G; z)(\mu - m)\end{aligned}$$

for $p \geqslant p'$ and $z \in G$. Taking the suprema over $z \in K$ shows that

$$\sup_K \log|f_p|^{\nu_p} \leqslant \mu' - (\mu - m)c_K$$

for any $\mu' > \mu$ if $p \geqslant p'$. Thus $\mu_K = \mu - (\mu - m)c_K$.

(*c*) If the sequence is eventually uniformly bounded on A itself by each $m' > m$, then $A_t = A$.

(*b*) If $m = -\infty$ then a limiting argument shows the result. ||

Application A5.1. Exponential order of entire functions. If u is holomorphic on a subset $D \times \mathbb{C}$ of $\mathbb{C}^n \times \mathbb{C}$ and $x \in D$ we define $u_x(z) = u(x, z)$, an entire function of z on $\mathbb{C}$. Of course u_x has a Taylor series representation

$$u_x(z) = \sum_{p=0}^{\infty} f_p(x) z^p = u(x, z).$$

In these terms the (exponential) order (Chapter I, §4) of the entire function u_x is

$$\rho(x) = \limsup_{p \to \infty} \frac{p \log p}{-\log |f_p(x)|}.$$

Proposition A5.1. Let D be a domain in $\mathbb{C}^n$, let u be holomorphic on $D \times \mathbb{C}$, and let u_x and $\rho(x)$ be as stated, for each $x \in D$. If $Z \subset D$ is the set on which $\rho(x) = 0$ and if $Z \cap \mathbb{R}^n$ has positive n-dimensional measure, no matter how small, then $\rho(x) \equiv 0$ on D.

The proof rests on Theorem A5.1. Let $G \times \mathcal{B}$ be the product of a subdomain $G \subset D$ and a disc about the origin in $\mathbb{C}$, with compact closure in $D \times \mathbb{C}$ and such that $A = G \cap Z \cap \mathbb{R}^n$ has some positive n-dimensional measure. Note that

$$-\frac{1}{\rho(x)} = \limsup_{p \to \infty} \log |f_p(x)|^{\nu_p}$$

where $\nu_p = 1/(p \log p)$. Thus, $m = -\infty$ in the application of Theorem A5.1. The uniform bound is taken from the Cauchy bounds, for there exist M and δ such that

$$|f_p(x)| \leq M\delta^{-p} \quad \text{if} \quad x \in G, \quad \text{all } p,$$

that is for each $\mu' > 0$

$$\log |f_p(x)|^{\nu_p} = \frac{\log |f_p(x)|}{p \log p} \leq \frac{\log M - p \log \delta}{p \log p} < \mu'$$

if $x \in G$ and p is large enough. Thus

$$\mu = \limsup_{p \to \infty} \left(\sup_G \log |f_p|^{\nu_p} \right) = 0.$$

By Theorem A5.1,

$$-\frac{1}{\rho(x)} = \limsup_{p \to \infty} \log |f_p(x)|^{\nu_p} = -\infty$$

for each $x \in G$. Since G contained any given point of D, $\rho(x) \equiv 0$ on D.

Application A5.2. Two lemmas of Hartogs type. For Hartogs' Lemma itself, see Bochner and Martin [19, p. 137].

Proposition A5.2. Let D be a domain in $\mathbb{C}^n$ *and let* $\{f_p\}$ *be a sequence of functions holomorphic on D. If the series*

$$\sum_{p=0}^{\infty} f_p(x)z^p = u_x(z) = u(x, z)$$

is uniformly absolutely convergent for (x, z) *on compact subsets of* $D \times \mathcal{B}_r(0)$ *for some positive r, no matter how small, and merely absolutely convergent on some product set* $A \times \mathbb{C}$ *where A is measurable in* $D \cap \mathbb{R}^n$ *with positive n-dimensional measure, no matter how small, then the series is uniformly absolutely convergent on compact subsets of* $D \times \mathbb{C}$.

Proof. For each $x \in D$, let $R(x)$ be the radius of convergence of the series

$$\sum_{p=0}^{\infty} f_p(x)z^p = u_x(z).$$

The following relationships are derived from taking the logarithm in the formula for $R(x)$, and they define for each compact K in D the quantities R_K and μ_K:

$$-\log R(x) = \limsup_{p\to\infty} \log |f_p(x)|^{1/p}$$

$$\leq \limsup_{p\to\infty}\left(\sup_K \log |f_p(x)|^{1/p}\right) = -\log R_K = \mu_K$$

if $x \in K$. If $R < R_K$ then there exists p_K such that $|z| \leq R$ implies that

$$\sum_{p=p_K}^{\infty} |f_p(x)z^p| \leq \sum_{p=p_K}^{\infty} (|z|/R)^p$$

so that the series is uniformly convergent on $K \times \mathcal{B}_R(0)$.

Let A' be the set of all $x \in D$ such that u_x is the restriction of an entire function, so that $R(x) = \infty$. By hypothesis, A' has positive n-dimensional measure. Let G and $\mathcal{B}$ be a subdomain of D and a concentric subdisc of $\mathcal{B}_r(0)$ respectively, each having compact closure in the larger set. Let G be so large that $A = A' \cap G$ has positive measure.

We now apply Theorem A5.1 with $\nu_p = 1/p$ and $m = -\infty$. The uniform bound μ is obtained from the Cauchy bounds as follows. For any δ less than the difference between r and the radius of $\mathcal{B}$, there exists a constant M such that

$$|f_p(x)| \leq M\delta^{-p} \quad \text{for each } p, \text{ each } x \in \bar{G}.$$

Thus

$$\limsup_{p\to\infty}\left(\sup_G \log|f_p|^{1/p}\right)\leqslant \limsup_{p\to\infty}\frac{1}{p}(\log M - p\log\delta) = -\log\delta.$$

Since $\delta > 0$ we apply Theorem A5.1, using $A = A'\cap G$ and $\mu = -\log\delta$, and obtaining $-\log R_K = \mu_K = -\infty$.

Thus, u_x is entire for $x\in G$ and the series is uniformly convergent on compact subsets of $G\times\mathbb{C}$. By taking unions we see that the same is true on $D\times\mathbb{C}$. ‖

For a generalization, let u be holomorphic on an arbitrary domain $D\subset\mathbb{C}^{n+1}$ and let $R(x, y)$ be the radius of convergence of the series for u_x:

$$u_x(z) = u(x, z) = \sum_{p=0}^{\infty} f_p(x, y)(z-y)^p, \qquad x\in\mathbb{C}^n, z\in\mathbb{C},$$

with expansion point $y\in\mathbb{C}$. Let

$$\mathcal{A} = \{(x, y)\in D : R(x, y) = \infty\}$$

be the set of points where the power series would give an entire function. Clearly, $\mathcal{A}$ is homogeneous in the last variable: at a given point $(x_0, y_0)\in D$ it happens that $R(x_0, y_0) = \infty$ if and only if $R(x_0, y) = \infty$ for all $(x_0, y)\in D$. Thus

$$\mathcal{A} = D\cap(\mathcal{A}_1\times\mathbb{C}),$$

where $\mathcal{A}_1$ is the projection of $\mathcal{A}$ into $\mathbb{C}^n$.

If $P = P_1\times\mathcal{B}_R(a)$ is any polydisc with compact closure in D, Proposition A5.2 shows the uniform convergence of the series on all compact subsets of $P_1\times\mathbb{C}$, provided only that $\mathcal{A}_1\cap P_1$ intersects some translate of $\mathbb{R}^n$ in a set having positive n-dimensional measure. Such a polydisc can be found if and only if $\mathcal{A}_1$, the projection of $\mathcal{A}$ on $\mathbb{C}^n$, itself intersects some translate of $\mathbb{R}^n$ in a set having positive n-dimensional measure.

Proposition A5.2 leads to the conclusion that P_1 is a subset of $\mathcal{A}_1$ and the series is uniformly convergent on compact subsets of $P_1\times\mathbb{C}$.

Under this condition, Theorem A5.1 can be applied, taking for A the intersection of the open set P with some translate of $\mathbb{R}^{n+1}$, for G any relatively compact subdomain of D containing P, and

$$\begin{aligned}-\log R(x, y) &= \limsup_{p\to\infty}\log|f_p(x, y)|^{1/p}\\ &\leqslant \limsup_{p\to\infty}\left(\sup_G\log|f_p|^{1/p}\right)\\ &\leqslant \limsup\frac{1}{p}(\log M - p\log\delta) = -\log\delta\end{aligned}$$

where $|f_p(x, y)| \leq M\delta^{-p}$ on $\bar{G}$, by the Cauchy bounds. Since $m = -\log R(x, y) = -\infty$ on A, the first conclusion of the theorem is that $G \subset \mathcal{A}$ and the series is convergent on $G_1 \times \mathbb{C}$ where G_1 is the projection of G into $\mathbb{C}^n$. However, the theorem also shows that the series is uniformly convergent on compact subsets of $G_1 \times \mathbb{C}$. Since G may be an arbitrarily large relatively compact subdomain of D, the conclusion holds for D itself. It is stated as follows.

Proposition A5.3. If u is holomorphic as a function of $(x_1, \ldots, x_n, y)$ *on a domain* $D \subset \mathbb{C}^{n+1}$, *if* u_x *is defined by* $u_x(y) = u(x, y)$, *and if the set*

$$\mathcal{A} = \{(x, y) \in D \colon u_x \text{ is an entire function of } y\}$$

has projection into $\mathbb{C}^n$ *which intersects some translate of* $\mathbb{R}^n$ *in a set having positive n-dimensional measure, then u has continuation as a holomorphic function on all of* $D_1 \times \mathbb{C}$, *where* D_1 *is the projection of D into* $\mathbb{C}^n$.

The two applications shown above have rested on estimates which can be written

$$\text{(A5.1)} \qquad -\log R(x) \leq -\log \delta - (\log r - \log \delta)\alpha(A, G; x)$$

where $R(x) \geq r$ is a relatively high pointwise lower bound holding on A while a constraint involving only $\log \delta$ applies to the suprema of the sequence on G. In these we have achieved the inequality $R(x) \geq r$ on G by restricting to the case $r = \infty$, provided that A is not too small.

The generalizations obtained by simply relaxing the requirement $r = \infty$ involve obtaining bounds which depend on the values $\alpha(A, G; x)$ and their infima on compact subsets of G. These are obvious and we shall not pursue them.

There are also generalizations obtained after weakening the constraint on the suprema. The results are somewhat analogous to Lemma A5.1 and Theorem A5.1 and are offered for the sake of completeness.

Lemma A5.2. Let G be an open set in $\mathbb{R}^n$ *and let* $\{g_p\}_{p=1}^{\infty}$ *be a sequence of functions continuous on G such that*

$$\limsup_{p \to \infty} g_p(x) < m \leq \infty \quad \textit{for each } x \in G.$$

There exists an open set G_0 *which is dense in G such that for each compact* $K \subset G_0$ *the number*

$$m_K = \limsup_{p \to \infty} \Big(\sup_K g_p \Big)$$

is strictly less than m.

Proof. Let $\{M_k\}$ be an increasing sequence of numbers with limit m, and observe that if $G_1 \subset G$ then

$$G_1 = \bigcup_{k=1}^{\infty} \bigcup_{p=1}^{\infty} \bigcap_{q \geq p} \{x \in G_1 : g_p(x) \leq M_K\}.$$

If $A_{k,p} = \bigcap_{q \geq p} \{x \in G_1 : g_p(x) \leq M_K\}$, closed in G_1, and G_1 has non-void interior, then the Baire Category Theorem implies that some $A_{k,p}$ also has non-void interior.

Now let $A^0_{k,p}$ be the interior of

$$\bigcap_{q \geq p} \{x \in G : g_p(x) \leq M_K\}.$$

Define

$$G_0 = \bigcup_{k=1}^{\infty} \bigcup_{p=1}^{\infty} A^0_{k,p}$$

and observe from the previous paragraph that G_0 has non-void intersection with every non-void open subset G_1 of G. Hence G_0 is dense. If K is compact in G_0 then it may be covered by finitely many of the $A^0_{k,p}$, in fact by one of them. Hence the result follows. ||

Theorem A5.2. Let G be an open set in $\mathbb{C}^n$. *Consider a sequence* $\{\log |f_p|^{\nu_p}\}$, *where each* f_p *is holomorphic on G and* $\{\nu_p\}$ *is a sequence of positive real numbers with limit zero, such that*

$$\limsup_{p \to \infty} \log |f_p(z)|^{\nu_p} < \infty \quad \textit{for each } z \in G.$$

Let G_0 *be the set constructed in Lemma A5.2 such that for each K compact in* G_0 *the number*

$$\limsup_{p \to \infty} \left(\sup_K \log |f_p|^{\nu_p} \right)$$

is finite, and let A be the set of $z \in G$ *such that*

$$\limsup_{p \to \infty} \log |f_p(z)|^{\nu_p} = -\infty.$$

If $A \cap \mathbb{R}^n$ *intersects some component* G_1 *of* G_0 *in a set with positive n-dimensional measure, then for each K compact in* G_1

$$\limsup_{p \to \infty} \left(\sup_K \log |f_p|^{\nu_p} \right) = -\infty.$$

Proof. Since G_1 can be exhausted by relatively compact subdomains intersecting $A \cap \mathbb{R}^n$ in sets with positive n-dimensional measure, to each of which Theorem A5.1 can be applied, the result follows. ||

For the last two estimates, of the form

$$-\log R(x) \leq -\log \delta - (\log r - \log \delta)\alpha(A, G; x)$$

as at (A5.1), instead of showing that $R(x) \geq r$ by restricting to the case $r = \infty$ we shall achieve the same result by showing that $\alpha(A, G; x) = 1$ at points of interest. For this, we call on the continuity properties of α, shown in the previous section.

Theorem A5.3. Let G be a domain in $\mathbb{C}^n$. Consider a sequence $\{\log |f_p|^{\nu_p}\}$ where each f_p is holomorphic on G and $\{\nu_p\}$ is a sequence of positive real numbers with limit zero such that

$$\limsup_{p\to\infty} \left(\sup_G \log |f_p|^{\nu_p} \right) = \mu < +\infty.$$

If A is open in $\mathbb{R}^n \cap G$ (or indeed if $A = D \backslash Z$ where D is open in $\mathbb{R}^n \cap G$ and $\sigma(Z) = 0$) and if

$$\limsup_{p\to\infty} \log |f_p(x)|^{\nu_p} \leq m \quad \text{for each } x \in A$$

then

$$\limsup_{p\to\infty} \log |f_p(x)|^{\nu_p} \leq \limsup_{p\to\infty} \left(\sup_K \log |f_p|^{\nu_p} \right) \leq m$$

for each x in each K compact in A (or compact in D, or in $\check{A}$, the union of all such D).

Proof. Since $\sigma(A) > 0$ for the only non-vacuous conclusions, the case $m = -\infty$ is simply Theorem A5.1. Therefore let $m > -\infty$ and let K be fixed. Since K is bounded it suffices to consider G to be bounded as well, so that $A \subset \mathbb{R}^n \cap G$ is bounded and therefore has finite measure.

Since there is nothing to prove if $m = \mu$ it is assumed that $m < \mu$. For each $\varepsilon > 0$ there exists p_G such that

$$\sup_G \log |f_p(z)|^{\nu_p} \leq \mu + \varepsilon \quad \text{for} \quad z \in G, p \geq p_G.$$

By Lemma A5.1, for each positive integer q there exists a set A_q such that $\sigma(A \backslash A_q) \leq \sigma(A)/q$ and

$$\limsup_{p\to\infty} \left(\sup_{A_q} \log |f_p|^{\nu_p} \right) \leq m.$$

By Corollary A4.4, there exists q so large that

$$\alpha(A_q, G; z) \geq 1 - \frac{\varepsilon}{\mu - m} \quad \text{for} \quad z \in K$$

since every point of A is a density point of A. Therefore, fix q.

There exists $p_\varepsilon \geq p_G$ such that

$$\sup_{A_q} \log |f_p|^{\nu_p} \leq m+\varepsilon \quad \text{for} \quad p \geq p_\varepsilon.$$

Under the same condition, if $z \in K$ then

$$\begin{aligned}\log |f_p(z)|^{\nu_p} &\leq \{1-\alpha(A_q, G; z)\}(\mu+\varepsilon)+\alpha(A_q, G; z)(m+\varepsilon)\\ &\leq \frac{\varepsilon}{\mu-m}(\mu+\varepsilon)+\left(1-\frac{\varepsilon}{\mu-m}\right)(m+\varepsilon)=m+\varepsilon,\end{aligned}$$

that is, for each ε, if p is large then

$$\sup_K \log |f_p|^{\nu_p} \leq m+\varepsilon. \quad \|$$

Theorem A5.3 leads to yet another version of Hartogs' Lemma.

Proposition A5.4. Let D be a domain in $\mathbb{C}^n$ and let $\{f_p\}$ be a sequence of functions holomorphic on D. If the series

$$\sum_{p=0}^{\infty} f_p(x) z^p$$

is uniformly absolutely convergent for (x, z) on compact subsets of $D \times \mathscr{B}_r(0)$ for some positive r, no matter how small, and merely absolutely convergent on some product set $A \times \mathscr{B}_R(0)$, where $R > r$ and A is an open subset of $D \cap \mathbb{R}^n$ such that $\sigma(D \cap \mathbb{R}^n \setminus A)=0$, then the series is also uniformly absolutely convergent on compact subsets of $(D \cap \mathbb{R}^n) \times \mathscr{B}_R(0)$.

Proof. For each $x \in D$ let $R(x)$ be the radius of convergence of the series

$$\sum_{p=0}^{\infty} f_p(x) z^p = u_x(z).$$

For each conditionally compact subset $K \subset D$ the following relationships define the quantity R_K:

$$\begin{aligned}-\log R(x) &= \limsup_{p\to\infty} \log |f_p(x)|^{\nu_p}\\ &\leq \limsup_{p\to\infty}\left(\sup_K \log |f_p(x)|^{\nu_p}\right) = -\log R_K \leq -\log r < \infty\end{aligned}$$

for $x \in K$. As in the proof of Proposition A5.2, the series is uniformly absolutely convergent on $K \times \mathscr{B}_t$ if $t < R_K$.

Now take G to be a conditionally compact subdomain of D. If $A' = A \cap G$ then $\sigma(\mathbb{R}^n \cap G \setminus A')=0$ and Theorem A5.3 applies, for $-\log R(x) \leq -\log R$ on A'. Thus

$$-\log R(x) \leq -\log R_K \leq -\log R \quad \text{for} \quad x \in K$$

where K is any compact subset of $\mathbb{R}^n \cap G$. By the arbitrariness of G, the same holds for each compact K in $\mathbb{R}^n \cap D$. As remarked above then, the series must be uniformly convergent on compact subsets of $K \times \mathcal{B}_{R_K}(0)$, and hence on compact subsets of $D \cap \mathbb{R}^n \times \mathcal{B}_R(0)$. ‖

Application A5.3. The upper semi-continuous envelope of the limit of a sequence of functions.

Any real-valued function L on a set D has an *upper semi-continuous envelope* $\hat{L}$ defined by

$$\hat{L}(z) = \inf\left\{\sup_U L \colon U \text{ is a neighbourhood of } z\right\}.$$

Naturally, $\hat{L}$ is *upper semi-continuous*, in that

$$\hat{L}(z) \geqslant \limsup_{x \to z} \hat{L}(x).$$

It is also the least semi-continuous function on D which is bounded below by L, that is if φ is an upper semi-continuous real function on D and $\varphi \geqslant L$ on D then $\varphi \geqslant \hat{L}$ on D.

Theorem A5.4. Let G be a domain in $\mathbb{C}^n$ and consider the sequence $\{|f_p|^{\nu_p}\}$ where each f_p is holomorphic on G and $\{\nu_p\}$ is a sequence of positive real numbers with limit zero. Define

$$L(z) = \limsup_{p \to \infty} |f_p(z)|^{\nu_p}, \qquad z \in G$$

and let L_r be the restriction of L to $G \cap \mathbb{R}^n$.

If

$$\limsup_{p \to \infty} \|f_p\|_G^{\nu_p} < \infty$$

then for every $x \in G \cap \mathbb{R}^n$ and every $\varepsilon > 0$ there exists a neighbourhood U of x relative to $G \cap \mathbb{R}^n$ such that

$$\text{(A5.2)} \qquad L_r(x) \leqslant \limsup_{p \to \infty} \|f_p\|_U^{\nu_p} \leqslant \hat{L}_r(x) + \varepsilon$$

where $\hat{L}_r$ is the upper semi-continuous envelope of L_r. $\hat{L}_r$ is the least function on $G \cap \mathbb{R}^n$ satisfying (A5.2) for each x.

Proof. Let $x \in G \cap \mathbb{R}^n$ and $\varepsilon' > 0$. Since $\hat{L}_r$ is the upper semi-continuous envelope of L_r, there exists a compact neighbourhood A of x in $G \cap \mathbb{R}^n$ on which

$$\log L(z) \leqslant \log \hat{L}_r(z) \leqslant \log \hat{L}_r(x) + \varepsilon'/3$$

for all $z \in A$. If U is any compact neighbourhood of x lying in the interior of A, it will be shown that there exists p_U such that

$$\log |f_p(z)|^{\nu_p} \leqslant \log \hat{L}_r(x) + \varepsilon'$$

if $z \in U$ and $p \geqslant p_U$.

Define

$$A_q = \left\{ z \in A : \log |f_p(z)|^{\nu_p} \leqslant \log \hat{L}_r(x) + \frac{2\varepsilon'}{3} \quad \text{for} \quad p \geqslant q \right\}$$

and note that the sequence $\{A_q\}$ is nested, non-decreasing with union A. To this is applied Corollary A4.4: if $\alpha_q(z) = \alpha(A_q, G; z)$ then

$$\lim_{q \to \infty} \|1 - \alpha_q\|_U = 0.$$

To use these, two estimates on $\{f_p\}$ are required. First, if ν is any number strictly larger than

$$\limsup_{p \to \infty} \log \|f_p\|_G^{\nu_p} \geqslant \log \|L\|_G \geqslant \log \|\hat{L}_r\|_{G \cap \mathbb{R}^n}$$

then there exists p_ν such that

$$\log \|f_p\|_G \leqslant \frac{1}{\nu_p} \nu \quad \text{if} \quad p \geqslant p_\nu.$$

Second, if $z \in A_q$ and $p \geqslant q$ then

$$\log |f_p(z)|^{\nu_p} \leqslant \log \hat{L}_r(x) + \frac{2\varepsilon'}{3}$$

and so

$$\log \|f_p\|_{A_q} \leqslant \frac{1}{\nu_p} \left\{ \log \hat{L}_r(x) + \frac{2\varepsilon'}{3} \right\}.$$

Thus, if z is any point of G and $p \geqslant q \geqslant p_\nu$, then

$$\begin{aligned} \log |f_p(z)| &\leqslant \frac{1}{\nu_p} \left\{ \log \hat{L}_r(x) + \frac{2\varepsilon'}{3} \right\} \alpha_q(z) + \frac{\nu}{\nu_p} \{1 - \alpha_q(z)\} \\ &= \frac{1}{\nu_p} \left[\log \hat{L}_r(x) + \frac{2\varepsilon'}{3} \alpha_q(z) + \{1 - \alpha_q(z)\}\{\nu - \log \hat{L}_r(x)\} \right] \end{aligned}$$

and if, further, $z \in U$ then

$$\begin{aligned} \log |f_p(z)|^{\nu_p} &= \nu_p \log |f_p(z)| \\ &\leqslant \log \hat{L}_r(x) + \frac{2\varepsilon'}{3} + \|1 - \alpha_q\|_U \{\nu - \log \hat{L}_r(x)\} \\ &\leqslant \log \hat{L}_r(x) + \varepsilon' \end{aligned}$$

if q is sufficiently large. For fixed x and small enough ε', the desired inequality follows.

The fact that $\hat{L}_r$ is least follows from the fact that if h satisfies (A5.2) then

$$h(x)+\varepsilon \geqslant L_r(z)$$

on a neighbourhood U of x, and the definition of $\hat{L}_r$ now shows that

$$h(x)+\varepsilon \geqslant \hat{L}_r(x), \quad \text{for every } \varepsilon>0. \quad \|$$

The property which $\hat{L}_r$ shows at (A5.2) will now be given a name to distinguish it more clearly from the related stronger property to appear in Corollary A5.1.

Definition A5.2. Let h be a real function on a set D, let $\{g_p\}$ be a sequence of real functions there, and let $x \in D$. If, for each $\varepsilon>0$, x has a neighbourhood $U(\varepsilon)$ such that

$$\limsup_{p\to\infty} \sup_{U(\varepsilon)} g_p \leqslant h(x)+\varepsilon,$$

then h provides the sequence with an *eventual uniform bound near x*. By contrast, the function h is a *local uniform majorant* for the sequence if each $x \in D$ has a neighbourhood U such that

$$\limsup_{p\to\infty} \sup_{U} (g_p - h) \leqslant 0.$$

In these terms, $\hat{L}_r$ provides $\{|f_p|^{\nu_p}\}$ with a uniform majorant near each $x \in \mathbb{R}^n \cap G$.

Corollary A5.1. Under the hypotheses of Theorem A5.4, any real function h on $G \cap \mathbb{R}^n$ which either

(1) *is continuous, with $h \geqslant L_r$, or*

(2) *is lower semi-continuous, with $h \geqslant \hat{L}_r$*

must be a local uniform majorant of $\{|f_p|^{\nu_p}\}$ on $G \cap \mathbb{R}^n$.

Proof. Since condition (1) implies condition (2), it must be shown that the latter implies the conclusion.

Let $x \in G \cap \mathbb{R}^n$ and $\varepsilon>0$. By the lower semi-continuity of h there exists a neighbourhood V of x such that $h(z)>h(x)-\varepsilon/2$ for $z \in V$. By Theorem A5.4 there exist a sub-neighbourhood U and p_ε such that if $z \in U$ and $p>p_\varepsilon$

$$\begin{aligned} |f_p(z)|^{\nu_p} &\leqslant \hat{L}_r(x)+\varepsilon/2 \\ &\leqslant h(x)+\varepsilon/2 \\ &\leqslant h(z)+\varepsilon. \end{aligned}$$

Thus

$$\sup_U(|f_p|^{\nu_p}-h)\leq\varepsilon \quad \text{for} \quad p>p_\varepsilon. \quad \|$$

Corollary A5.2. Under the hypotheses of Theorem A5.4, let $\hat{L}$ be the upper semi-continuous envelope of L on G. If $\hat{L}$ is continuous on G then it is a local uniform majorant of $\{|f_p|^{\nu_p}\}$ on G.

Proof. Let $x\in\mathbb{R}^n\cap G$. Since $\hat{L}$ is lower semi-continuous, for any $\varepsilon>0$ there is a neighbourhood U of x in G on which

$$\text{(A5.3)} \qquad \hat{L}(z)>\hat{L}(x)-\varepsilon/4, \quad \text{for all } z\in U.$$

There is also a neighbourhood of x in $G\cap\mathbb{R}^n$ on which the conclusion of Theorem A5.4 holds. By intersection we take this to be $A=U\cap\mathbb{R}^n$ and we can assume it to be a compact set:

$$\text{(A5.4)} \qquad \begin{aligned}\limsup_{p\to\infty}\|f_p\|_A^{\nu_p}&\leq\hat{L}_r(x)+\varepsilon/4\\ &\leq\hat{L}(x)+\varepsilon/4.\end{aligned}$$

The third and last contribution to the desired inequality begins

$$\begin{aligned}\nu_p\log|f_p(z)|&\leq\nu_p\log\|f_p\|_A\alpha(A,G;z)+\nu_p\log\|f_p\|_G\{1-\alpha(A,G;z)\}\\ &\leq\nu_p\log\|f_p\|_A\delta_K+\nu_p\log\|f_p\|_G(1-\delta_K)\end{aligned}$$

if z lies in some set K and

$$\delta_K=\inf_{y\in K}\alpha(A,G;y).$$

Thus, if $p\geq p_\varepsilon$ then (A5.4) implies that

$$\|f_p\|_K^{\nu_p}\leq\{\hat{L}(x)+2\varepsilon/4\}^{\delta_K}\left(\limsup_{p\to\infty}\|f_p\|_G^{\nu_p}\right)^{1-\delta_K}.$$

Corollary A4.3 shows that $\alpha(A,G;\cdot)$ is continuous at x. If K is a sufficiently small neighbourhood of x then δ_K is arbitrarily near unity. Therefore

$$\begin{aligned}|f_p(z)|^{\nu_p}\leq\|f_p\|^{\nu_p}&\leq\hat{L}(x)+3\varepsilon/4\\ &\leq\hat{L}(z)+\varepsilon \quad \text{if} \quad z\in K\cap U\end{aligned}$$

where the last inequality is (A5.3). $\|$

The strength of the condition of eventual uniform boundedness is now illustrated in a way which relates it, although weakly, with the continuity properties of the limit supremum.

Theorem A5.5. Let G be a domain in $\mathbb{C}^n$ and consider the function

$$L(z)=\limsup_{p\to\infty}|f_p(z)|^{\nu_p}, \qquad z\in G$$

where each f_p is holomorphic on G and $\{\nu_p\}$ is a sequence of positive real numbers with limit zero. For every $\varepsilon>0$ and $x\in D=G\cap\mathbb{R}^n$ define

$$A(x,\varepsilon)=\{z\in D\colon L(z)<L(x)-\varepsilon\}.$$

If

$$\limsup_{p\to\infty}\|f_p\|_G^{\nu_p}<\infty$$

then a point $x\in D$ is not a density point of $A(x,\varepsilon)$, and in fact

$$\liminf_{R\to 0}\Theta_R(x,A(x,\varepsilon))\geq\left[\frac{\varepsilon/c_n}{\varepsilon+L(x)\log\{\hat{L}(x)/L(x)\}}\right]^{n+1}>0$$

where $c_n=(n+1)(48/\pi^2)^{n/(n+1)}$ and $\hat{L}$ is the upper semi-continuous envelope of L on G.

Proof. If $A(x,\varepsilon)$ has zero measure there is nothing to prove. Therefore, fix x and $\varepsilon>0$ such that $L(x)>\varepsilon$ and $\sigma(A(x,\varepsilon))>0$. Also let

$$\nu>\mu=\limsup_{p\to\infty}\|f_p\|^{\nu_p}.$$

Of course

$$A(x,\varepsilon)=\bigcup_{\xi>0}A(x,\varepsilon+\xi).$$

Accordingly, for each $\eta>0$ there exists $\xi>0$ such that $\sigma(A(x,\varepsilon)\backslash A(x,\varepsilon+\xi))<\eta/2$. Thus, by Lemma A5.1 there exists a set $A_\eta\subset A(x,\varepsilon+\xi)$ with the properties

(a) $\sigma(A(x,\varepsilon)\backslash A_\eta)\leq\sigma(A(x,\varepsilon)\backslash A(x,\varepsilon+\xi))+\sigma(A(x,\varepsilon+\xi)\backslash A_\eta)<\eta$

and

(b) there exists p_ε such that if $p\geq p_\varepsilon$ and $z\in A_\eta$ then

$$|f_p(z)|^{\nu_p}<L(x)-(\varepsilon+\eta)<L(x)-\varepsilon.$$

Naturally, p_ε can also be taken so large that

$$\|f_p\|_G^{\nu_p}<\nu\quad\text{if}\quad p\geq p_\varepsilon.$$

Therefore

$$\log|f_p(z)|^{\nu_p}\leq\log\{L(x)-\varepsilon\}\alpha(A_\eta,G;z)+\log\nu\{1-\alpha(A_\eta,G;z)\}$$

for all $z\in G$, and in particular for $z=x$. Taking the limit supremum shows that

$$\log L(x)\leq\log\{L(x)-\varepsilon\}\alpha(A_\eta,G;x)+\log\nu\{1-\alpha(A_\eta,G;x)\}$$
$$\leq\left\{\log L(x)-\frac{\varepsilon}{L(x)}\right\}\alpha(A_\eta,G;x)+\log\nu\{1-\alpha(A_\eta,G;x)\}$$

so that

$$\frac{\varepsilon}{L(x)} \leqslant \left\{\log \nu - \log L(x) + \frac{\varepsilon}{L(x)}\right\}\{1-\alpha(A_\eta, G; x)\}$$

or

$$\frac{\varepsilon}{\varepsilon + L(x)\log\{\nu/L(x)\}} \leqslant 1-\alpha(A_\eta, G; x).$$

If R is sufficiently small that the polydisc $P_R(x)$ lies in G, then Theorem A4.1 implies that

$$\begin{aligned} 1-\alpha(A_\eta, G; x) &\leqslant c_n(\Theta_R(x, \bar{A}_\eta))^{1/(n+1)} \\ &\leqslant c_n(\Theta_R(x, A_\eta))^{1/(n+1)} \end{aligned}$$

and so

$$\Theta_R(x, A_\eta) > \left[\frac{\varepsilon/c_n}{\varepsilon + L(x)\log\{\nu/L(x)\}}\right]^{n+1}.$$

Of course this is true for all η, and hence also for $A(x, \varepsilon) = \cup A_\eta$, also for all $\nu > \mu$, and hence for μ as well. Since it is also true for all small enough $R > 0$, and since G may be restricted to be $P_R(x)$, it follows that μ may be arbitrarily close to $\hat{L}(x) = \inf\{\|f_p\|_U^{\nu_p}: U$ is a neighbourhood of x in $G\}$. Hence the result follows, since

$$\lim_{R\to 0} \Theta_R(x, A) = 0$$

if x is a density point of $A(x, \varepsilon)$. ‖

A6. Completeness

In Proposition A3.1 we introduced the set $A_1 = \{z \in G: \alpha(A, G; z) = 1\}$. We saw that $A \subset A_1$, A_1 is relatively closed in G, is the largest subset of G on which $\alpha(A, G; z) \equiv 1$, and is the largest subset of G with the property that $\alpha(A, G; z) \equiv \alpha(A_1, G; z)$.

Definition A6.1. Let G be a domain in $\mathbb{C}^n$ and let $A \subset G$.

(a) The set A is called *complete in* G, or simply G*-complete* if $A = \{z \in G: \alpha(A, G; z) = 1\}$.

(b) The set $A_1 = \{z \in G: \alpha(A, G; z) = 1\}$ is called the *completion of* A *in* G, or simply the G*-completion of* A.

It is clear that if A is G-complete then it is its own G-completion, that is $A = A_1$, and in all cases A_1 is G-complete. In addition, a G-complete set is always relatively closed in G. We shall give conditions under which a set A is G-complete, give several examples, and then discuss the connection with domains of holomorphy.

Theorem A6.1. A necessary and sufficient condition that A be G-complete is that for every $z \in G\backslash A$ *there exists a function* $\varphi \in \Phi(A, G)$ *such that* $|\varphi(z)| > \|\varphi\|_A$.

Proof. The set A is G-complete if and only if $\alpha(A, G; z) < 1$ for all $z \in G\backslash A$, and this holds if and only if for each $z \in G\backslash A$ there exists $\varphi \in \Phi(A, G)$ with

$$\alpha(A, G; z) \leqslant \frac{\log |\varphi(z)|}{\log \|\varphi\|_A} < 1. \quad \|$$

Theorem A6.2. A necessary and sufficient condition that A be G-complete is that there exists a (finite or infinite) class of functions $\Psi \subset \Phi(A, G)$ *and for each* $\psi \in \Psi$ *there exists a number* $\rho_\psi > 0$ *such that* $A = \{z \in G : |\psi(z)| \leqslant \rho_\psi$, for all $\psi \in \Psi\}$, *that is*

$$A = \bigcap_{\psi \in \Psi} \{z \in G : |\psi(z)| \leqslant \rho_\psi\}.$$

Proof. If the condition holds, then for each $z \in G\backslash A$ there exists $\psi_0 \in \Psi$ such that $|\psi_0(z)| > \rho_{\psi_0}$. Since $\|\psi_0\|_A \leqslant \rho_{\psi_0}$ the G-completeness follows from Theorem A6.1. Conversely, if A is G-complete, take $\Psi = \Phi(A, G)$ and $\rho_\psi = \|\psi\|_A$. $\|$

Example A6.1. Let A be the closed polydisc

$$\bar{P}_r(a) = \prod_{j=1}^{n} \overline{\mathcal{B}_{r_j}(a_j)}$$

and let G be the open polydisc

$$P_R(a) = \prod_{j=1}^{n} \mathcal{B}_{R_j}(a_j)$$

with $0 < r_j < R_j$. That A is G-complete can be seen either from Theorem A6.1 or Theorem A6.2 (or from Ex. A2.4 where α is given explicitly). Let $z \in G\backslash A$. There exists j, $1 \leqslant j \leqslant n$, such that $|z_j - a_j| > r_j$. Taking

$$\varphi(x) = \frac{1}{R_j}(x_j - a_j)$$

we have $\varphi \in \Phi(A, G)$ and

$$|\varphi(z)| = \frac{1}{R_j}|z_j - a_j| > \frac{r_j}{R_j} = \|\varphi\|_A,$$

so that Theorem A6.1 yields the conclusion. Alternatively, noting that a point

$z \in A$ if and only if

$$\frac{1}{R_j}|z_j - a_j| \leq \frac{r_j}{R_j} \quad \text{for all } j = 1, \ldots, n$$

and the functions

$$\psi_j(x) = \frac{1}{R_j}(x_j - a_j)$$

all belong to $\Phi(A, G)$, we can take $\Psi = \{\psi_j\}$, $\rho_{\psi_j} = r_j$, and apply Theorem A6.2. A slight modification of this argument shows that $\bar{P}_r(a)$ is G-complete in any domain G which has non-constant bounded holomorphic functions.

Example A6.2. Let G be the open polydisc

$$P_R(a) = \prod_{j=1}^{n} \mathcal{B}_{R_j}(a_j)$$

and let

$$A = \prod_{j=1}^{n} \partial\mathcal{B}_{r_j}(a_j),$$

the distinguished boundary of

$$P_r(a) = \prod_{j=1}^{n} \mathcal{B}_{r_j}(a_j), \qquad 0 < r_j < R_j.$$

The set A is not G-complete since if $\varphi \in \Phi(A, G)$ then for all $z \in P_r(a)$ we have $|\varphi(z)| \leq \|\varphi\|_A$. Thus $A_1 \supset P_r(a)$ and by Ex. A6.1 we have that $\bar{P}_r(a)$ is G-complete. Hence $A_1 = \bar{P}_r(a)$. This example is actually a special case of the more general result (whose proof is the same): if K is a compact subset of a domain G, if K is G-complete, and if A is the Šilov boundary of K, then $A_1 = K$.

Example A6.3. Let $G = P_3 \backslash \bar{P}_1$ where

$$P_r = \prod_{j=1}^{n} \mathcal{B}_r$$

and $n > 1$. Let A be the distinguished boundary $\prod_{j=1}^{n} \partial\mathcal{B}_2$ of P_2. It is well-known (e.g. Hormander [50, Theorem 2.3.2]) that every holomorphic function on G has a unique analytic extension to all of P_3. Thus if $\varphi \in \Phi(A, G)$, then for every $z \in P_2 \cap G$ we have $|\varphi(z)| \leq \|\varphi\|_A$. By Ex. A6.2 it follows that the G-completion is $A_1 = \bar{P}_2 \backslash \bar{P}_1 = \bar{P}_2 \cap G$. Note that A_1 is relatively closed in G but is not compact. (For $n = 1$, A is G-complete as the next example shows.)

Example A6.4. Let $R_j = \{z_j \in \mathbb{C}^1 : 1 < |z_j| < 3\}$ and let $G = \prod_{j=1}^{n} R_j$, that is G is a product of annular regions. Take $A = \prod_{j=1}^{n} \partial\mathcal{B}_2$, the distinguished boundary of the polydisc P_2. Using theorem A6.2 we can show that A is G-complete (compare Exx. A6.2 and A6.3). Indeed, let $\psi_j(z) = z_j/3$, $1 \leq j \leq n$, and let $\tilde{\psi}_j(z) = 1/z_j$, $1 \leq j \leq n$. The collection $\Psi = \{\psi_j, \tilde{\psi}_j\}_{1 \leq j \leq n} \subset \Phi(A, G)$ and $z \in A$ if and only if $|\psi_j(z)| \leq \frac{2}{3}$, $|\tilde{\psi}_j(z)| \leq \frac{1}{2}$, $1 \leq j \leq n$.

Exx. A6.2 and A6.4 show that if G' and G are domains in $\mathbb{C}^n$, and if

$A \subset G' \subset G$, then A may be G'-complete but not G-complete. However, we have the following.

Theorem A6.3. If A is G-complete then A is G'-complete for every $G' \subset G$ with $A \subset G'$.

Proof. By Proposition A1.2, $\alpha(A, G'; z) \leq \alpha(A, G; z)$ for all $z \in G'$. Since A is G-complete, for every $z \in G \setminus A$ we have $\alpha(A, G; z) < 1$. Thus if $z \in G' \setminus A$ then $\alpha(A, G'; z) < 1$ so that A is G'-complete. ‖

The next two theorems give set-theoretic and topological conditions for completeness of A.

Theorem A6.4. Let $\{A_k\}$ be a decreasing sequence of subsets of a domain G and let

$$A = \bigcap_k A_k.$$

If each A_k is G-complete and $A \neq \phi$, then A is G-complete.

Proof. If $A = G$ there is nothing to prove, so we can assume that $A_1 \neq G$. Let $z \in G \setminus A$. There exists k_0 such that $z \in G \setminus A_{k_0}$, and, since A_{k_0} is G-complete, by Theorem A6.1 there is a function $\varphi_0 \in \Phi(A_{k_0}, G)$ with $|\varphi_0(z)| > \|\varphi_0\|_{A_{k_0}}$. If $\|\varphi_0\|_A > 0$, the theorem is proved. If not, let $\varphi(x) = \frac{1}{2}\varphi_0(x) + \frac{1}{4}\varphi_0(z)$. The function φ is in $\Phi(A, G)$ because $|\varphi(x)| < \frac{3}{4}$ for all $x \in G$ and $\|\varphi\|_A = \frac{1}{4}|\varphi_0(z)| > 0$. Moreover we have $|\varphi(z)| = \frac{3}{4}|\varphi_0(z)| > \frac{1}{4}|\varphi_0(z)| = \|\varphi\|_A$. Thus A is G-complete. ‖

Remark A6.1. We included $A \neq \phi$ in the hypothesis because it can happen that

$$\bigcap_k A_k = \phi.$$

For instance, in $\mathbb{C}^1$ take G to be the unit ball and $A_k = \{iy : 1 - 1/k \leq y < 1\}$.

Let L be a k-dimensional complex linear affine subspace (that is a translate of a subspace) of $\mathbb{C}^n$. If $H \subset L$, denote by $\partial_L H$ the boundary of H when H is considered as a subset of L.

Theorem A6.5. Let A be G-complete. If L is any k-dimensional complex linear affine subspace of $\mathbb{C}^n$, $1 \leq k \leq n$, if $H \subset L$ is a bounded open subset of L with $\bar{H} \subset G$, and if $\partial_L H \subset A$, then $\bar{H} \subset A$.

Proof. Under the conditions of the theorem, every function $\varphi \in \Phi(A, G)$ is holomorphic as a function restricted to $G \cap L$. Thus $\|\varphi\|_H \leq \|\varphi\|_{\partial_L H} \leq \|\varphi\|_A$, so by Theorem A6.1 every $z \in H$ belongs to the G-completion of A, that is to A. ‖

Example A6.5. Let G be the polydisc $P_2 = \prod_{i=1}^n \mathcal{B}_2$ and let $A = \{z \in \mathbb{C}^n : |z_1| = 1, z_2 = \ldots = z_n = 0\}$. By Theorem A6.5, A cannot be G-complete since it would have to contain $\bar{H}=\{z\in\mathbb{C}^n: |z_1|\leqslant 1, z_2=\ldots=z_n=0\}$. However, $\bar{H}$ is the G-completion of A. To see this, let $\bar{H}_k$ be the closed polydisc $\mathscr{B}_1\times\prod_{j=2}^n \mathscr{B}_{1/k}$, $k=1,2,\ldots$. By Ex. A6.1, $\bar{H}_k$ is G-complete and since

$$\bar{H}=\bigcap_k \bar{H}_k$$

by Theorem A6.4 we have that $\bar{H}$ is G-complete.

In Ex. A6.3, with $G=P_3\backslash\bar{P}_1$ and A the distinguished boundary of P_2, we saw that the G-completion of A was not compact in G even though A was compact. The fact that G is not a domain of holomorphy led to our result. We shall show that this is the general situation.

Definition A6.2. Let G be open in $\mathbb{C}^n$ and let K be a compact subset of G. Define $\hat{K}_G\equiv\{z\in G: |f(z)|\leqslant\|f\|_K$ for all f holomorphic on $G\}$.

Recall that K_1 denotes the G-completion of K.

Lemma A6.1. If K is a compact subset of G then $K_1=\{z\in G: |f(z)|\leqslant\|f\|_K$ for all bounded holomorphic f on $G\}$.

Proof. If f is bounded and holomorphic on G with $\|f\|_K>0$ we have

$$|f(z)|\leqslant\|f\|_K^{\alpha(K,G;z)}\,\|f\|_G^{1-\alpha(K,G;z)}, \qquad z\in G.$$

Thus $z\in K_1$ if and only if $|f(z)|\leqslant\|f\|_K$ for such f. Therefore the set described in the statement of the lemma is a subset of K_1. Now if f is bounded and holomorphic on G but $\|f\|_K=0$, for each $\varepsilon>0$ define $g_\varepsilon(z)=f(z)+\varepsilon$. Then $\|g_\varepsilon\|_K=\varepsilon>0$, and so if $z\in K_1$ we have $|g_\varepsilon(z)|\leqslant\varepsilon$, or $|f(z)+\varepsilon|\leqslant\varepsilon$. Letting ε approach zero we obtain $|f(z)|\leqslant 0=\|f\|_K$ if $z\in K_1$. This shows the reverse inclusion and completes the proof. ||

Corollary A6.1. If K is a compact subset of G then $\hat{K}_G\subset K_1$.

Corollary A6.2. If K is a compact subset of G and if K is G-complete then $\hat{K}_G=K$.

Lemma A6.2. A domain G in $\mathbb{C}^n$ is a domain of holomorphy if and only if for every compact $K\subset G$, $\hat{K}_G$ is compact in G.

For a proof see Hörmander [50, Theorem 2.5.5].

Theorem A6.6. Let G be a domain in $\mathbb{C}^n$. If K_1 is compact in G for every compact subset K of G then G is a domain of holomorphy.

Proof. Let K be a compact subset of G. By Corollary A6.1, $\hat{K}_G\subset K_1$, and by hypothesis K_1 is compact in G. Since $\hat{K}_G$ is closed in G it follows that $\hat{K}_G$ is compact in G. By Lemma A6.2, G is a domain of holomorphy. ||

The converse of Theorem A6.6 is not true. In standard terminology, a domain $G\subset\mathbb{C}^n$ is *convex with respect to a family $\mathscr{F}$ of functions* defined on G if for every set $K \subset G$ with compact closure in G the set $\{z \in G: |f(z)| \leq \|f\|_K \text{ for all } f \in \mathscr{F}\}$ has compact closure in G. In Lemma A6.2, domains of holomorphy are characterized by being convex with respect to the family of *all* holomorphic functions on G, but the set K_1 refers to the family of *bounded* holomorphic functions on G. In order for the converse of Theorem A6.2 to be true it would be necessary that convexity with respect to bounded holomorphic functions follow from convexity with respect to all holomorphic functions, which is not true in general. This is easy to see in $\mathbb{C}^1$. If $G = \mathscr{B}_1(0) \backslash \{0\}$ then G is a domain of holomorphy (all domains in $\mathbb{C}^1$ are domains of holomorphy). However, any bounded holomorphic function on G has a unique analytic continuation to $\mathscr{B}_1(0)$. Thus if $K = \{z \in \mathbb{C}^1: |z| = \frac{1}{2}\}$, it follows readily that $K_1 = \overline{\mathscr{B}_{\frac{1}{2}}(0)} \backslash \{0\}$, which is not compact in G. Sibony [80] has given an example in $\mathbb{C}^2$ which shows that $\hat{K}_G$ can be a proper subset of K_1, so that the converse of Theorem A6.6 cannot be true for $n > 1$.

A7. Holomorphic capacity on manifolds

In this section we show how the notion of capacity can be taken over to complex manifolds and indicate how the results of the preceding sections can be formulated on these manifolds. We shall assume that the reader is familiar with the definition of complex manifold and submanifold, and in keeping with our terminology we shall refer to 'holomorphic' manifolds, which are called 'complex analytic' in much of the literature.

We consider $\mathbb{C}^n$ as both a complex n-dimensional vector space and a real $2n$-dimensional vector space. Let S be a subspace of the real $2n$-dimensional vector space and let $S_c = \{v \in S: iv \in S\}$. Clearly $S_c \subset S$ and S_c is a complex subspace of $\mathbb{C}^n$. We denote by k the dimension of S as a real subspace and by k_c the dimension of S_c as a complex subspace (so that the real dimension is $2k_c$). The real dimension of a complement of S_c in S is $k - 2k_c$.

Definition A7.1. If $k_c = 0$ the subspace S is said to be of *real type.*

For example, $\mathbb{R}^k$ and $i\mathbb{R}^k$ are of real type for each k, $1 \leq k \leq n$.

Definition A7.2. The *dimension* (of S) *of complex type* is k_c, and the *dimension* (of S) *of real type* is $k - 2k_c$.

Proposition A7.1. If two real subspaces S and S' have the same dimensions of complex and real types, then there exists a holomorphic linear homeomorphism of S onto S'.

The manifolds and submanifolds which we consider will be assumed to be connected. Since a complex n-dimensional holomorphic manifold can be viewed as a real $2n$-dimensional manifold, we shall use the symbols $\mathcal{M}_h^n$ and $\mathcal{M}_r^{2n}$ to distinguish these two structures.

Definition A7.3. A real submanifold $\mathcal{N}_r^k$ of $\mathcal{M}_r^{2n}$ is said to be *compatible with the holomorphic structure* of $\mathcal{M}_h^n$ if for every $z \in \mathcal{N}_r^k$ there exists a chart $[U, F]$ of z in $\mathcal{M}_h^n$ such that $F(U \cap \mathcal{N}_r^k) \subset S^k \subset \mathbb{C}^n$ where S^k is a real k-dimensional subspace of $\mathbb{C}^n$.

The dimension k decomposes into $k = k_r + 2k_c$ where k_c is the dimension of complex type and k_r the dimension of real type.

Proposition A7.2. If $\mathcal{N}_r^k$ is compatible with the holomorphic structure of $\mathcal{M}_h^n$ then the dimensions of complex type and real type are constant on $\mathcal{N}_r^k$.

Definition A7.4.

(a) If the dimension of real type of $\mathcal{N}_r^k$ is zero, $\mathcal{N}_r^k = \mathcal{N}_h^{k_c}$ is a *holomorphic submanifold* of $\mathcal{M}_h^n$.

(b) If the dimension of complex type of $\mathcal{N}_r^k$ is zero, $\mathcal{N}_r^k$ is a *k-dimensional submanifold of real type* of $\mathcal{M}_h^n$.

In the case $\mathcal{M}_h^n = \mathbb{C}^n$, we have already encountered submanifolds of real type (§§A3, A4 and A5).

The definitions in the preceding sections can now be transferred to manifolds. Let $\mathcal{M}_h^n$ be a holomorphic manifold and let A be a subset of $\mathcal{M}_h^n$. The class of functions $\Phi(A, \mathcal{M}_h^n)$ is defined as before with G replaced by $\mathcal{M}_h^n$ and the holomorphic capacity $\alpha(A, \mathcal{M}_h^n; z)$ is defined. As before, A is polar if $\alpha(A, \mathcal{M}_h^n; z) = 0$ for some $z \in \mathcal{M}_h^n$; otherwise A is non-polar.

All of the results of §A1 transfer immediately to the case of manifolds. In §A3 the results hold on manifolds by making the appropriate local interpretations in the proofs. In particular, Theorem A3.1(b) is stated as follows.

Theorem A7.1. If $\mathcal{M}_h^n$ is a holomorphic manifold and $\mathcal{N}_r^n$ is a submanifold (compatible with the holomorphic structure) of dimension n of real type, and if A is a subset of $\mathcal{M}_h^n$ whose closure intersects $\mathcal{N}_r^n$ in a set of positive n-dimensional Lebesgue measure on $\mathcal{N}_r^n$, then A is non-polar.

Similarly most of the results of §§A4 and A5 are valid on manifolds with the appropriate interpretation. In those statements which involve n-dimensional Lebesgue measure of a set $A \subset \mathbb{R}^n \subset G$, the change should be made as in the formulation of Theorem A7.1.

REFERENCES

The numbers in parentheses refer to the section(s) in which the reference is cited.

1. Agmon, S., Douglis, A., and Nirenberg, L. (1959). Estimates near the boundary for solutions of elliptic partial differential equations satisfying general boundary conditions, I. *Commun. Pure Appl. Math.* **12,** 623–727.
   Agmon, S., Douglis, A., and Nirenberg, L. (1964). Estimates near the boundary for solutions of elliptic partial differential equations satisfying general boundary conditions, II. *Commun. Pure Appl. Math.* **17,** 35–92. (§6, Chapter VI.)
2. Aleksandrov, P. S. (1960). *Combinatorial topology*, Vol. 3. Graylock Press, Albany, NY. (§4, Chapter II.)
3. Almansi, E. (1898). Sulle integrazione dell' equazione differenziale $\Delta^{2m}u = 0$. *Ann. Mat. Pura Appl., Suppl.* 3, **2.** (§1, Chapter I.)
4. Armitage, D. H. (1978). Reflection principles for harmonic and polyharmonic functions. *J. Math. Anal. Appl.* **65,** 44–55. (§1, Chapter I.)
5. Armitage, D. H. (1979). On the derivatives at the origin of entire harmonic functions, *Glasgow Math. J.* **20,** 147–154. (§2, Chapter VI.)
6. Aronszajn, N. (1935). Sur les décomposition des fonctions analytiques uniformes et sur leurs applications. *Acta Math.* **65,** 1–156. (§1, Chapter II; §2, Chapter II; §2, Chapter III.)
7. Aronszajn, N. (1937). Sur un théorème de la théorie des fonctions analytiques de plusieurs variables complexes. *C. R. Acad. Sci. Paris* **205,** 16–18. (§1, Chapter II; §1, Chapter V.)
8. Aronszajn, N. (1938). Un théorème sur les fonctions analytiques de plusieurs variables complexes. *Bull. Sci. Math., Ser. II* **62,** 149–160. (§A1, Appendix.)
9. Aronszajn, N. (1951). Green's functions and reproducing kernels. *Proc. Symp. on Spectral Theory and Differential Problems, Stillwater,* OK, 1951, pp. 355–412. (§6, Chapter VI.)
10. Aronszajn, N. (1973). Preliminary notes for the talk on 'Traces of analytic solutions of the heat equation'. *Colloq. Int. C.N.R.S., Soc. Math. France, astérisque* 2 *et* 3, pp. 5–34. (§7, Chapter VI.)
11. Aronszajn, N. (1973). Traces of analytic solutions of the heat equation. *Colloq. Int. C.N.R.S., Soc. Math. France, astérisque* 2 *et* 3, pp. 35–68. (§7, Chapter VI.)
12. Aronszajn, N. (1977). Calculus of residues and general Cauchy formulas in $\mathbb{C}^n$. *Bull. Sci. Math.* 2 **101** (4), 319–352. (§1, Chapter V.)
13. Aronszajn, N. and Milgram, A. N. (1953). Differential operators on Riemannian manifolds, *Rend. Circ. Mat. Palermo, Ser.* 2 **2,** 266–325. (§6, Chapter VI.)
14. Avanissian, V. (1977). Distributions harmoniques d'order infini et l'analyticité (réelle) liée à l'opérateur laplacien itéré, *Lect. Notes Math.* **581,** 1–20. (§1, Chapter II.)
15. Avanissian, V. and Fernique, X. (1968). Sur l'analyticité des distributions 'harmoniques d'order infini', *Ann. Inst. Fourier* (*Grenoble*) **18,** 1–9. (§1, Chapter II.)
16. Baouendi, M. S. (1976). Solvability of partial differential equations in the traces of analytic solutions of the heat equation, *Amer. J. Math.* **97,** 983–1005. (§7, Chapter VI.)

17. Baouendi, M. S., Goulaouic, C., and Lipkin, L. J. (1974). On the operator $\Delta r^2 + \mu(\partial/\partial r)r + \lambda$. *J. Diff. Eqn.* **15,** 499–509. (§1, Chapter V.)
18. Boas, R. P., Jr. (1954). *Entire Functions.* Academic Press, New York. (§4, Chapter I; §§1, 7 Chapter VI.)
19. Bochner, S. and Martin, W. T. (1948). *Several complex variables.* Princeton University Press, Princeton, NJ. (§3, Chapter II; §A5, Appendix.)
20. Boggio, T. (1905). Sulle funzioni di Green d'ordine *m*. *Rend. Circ. Math. Palermo* **20,** 97–135. (§2, Chapter II.)
21. Bramble, J. H. (1958). Continuation of biharmonic functions across circular arcs. *J. Math. Mech.* **7,** 905–924. (§1, Chapter I.)
22. Bramble, J. H. and Payne, L. E. (1966). Mean value theorems for polyharmonic functions. *Amer. Math. Mon.*, **73** (part II), 124–127. (§1, Chapter I.)
23. Bremekampt, H. (1945). Properties of the solutions of $\Delta^k u = 0$. *Indag. Math.* **7,** 34–41. (§1, Chapter I.)
24. Cheng, M.-T. (1951). On a theorem of Nicolesu and generalized Laplace operators. *Proc. Amer. Math. Soc.* **2,** 77–86. (§1, Chapter I.)
25. Ciorănescu, N. (1936). Sur les fonctions analytiques de deux variables réelles. *Bull. Math. Soc. Roum. Sci.* **38,** 63–70. (§2, Chapter VI).
26. Ciorănescu, N. (1937). Sur la représentation des fonctions analytiques de plusieurs variables réelles. *Bull. Soc. Math. Fr.* **65,** 41–52. (§1, Chapter V.)
27. Ciorănescu, N. (1944). Sur la détermination des fonctions harmoniques par leurs valeurs sur des variétés caractéristiques *Math. Timişoara* **20,** 137–147. (§1, Chapter V.)
28. Colucci, A. (1937). Sulla rappresentazione delle funzioni iperarmoniche a mezzo di armoniche. *Acad. Sci. Fis. Mat. Napoli, Rend.* **4**(Suppl. 7), 55–62. (§1, Chapter I.)
29. Duffin, R. J. (1955). Continuation of biharmonic functions by reflection. *Duke Math. J.* **22,** 313–324. (§1, Chapter I.)
30. Duffin, R. J. and Nehari, Z. (1961). Note on polyharmonic functions. *Proc. Amer. Math. Soc.* **12,** 110–115. (§1, Chapter I.)
31. Fichera, G. (1941). Sviluppi in serie e teoremi di decomposizione in somma per le funzioni iperarmoniche. *Rend. Circ. Math. Palermo* **63,** 24. (§1, Chapter I.)
32. Fichera, G. (1942). Un teorema generale sulla struttura delle funzioni iperàrmoniche. *Atti Accad. Ital., Rend. Cl. Sci. Fis. Mat. Nat.* **3** (7), 511–523. (§1, Chapter I.)
33. Friedman, A. (1957). On *n*-metaharmonic functions and harmonic functions of infinite order. *Proc. Amer. Math. Soc.* **8,** 223–229. (§§3, 5, Chapter VI.)
34. Friedman, A. (1969). *Partial differential equations.* Holt, Rinehart, and Winston, New York, 1969. (§§3, 5, Chapter VI.)
35. Friedman, A. (1957). Mean values and polyharmonic polynomials. *Michigan Math. J.* **4,** 67–74. (§1, Chapter I.)
36. Fryant, A. J. (1978). Growth of entire harmonic functions in $\mathbb{R}^3$. *J. Math. Anal. Appl.* **66,** 599–605. (§2, Chapter VI.)
37. Fugard, T. B. (1980). Growth of entire harmonic functions in $\mathbb{R}^n$, $n \geq 2$. *J. Math. Anal. Appl.* **74,** 286–291. (§2, Chapter VI.)
38. Fuks, B. A. (1963). *Introduction to the theory of analytic functions of several complex variables, Translations of Mathematical Monographs,* Vol. 8, American Mathematical Society, Providence, RI, 1963. (§4, Chapter I.)
39. Gagaeff, B. (1937). Sur les familles normales de fonctions polyharmoniques. *Rec. Math. Moscow, New Ser.* **2,** 759–767. (§1, Chapter I.)
40. Ghermanescu, M. (1937). Sur un théorème de M. Mauro Picone. *Atti. Accad. Naz. Lincei, Rend.* **6** (Suppl. 25), 553–557. (§1, Chapter I.)

41. GHERMANESCU, M. (1944). Sur les valeurs moyennes des fonctions. *Math. Ann.* **119,** 288–320. (§5, Chapter VI.)
42. GHERMANESCU, M. (1945). Sur les valeurs moyennes des fonctions. *Bull Sci. Ec. Polytech. Timişoara* **12,** 17–48. (§5, Chapter VI.)
43. GHERMANESCU, M. (1956). Sur les fonctiones n-harmoniques. *Acad. Repub. Pop. Rom. Bul. Stiint. Sect. Stiinte. Mat. Fiz.* **8,** 529–536. (§1, Chapter I.)
44. GILLMAN, L. and JERISON, M. (1960). *Rings of continuous functions*, Van Nostrand, New York. (§4, Chapter II.)
45. GRAUERT, H. and FRITZSCHE, K. (1976). *Several complex variables.* Springer Verlag, New York. (§4, Chapter I.)
46. HARDY, G. H. and WRIGHT, E. M. (1960). *An introduction to the theory of numbers*, 4th edn. Clarendon Press, Oxford. (§A1, Appendix.)
47. HELMS, L. L. (1969). *Introduction to potential theory.* Wiley-Interscience, New York. (§2, Chapter I.)
48. HERVÉ, R.-M. (1962). Recherches axiomatiques sur la théorie des fonctions surharmonique et du potential. *Ann. Inst. Fourier (Grenoble)* **12,** 415–571. (§5, Chapter VI.)
49. HILLE, E. (1959). *Analytic function theory*, Vol. 2. Ginn, Boston. (§2, Chapter IV.)
50. HÖRMANDER, L. (1966). *An introduction to complex analysis in several variables.* Van Nostrand, Princeton, NJ. (§4, Chapter I; §A6, Appendix.)
51. HUA, L.-K. (1963). *Harmonic analysis of functions of several complex variables in the classical domains, Translations of Mathematical Monographs*, Vol. 6. American Mathematical Society, Providence, RI, 1963. (§3, Chapter II; §1, Chapter V; §A.2, Appendix.)
52. HUBER, A. (1956). On the reflection principle for polyharmonic functions. *Commun. Pure Appl. Math.* **9,** 471–478. (§1, Chapter I.)
53. JARNICKI, M. (1975). Analytic continuation of harmonic functions. *Zesz. Nauk. UJ, Pr. Mat.* **17,** 93–104. (§2, Chapter II.)
54. JARNICKI, M. (1977). Analytic continuation of pluriharmonic functions, *Zesz. Nauk. UJ, Pr. Mat.* **18,** 45–51. (§2, Chapter II.)
55. KAMPÉ DE FÉRIET, J. (1948). Sur la moyenne polyharmonique d'une fonction. *C. R. Acad. Sci. Paris* **226,** 621–623. (§1, Chapter I.)
56. KISELMAN, C. O. (1969). Prolongement des solutions d'une équation aux dérivées partielles a coefficients constant. *Bull. Soc. Math. Fr.* **97,** 329–356. (§2, Chapter II.)
57. KRAFT, R. (1968). Reflection of polyharmonic functions, *J. Math. Anal. Appl.* **22,** 670–678. (§1, Chapter I.)
58. KRASOVSKII, JU. P. (1967). Singularities of the Green's function. *Izv. Akad. Nauk USSR, Ser. Mat.* **31,** 977–1010. (§6, Chapter VI.)
59. LELONG, P. (1946). Sur la définition des fonctions harmoniques d'order infini. *C. R. Acad. Sci. Paris* **223,** 372–374. (§1, Chapter II.)
60. LELONG, P. (1947). Sur les fonctions indéfiniment dérivables de plusieurs variables dont les laplaciens successifs ont des signes alternés. *Duke Math. J.* **14,** 143–149. (§1, Chapter II.)
61. LELONG, P. (1948). Sur l'approximation des fonctions de plusieurs variables au moyens des fonctions polyharmoniques d'orders croissants. *C. R. Acad. Sci. Paris* **227,** 26–28. (§2, Chapter II.)
62. LELONG, P. (1951). Sur les singularités complexes d'une fonction harmonique. *C. R. Acad. Sci. Paris* **232,** 1895–1897. (§2, Chapter II.)
63. LELONG, P. (1954–1955). Prolongement analytique et singularités complexes des fonctions harmoniques. *Bull. Soc. Math. Belg.* **7,** 10–23. (§2, Chapter II.)
64. LEVIN, JA. B. (1964). *Distribution of zeros of entire functions, Translations of*

Mathematical Monographs, Vol. 5. American Mathematical Society, Providence, RI. (§4, Chapter I; §7, Chapter VI.)

65. Lions, J. L. and Magenes, E. (1973). *Non-homogeneous boundary value problems and applications*, Vol. 3. Springer Verlag, Berlin. 1973. (§§4, 6, Chapter VI.)
66. Morrey, Jr, C. B. (1966). *Multiple integrals in the calculus of variations.* Springer Verlag, New York. (§6, Chapter VI.)
67. Narasimhan, R. (1971). *Several complex variables.* University of Chicago Press, Chicago. (§4, Chapter I).
68. Nicolescu, M. (1936). *Les fonctions polyharmoniques.* Hermann, Paris. (§1, Chapter I: §§1, 2, Chapter II).
69. Nicolescu, M. (1936). Nouvelles contributions dans la théorie des fonctions polyharmoniques. *Bull. Math. Soc. Roum. Sci.* **37** (2), 79–95. (§1, Chapter I.)
70. Nicolescu, M (1960). Sur les suites convergentes de fonctions polyharmoniques. *Acad. Repub. Pop. Rom. Stud. Cerc. Math.* **11,** 287–291. (§1, Chapter I.)
71. Nicolescu, M. (1960). Convergent sequences of polyharmonic functions (in Russian). *Rev. Math. Pures Appl.* **5,** 467–471. (§1, Chapter I.)
72. Nicolescu, M. (1966). Sur quelques propriétés de structure des solutions des équations métaharmonique et métacalorique. *Rev. Roum. Math. Pures Appl.* **11,** 609–620. (§5, Chapter VI.)
73. Ovčarenko, I. E. (1961). On multiply superharmonic functions (in Russian). *Usp. Mat. Nauk* **16** (3) (99), 197–200. (§1, Chapter I.)
74. Pascali, D. (1960). Les polynômes aréolaires et le développement d'Almansi dans le plan. *Commun. Acad. Repub. Pop. Rom.* **10,** 257–262. (§1, Chapter I.)
75. Pčelin, B. K. (1969). On the general theory of polyharmonic functions. *Sov. Math. Dokl.* **10** (4), 983–984. (§1, Chapter I.)
76. Picone, M. (1936). Sulla convergenza delle successioni di funzioni iperarmoniche. *Bull. Math. Soc. Roum. Sci.* **38,** 105–112. (§1, Chapter I.)
77. Privaloff, I. and Pčelin, B. (1937). Sur la théorie générale des fonctions polyharmoniques. *C. R. Acad. Sci. Paris* **204,** 328–330. (§1, Chapter I.)
78. Privaloff, I. and Pčelin, B. (1937). Sur la théorie générale des fonctions polyharmoniques. *Rec. Math. Moscow, New Ser.* **2,** 745–757. (§1, Chapter I.)
79. Reade, M. O. (1957). Remarks on a paper by A. Friedman. *Michigan Math. J.* **4,** 75–76. (§1, Chapter I.)
80. Sibony, N. (1975). Prolongement des fonctions holomorphies bornées et métrique de Caratheodory. *Inventiones Math.* **29,** 205–30. (§A.6, Appendix.)
81. Siciak, J. (1974). Holomorphic continuation of harmonic functions. *Ann. Polon. Math.* **29,** 67–73. (§2, Chapter II; §A.2, Appendix.)
82. Spanier, E. (1966). *Algebraic topology.* McGraw-Hill, New York. (§4, Chapter II.)
83. Svešnikov, A. G. (1952). The principle of limiting absorption for the metaharmonic equation (in Russian). *Dokl. Akad. Nauk USSR* (N.S.) **86,** 231–234. (§5, Chapter VI.)
84. Teissier du Cros, F. (1952). Sur le lien entre les notions 'champ réel autonome' et 'cellule d'harmonicité'. *C. R. Acad. Sci. Paris* **235,** 600–601. (§2, Chapter II.)
85. Teleman, S. (1953). Une propriété des fonctions du développement d'Almansi d'une fonction polyharmonique. *Acad. Repub. Pop. Rom. Bul. Stiint. Sect. Stiinte. Mat. Fiz.* **5,** 385–391. (§1, Chapter I.)
86. Titchmarch, E. C. (1958). *The theory of functions*, 2nd edn. Claredon Press, Oxford, 1958. (§7, Chapter VI.)
87. Tolotti, C. (1946). Sulla struttura delle funzione bi-iperarmoniche in tre

variabili indipendenti. *Atti Accad. Naz. Lincei Rend. Cl. Sci. Fis. Mat. Nat.* **8** (1), 359–363. (§1, Chapter I.)
88. TOLOTTI, C. (1946). Sulla struttura delle funzioni iperarmoniche in puì variabili indipendenti. *Ric. Sci.* **16,** 315–317. (§1, Chapter I.)
89. TOLOTTI, C. (1947). Sulla struttura delle funzioni iperarmoniche in puì variabili indipendenti. *G. Math. Battaglini* **4** (1) (77), 61–117. (§1, Chapter I.)
90. VEKUA, I. N. (1967). *New methods for solving elliptic equations.* North-Holland, Amsterdam. (§5, Chapter VI.)

NON-ALPHABETIC SYMBOLS

The page reference is to the place(s) where the symbol is introduced.

D':	domain of canonical extension from D into $\mathbb{C}^n$ 20
$\|$:	end of proof 2
U_e; U_0:	even extension of U from $\mathbb{R}^n$ to $\mathbb{R}^{n+1}$; odd extension 154
$\tilde{G}$; $\hat{G}$:	harmonicity hull of G; reduced harmonicity hull 42
(x, y):	inner product in $\mathbb{R}^n$ 37
$(x, y), (x; y)$:	point in $\mathbb{R}^n \times \mathbb{R}^m$ 1, 235

$E_\tau = \bigcap_{z \in \bar{D}} B_{1/\tau}(z) \setminus \bar{D}$; G_τ, $\mathscr{D}_\tau$: intersection of balls 35, 46

$\|x\|$:	modulus, in $\mathbb{C}^n$ 2
$\|q\|$, $q!$, x^q, D^q:	multi-index q, attributes of 1
D_δ; T_δ:	neighborhood of ∂D; mapping of D_δ 184
$\hat{x}$:	point on Weierstrassian manifold 19, 21
x':	projection of x on ∂D 181

$\xi + i\eta \in \mathbb{C}^n$, where ξ, $\eta \in \mathbb{R}^n$: real and imaginary parts 41

x^*, S^*:	reflections in unit sphere 11, 64, 86
l_r:	restriction of l to $\mathbb{R}^n$ 100
$\|f\|$, $\|f\|_D$:	supremum of modulus 13

$K(R) = \{z : |z - x| < R$ for some $x \in K\}$: union of balls 29

INDEX

For symbols the reference is to the place(s) where introduced. Non-alphabetic symbols such as diacritical marks are listed on page 257.

A_ε: set enclosed by level sets for holomorphic capacity 223
A_ι: set on which sequence is eventually uniformly bounded 231
A_u, $A(u, K, z)$: analytic part of u 78–9
$\mathscr{A}$: space of germs of analytic functions 130–1
$\alpha(z)$, $\alpha(A, G; z)$: holomorphic capacity function 208
Alexander duality 70, 74
algebraic topology, facts from 69–70
Almansi expansion 4–6, 122–53
  for analytic functions, by spherical harmonics 125
  coefficients
    conditions implying Laplacian order and type 131–2
    formulas for 126, 135
    and growth of functions f_K 147
    Laplacian order and type imply conditions 142
    necessary and sufficient conditions for Laplacian order and type 145
  finite, for polyharmonic functions of finite degree 4–6
  for holomorphic functions 122
  for polyharmonic functions 135
analytic continuation
  across surface of integration 50–1
  along path 19
  Borel's method 203–7
  defines holomorphic function on reduced harmonicity hull 50–1
  of polyharmonic function is polyharmonic 77
  or principal and analytic parts 81
  unlimited 22, 69
    from G to $\hat{G}$ and $\tilde{G}$ 44, 51
analytic functions 2, 17
  necessary and sufficient condition: $\rho=\infty$ and $\tau=\infty$ 36
  not polyharmonic
    see also examples
    failure of maximum principle 120
    local serio-integral representation of 52
analytic part, A_u, of function u, *see* principal part, R_u, of function u
auxiliary functions $\imath_l$ 8, 35
  analytic continuation of 44–5, 62, 65
  are their own principal parts 80–1, 102
  see also estimates

B^+, B^-: neighborhoods of sides of surface 50
B_k: operator, extension of $\check{B}_k$ 180
$\check{B}_k$; $\check{B}_k^0$; $\check{B}_k^0(x, \xi)$: boundary operator; principal part; characteristic polynomial 180
$\{\check{B}_k^{(i)}\}$: sequence of boundary operators for iterated system 193
B_R, $B_R(x)$, $B_R^n(x)$: balls in $\mathbb{R}^n$ 1
$\tilde{B}_R$; $\hat{B}_R$: harmonicity hull of ball; reduced harmonicity hull 59
$\mathscr{B}_R$, $\mathscr{B}_R(x)$, $\mathscr{B}_R^n(x)$: balls in $\mathbb{C}^n$ 1
Baire category 237
ball, B
  convexity of local Laplacian order and type on 116
  harmonicity hull, $\tilde{B}=\hat{B}$ 58–63
    and Cartesian products 127–8
    as classical domain $\mathscr{R}_{IV}$ of E. Cartan (Lie ball) 59, 126, 214
    convergence of series on 62–3
    convexity of local Laplacian order and type on 113
    defined by norm, q 59
    examples of local Laplacian order and type on 153
    expreme points, Λ; Šilov boundary 60–2, 64
    separation properties 59, 65
beta function 48

Borel
method of continuation, *see* analytic continuation
transform, φ 202–7
boundary operators
normal; normal sequence of 181
'complementing condition' on ∂D 181
extension of 181, 184–97
distinguished 186, 188, 193
see also operators, linear differential
boundary value problems 180–97
finite order 180–90
existence problem 186–90
existence theorem 190
formulas for solutions 183, 185, 191
uniqueness for $\{\Delta^m, \{\check{B}_k\}\}$ 185
infinite order 180, 190–7
existence theorem 194–5
periodic relative to Δ 193
periodically uniform conditions, (U_∞, m, τ), (U, m, τ), (U_1, m, τ) 194–7
uniqueness theorem 196
well-posed 197
bounded polyharmonic functions 169–73

c_l: coefficient in auxiliary function $\imath_{l+n/2}$ 8
C_R, $C_R(0) = \mathbb{R}^n \backslash \bar{B}_R(0)$: comball 63
$\{\check{C}'_k\}$: adjoint system of operators 184
$\mathbb{C}^n$: complex n-dimensional space 1
$\mathscr{C}^*(X)$: Space of Borel measures on X 170
$\mathscr{C}^k, \mathscr{C}^\infty, \mathscr{C}^k(D)$: spaces of differentiable functions 1
Cauchy
bounds 18
integral formula 17
theorem 17
characteristic solution of $\Delta^l u = 0$ 7–8
χ_A: characteristic function of A 216
circular domain, complete 59
classical domain $\mathscr{R}_{\mathrm{IV}}$ of E. Cartan 59, 214
Cauchy kernel; volume element 126
Co(k): convexity in dimension k 66
cohomology group (Cech), $H^0_c(F, \mathscr{G})$ 69
comball, C_R 63–5
convexity of local Laplacian order and type on 116
harmonicity hull, $\tilde{C}_R$ 63–5
convexity of local Laplacian order and type on 113
defined by homogeneous function, s 64
separation properties 64–5
complementary system of operators, $\{\check{\Phi}_k\}$ 183
completeness (completion) of sets in a domain 245–9
compared with domains of holomorphy 249
examples 246–9
intersection with affine subspaces 248
cone $V(x)$, *see* isotropic cone, $V(x)$
constancy
of bounded polyharmonic functions 169–73
of local Laplacian order and type 106–7
conventions
A is a measurable set 224
centre of B_R is the origin 122
for $h(\theta)$, if f has exponential growth 202
dimension $n \geq 2$ 164
domains, attention restricted to 55
convex domain 56
harmonicity hull of 56–8
convexity of functions
logarithmic 112
of local Laplacian order and type 110–21
examples 110–21, 150–3
on $\mathbb{C}^n$ 113–14
see also Laplacian order and type
trigonometric 201
convexity in dimension k: Co(k) 66–8

$d(y, A)$: minimum distance 13, 31
D_{t_0}: domain of section at t_0 198
$D^q = \partial^{|q|}/\partial x_1^{q_1} \dots \partial x_n^{q_n}$ 1
$\mathscr{D}(x, R)$: polydisc 17
∂D: boundary 9
$\partial/\partial\nu$, $\partial/\partial\nu_x$; $\partial/\partial\tau$: inner normal derivative; directional derivative 9, 11, 12
Δ; Δ^p: Laplace operator; pth iterate of Δ 1
Δ_S: Laplace–Beltrami operator on sphere 3
$\{\Delta^m, \{\check{B}_k\}\}, \{A^{(j)}, \{\check{B}_k^{(j)}\}\}$: differential systems 181
density 224
diam $A = \sup\{|x - y|: x, y \in A\}$ 91
differential operators, *see* operators, linear differential; boundary operators
differential system 181–97

differential system (*contd*)
  adjoint system 184
  complementary system 183
  iterated: $\{\Delta^{mp}, \{\check{B}_k^{(i)}\}\}$ 190–7
  satisfying conditions (1), (2), (3) and (A), (B), (C) 181–3
  self-adjoint, positive definite 182
dimension $n = 1$, special case
  Almansi expansion 125
  applications in $\mathbb{R}^{n+1}$ 101–2, 155–8
  harmonicity hull, reduced harmonicity hull 42, 52, 119
  holomorphic capacity 211, 217–20
  Laplacian order and type 29
  local 108
    convexity 116, 119
    constancy 107, 152
    example 52
distinguished boundary 246–7
distributions 'harmonic of infinite order' 34
domain, defined 1
domains of holomorphy 249

$E(x, t)$: fundamental solution of heat equation 197
elliptic iterates, theorem on 194
entire functions 18, 22–4, 28–9, 101, 164–7
  exponential growth 202
  exponential order and type 22–4, 29–30, 115, 202, 233
  multiplier theorem 166
  are polyharmonic 28–9, 164
estimates
  of $D^q r^k$, $D^q r^k \log r$, $D^q \mathfrak{r}_l$ 15–16
  of $\Delta^p(r^{2k} h_k)$ 142
  of derivatives 12–14, 16
  see also holomorphic capacity
eventual uniform boundedness of sequences 231, 242
examples 26–9, 52–3, 58–9, 63–8, 75, 79–80, 83, 101–2, 104, 110, 148–53, 159–62, 177–9
  analytic functions
    have Laplacian order ∞, type ∞ 26
    not polyharmonic 104–5, 120, 159–62
      failure of maximum principle 160
  harmonicity hull; reduced harmonicity hull
    ball 58–63
      see also ball, B
    comball 63–5
      see also comball, C_R
    half-space and convex domains 58
    intersections and products 65, 68
    non-topological 65–8
    null-homotopic 66
  holomorphic capacity 210–20
    see also holomorphic capacity
  Laplacian order and type related to exponential order and type 147
  local Laplacian order and type 99, 101–2, 104, 110
    on $\hat{B}_R$ 153
    constant 148
    local type discontinuous 150
    local type not smooth 102
    logarithmically convex in $r(x)$ 150–3
    of continuations into $\mathbb{C}^n$ 152–3
    power of $r(x)$ 149–50
  local nature of Laplacian order and type 33
  local representation of function which is not polyharmonic 52
  maximum principle for local type requires polyharmonicity 104
  minima for local Laplacian type 110
  monodromy group 75–6
  non-analytic solutions of heat equation 198
  polyharmonic functions satisfying $\Delta^m u = Cu$ 27
  polyharmonic function satisfying $\Delta^m u + Au = 0$ 177–9
  principal part and analytic part of function 83, 79–81, 102
  r^k, $r^k \log r$ 27
exponential growth 170
exponential order and type 101
  compared with Laplacian order and type 115, 164–9
  inclusive 170
  related to pointwise Laplacian order and type 155–8
  see also entire functions
exponential type (or growth) in sector 202–7
extension of function
  analytic, complete 21
  branch of 21

holomorphic, canonical 20
see also analytic continuation
extension of boundary operator 181
extremal principles, *see* maximum principle for local Laplacian order and type; zero minimum principle; *see also* Laplacian order and type, local

$f(x)=\exp\{1-l(x)^{-1}\}$ 99
$f_K=\sum\|h_k\|_K z^{2k}$ 147
$F(x)=\exp\{1-\lambda(x)^{-1}\}$ 100
convexity results on 114, 116
examples 149–52
$\mathscr{F}$: class of solutions of partial differential equations 180
finite degree, functions polyharmonic of 2
in separation of singularities 90
constructed in proofs 85–9
finiteness conditions for local type, μ 103–4, 110
full cone, $V'(x)$ 40–2
see also isotropic cone, $V(x)$
function classes $\mathscr{P}(D)$, $\mathscr{P}^{\rho}(D)$, $\mathscr{P}^{\rho,\tau}(D)$ defined 28, 45
fundamental group, first, Π_{x^0} 69
fundamental solution of $\Delta^l u=0$ 8, 27

$g(z, y)$: Green function for half disc 225
g_k: boundary data for partial differential equation 180
$\hat{G}$; $\tilde{G}$: reduced harmonicity hull of G; harmonicity hull 42
$\mathscr{G}$: class of boundary data 180
γ_l, γ_l': coefficient in auxiliary function $\imath_l$ 7–8
Γ, $\Gamma(t)$: path, arc in $\mathbb{C}^n$ 18, 42
Γ-function, *see* gamma function
gamma function, formulae on 14
G-complete, G-completion 245
germs of functions 129–31
Gevrey classes 173–4
Green formulae
for ball 11
first and second 9–10
Green function
for ball 11
for boundary problem of infinite order 180
for $\{\Delta^m, \{\check{B}_k\}\}$ 183, 190
growth (ρ, τ) 170

h: homogeneous function on $\mathbb{C}^n$ 112
$h(\theta)$: (Phragmén–Lindelöf) indicator 201
h_k: Almansi coefficient 122
$H(\theta)$: indicator function 201
$H_{n-1}(D, \mathscr{G})$: singular homology group 70
$H_R(a)$: upper half disc in $\mathbb{C}^1$ 225
$\check{H}^0_c(F, \mathscr{G})$: Cech cohomology group 69–70
$\mathscr{H}$, $\mathscr{H}(B_R)$: space of germs of harmonic functions 130–1, 170–2
Hadamard
factorization 101
multiplication theorem 200, 206
product 205–6
three-circle theorem 112
generalized to three-shell lemma 112
half disc $H_R(a)\subset\mathbb{C}^1$ 225
harmonic functions
integral formula for 126
of infinite order, *see* out of date terminology
on $\mathbb{R}^{n+1}$, extensions from functions on $\mathbb{R}^n$ 154–64
harmonic measure, ω 117–18
harmonicity hull, $\tilde{G}$, of G 39–40, 42–3
characterized by unlimited continuation from G 51–2
construction of monodromy group 70–4
and reduced harmonicity hull $\hat{G}$: properties 43, 55–69
harmonicity hull of ball 58
and Cartesian products 127–8
as classical domain $\mathscr{R}_{IV}$ of E. Cartan (Lie ball) 59, 126, 214
see also ball, B, harmonicity hull
Harnack's theorem 180
Hartogs' lemma
related results 156–9
similar results 234–40
heat equation, solution of 197–207
analytic 197–207
if and only if Laplacian order 2, type $\tau\leq\infty$ 198
section of, if and only if Laplacian order 2, type $\tau<\infty$ 199
non-analytic, example 198
(space) section of 198
traces of 197
historical notes, *see* notes on history, background
holomorphic capacity 110, 208–51
applications
to polyharmonic functions 77, 100,

holomorphic capacity, applications, to polyharmonic functions (*contd*)
103, 109–11, 153, 156–7
to sequences of functions 231–45
convergence properties 234–40
exponential order of entire functions 233
upper semi-continuous envelopes 240–5
approximate continuity is continuity 229–30
basic properties 208–10
continuity at density points in $\mathbb{R}^n$ 228–9
estimates
on polydiscs 223–4
on sequences of sets 229
preliminary 226
two-step 227
examples 210–20
concentric polydiscs in $\mathbb{C}^n$ 213
infinite strip 218
Lie ball (harmonicity hull of ball) 214–16
nested discs in $\mathbb{C}$ 211, 217–18
nested half planes 217–18
real subsets of disc 218–19
regions bounded by level sets 211
single point: characteristic function 216
inequalities on Cartesian products 212
invariance under holomorphic mapping 210
and measure 219–20
monotonicity 208
on manifolds 250–1
polar, non-polar sets: positivity results 220–3, 251
positivity lemma 220–1
subadditivity 209
supermultiplicativity 209
upper semi-continuity 210
holomorphic function: terminology 2, 17
homogeneous polynomials 15, 62, 87–8, 123, 125
homology groups $H_{n-1}(D, \mathcal{G})$ 69–71

I^+, I^-: continuations of I across surface of integration 51
$\bar{I}_f$: conjugate diagram of f 202
I_k, $I_k(S, u; y)$: term of serio-integral representation 39
$\mathcal{I}m\, z$: imaginary part of $z \in \mathbb{C}^1$ 117
index, $w(p; C)$ 70–3
indicator, Phragmén–Lindelöf, $h(\theta)$ 201–7
isotropic cone, $V(x)$ 40–2, 113, 174
functions analytic or harmonic on, and their germs 127–31

$k(\theta)$: support function for convex set 201
$\hat{K}_G = \{z \in G : |f(z)| \leq \|f\|_K, f$ holomorphic on $G\}$ 249
Kelvin transformation 5–7, 64
extended 6–7
Krasovskiĭ's inequalities 187
Kronecker's theorem 209

$l(x)$: pointwise Laplacian order 99
L, $L(z)$; L_r: limit supremum; restriction of L to $\mathbb{R}^n$ 100, 240
$\hat{L}$: upper semi-continuous envelope of L 240
λ: least Laplacian order 28
$\lambda(x)$: local Laplacian order 33
λ_l: eigenvalue 182
λ_l^p: eigenvalue of pth iterated system 191
λ_U: least Laplacian order on U 98
Λ: Šilov boundary of $\hat{B}_1(0)$ 126
Laplace operator, defined 1, 3
Laplacian order and type 25–34, 45
in $\mathbb{C}^n$ 45
compared with exponential order and type 164–9
on harmonicity hull, reduced harmonicity hull 45–9, 51
least 28, 30, 45, 98
local 33–4, 45
consistency of definitions in $\mathbb{R}^n$, $\mathbb{C}^n$ 53
constancy on $\mathbb{R}^n$, $n \geq 2$ 120
convexity 110–21
for polyharmonic functions 116
equal to limit supremum 105
equality for bounds on G, $\tilde{G}$, ∂G, $\partial\tilde{G}$ 53–4
maximum principle 102
see also maximum principle for local Laplacian order and type
minima of 110
properties of the sets $P^\rho(u)$, $P^{\rho,\tau}(u)$ 109–10
and the radius of convergence of power series 108
restriction from $\mathbb{C}^n$ to $\mathbb{R}^n$ 100

upper semi-continuity of 33–4
upper semi-continuous envelopes of pointwise Laplacian order and type 99–100
for meta-harmonic functions 174–9
multiplier theorem 166
for non-polyharmonic functions 154–64
pointwise 99, 103, 106
compared with exponential order and type 155–8
constancy of, *see* constancy of local Laplacian order and type
maximum principle and 103
see also maximum principle
related to convergence of power series 155–6
related to exponential order and type 101, 115, 147
related to Gevrey classes 173–4
in separation of singularities 90, 163–4
for solutions of boundary value problems 191–3
for solutions of $\Delta^m u + Au = 0$ 176–9
unattainable 28
see also Almansi expansion, coefficients
Laplacians, Δ^p, sequence of 25
Legendre's formula 14
Lie ball, *see* ball, B, harmonicity hull
Lipschitzian graph manifold 8–9
Liouville's theorem 173, 180

$m(x)$: pointwise type 99
$M(u; R)$: maximum modulus of function u 22
$\mathcal{M}$: Lipschitzian graph manifold 8
μ: least type 28
$\mu(x)$: local type 33
μ_U: least type on U 98
majorant, local uniform 242–3
manifolds 8
by analytic continuation 19, 21
bearing holomorphic capacity 250–1
Laplacian order and type on 45
maximum principle for local Laplacian order and type 102
applications 105–6, 108, 114–15, 118, 120–1
meromorphic function 18
meta-harmonic equations and related equations 173–8
monodromy group
of functions, $\Pi_{x^0}/\Pi^0_{x^0}$ 69
of $\tilde{G}$ 74–6
multi-index, defined 1
multiplier theorem for entire functions and polyharmonic functions 166

ν: unit inner normal 9
$n = 1$, *see* dimension $n = 1$
normal order of operator 181
notes on history, background v, 7, 34, 54, 131, 173, 179

ω: harmonic measure 117–18
ω_n, Ω_n: area of sphere 10
$\Omega_{\rho,\tau}$: a collection of sequences of integers 171–3
open questions 109–10, 121, 158, 186
operators, linear differential 1–3, 31–3, 174–6, 180–97
effect on finite degree 31–2
effect on Laplacian order and type 31–3
see also boundary operators
out-of-date terminology i, 25, 34
order
of operator: normal, total 181
see also exponential order and type; Laplacian order and type; out-of-date terminology

$P(\Delta)$: differential operator, P a polynomial 174
$P(x; R)$: polydisc 17
$P^\rho(u)$, $P^{\rho,\tau}(u)$: open sets determined by Laplacian order (and type) of u 109
$\mathcal{P}^\rho$, $\mathcal{P}^{\rho,\tau}$, $\mathcal{P}^\rho(D)$, $\mathcal{P}^{\rho,\tau}(D)$: function classes 28, 131
$\varphi(z)$: Borel transform 202
$\Phi(A, G)$: a collection of functions holomorphic on G 208
$\check{\Phi}_k$: boundary operators corresponding to $\{\Delta^m, \{\check{B}_k\}\}$ 183
$\check{\Phi}A|_{\partial D}$ 191
$\Pi(z)$: plane in $\mathbb{R}^n$ determined by $\Sigma(z)$, $\Sigma'(z)$, for $z \in \mathbb{C}^n$ 41–2
$\Pi_q(l)$: polynomial 16
Π_{x^0}: first fundamental group 69
$\Pi^0_{x^0}$: class of paths 69
$P-0, \ldots, P$-7: properties 186–7
path, Γ, admissible 18
periodically uniform inequalities 193

permanence of functional equations, generalized 107
Phragmén–Lindelöf
  indicator, $h(\theta)$ 201–7
  theorem 171
piecewise flat; piecewise smooth 9
Poisson formula and kernel for ball $\bar{B}_R$ 12, 132
Polya's theorem 202
poly-half-disc 224
polyharmonic function, definition and terminology 25
  Lelong's 34
  finite degree, least degree 2
    see also finite degree
  order *m*, *see* out-of-date terminology
positive (or zero) Lebesgue measure 98, 103, 105–6, 108, 155–8, 233
  and eventual uniform boundedness 231
  and uniform convergence of power series 234–40
power series, uniform convergence 234–40
principal part R_u, of function *u* 78–85, 120
  local, for some analytic functions 84–5
  properties: existence, uniqueness, Laplacian order and type 78–80
products, Cartesian 67–8, 128, 154–64, 197–200, 212–17, 220–1, 234–9
properties *P* 186–7
  relationships among 188
ψ_l: eigenfunctions 182

q: norm defining $\tilde{B}_R(0)$ 59
  and convexity results 113–21
Q_R, $Q_R(z)$: cube in $\mathbb{R}^n$, real cube in $\mathbb{C}^n$ 224

$r(x)=r(x_1,\dots,x_n)=(\Sigma x_k^2)^{\frac{1}{2}}$ 2
  continuation into $\mathbb{C}^n$ 21–2
  related to $q(x)$, $s(x)$ 64
R_u, $R(u,k;x)$: principal part of function 78–9
$\mathbb{R}^n$: real n-dimensional space 1
$\imath_l$: auxiliary function 8, 35
$\mathcal{R}_{IV}$: classical domain of E. Cartan (Lie ball) 59
$\mathcal{Re}\, z$: real part of $z\in\mathbb{C}^1$ 37
ρ: exponential order 22
ρ: Laplacian order 25–6
(ρ,τ): exponential growth 170

rational functions 18
reciprocal radii, *see* reflection in unit sphere
reduced harmonicity hull, $\hat{G}$, 42–3
  see also harmonicity hull $\tilde{G}$
reflection in unit sphere 11–12, 64, 86–7
regularity at infinity 6–7
representation of $z\in\mathbb{C}^n$ as oriented sphere in $\mathbb{R}^n$ 41–2
reproducing kernel 183
Rheinhardt domain, complete 63
Runge's theorem, lemma reminscent of 85, 88–9

s: homogeneous function defining $\tilde{C}_R(0)$ 64
  and convexity results 113–21
s: winding number in $\mathbb{C}^1$ 44
S; S_C; $S^+=\partial B^+\backslash S$, S^-: surfaces of integration 39, 50
σ, $\sigma_{\mathcal{M}}$, $\sigma(x)$: measures 9
σ_R: Lebesgue measure on ∂B_R 123
$\Sigma(x)=V(x)\cap\mathbb{R}^n$, $\Sigma'(x)=V'(x)\cap\mathbb{R}^n$: representation in $\mathbb{R}^n$ of $x\in\mathbb{C}^n$ 40–2, 71–2
separation of singularities 85–97
  for analytic functions 162
  for polyharmonic functions 90
series domination, technique for 134
serio-integral construction 45–9
  local 45–6
serio-integral representation
  and unlimited continuation 72–3
  local representation in $\mathbb{R}^n$
    of analytic functions 35, 37
    of functions with $\rho=\infty$, $\tau=\infty$ 35
  of polyharmonic functions, summarized 51
  properties of terms I_k 39–40, 43–5
Šilov boundary 247
  of $\hat{B}_1(0)$ 126
singular point, singular set: singularity 77, 85
  polar 18
  removable, non-removable 18, 77
star domain 4, 55
  harmonicity hull of 56
Stirling's formula 14, 134
submanifolds of real type 222, 250–1
  n-dimensional measure on 222
support function $k(\theta)$ for convex set 201–6

symbolic integrals 200
τ: exponential type 23
τ: Laplacian type 25–6
Taylor series, Taylor development 18–20
total order of operator 181
tube $\Sigma \circ \Gamma$ 42–3, 55
full tube $\Sigma' \circ \Gamma$ 42–3
type, *see* exponential order and type; Laplacian order and type
θ: parameter in convexity 112, 114, 116–18
θ: point on sphere in $\mathbb{R}^n$ 59, 60, 123
$\Theta_R(a, A)$: density of complement of A 224

$u(x, t_0)$: section of u at t_0 198
$u_x(z) = u(x, z)$: domain restriction 235
$(U_\infty, m, \tau), (U_1, m, \tau), (U, m, \tau)$: periodically uniform conditions 194–7
upper semi-continuity
of holomorphic capacity 240–5
of local Laplacian order and type 33, 99–100

$V(x)$; $V'(x)$: isotropic cone; full cone 40–1, 70–2

$w(p; C)$: index of p with respect to C 70
Weierstrassian manifold of u 19, 69
winding number, s, in $\mathbb{C}^1$ 44, 70–1

zero Lebesgue measure, *see* positive (or zero) Lebesgue measure
zero minimum principle 106–7
see also Laplacian order and type, local